Two Manuals Combined into One Book

Military Police Operations

And

Military Intelligence

Department of Defense

Pentagon
Publishing

***FM 3-19.1** (FM 19-1)

Field Manual
No. 3-19.1

Headquarters
Department of the Army
Washington, DC, 22 March 2001

Military Police Operations

Contents

Distribution Restriction: Approved for public release; distribution is unlimited.

*This publication supersedes FM 19-1, 23 May 1988.

Preface

The Military Police (MP) Corps supports the commander across the full spectrum of military operations. This manual is the foundation for all MP doctrine as it relates to this support. It communicates to all levels of leadership and staffs how the MP provide a flexible and lethal force capable of operating across this full spectrum. As the keystone manual, it identifies what the MP train on and how their forces are organized and equipped in support of all Army echelons. Additionally, this manual provides guidance that can be used by United States (US) Army service schools, students, sister services, and federal agencies.

This manual is based on the purpose, organization, responsibilities, and goals of the US Army as set forth in Field Manuals (FMs) 100-1 and 3-0, as well as corps, division, and brigade manuals. Additionally, this manual is fully compatible with current joint, multinational, and interagency doctrine.

Appendix A contains a metric conversion chart.

The proponent of this publication is HQ TRADOC. Send comments and recommendations on Department of the Army (DA) Form 2028 directly to Commandant, US Army Military Police School (USAMPS), ATTN: ATSJ-MP-TD, 401 Engineer Loop, Suite 2060, Fort Leonard Wood, Missouri 65473-8926.

Unless this publication states otherwise, masculine nouns or pronouns do not refer exclusively to men.

Chapter 1

Introduction

The MP Corps has a strong history evolving over the past five decades. We, as a corps, continue to transform our organizations and doctrine as we have in the past to support the Army in the active defense of the 1970s, the AirLand battle of the 1980s, and now the force-projection doctrine of the 1990s. Our five MP functions clearly articulate the diverse role the MP play across the full spectrum of military operations. We cannot bask in our successes, nor reflect or celebrate. Our charter is to continue our legacy of stellar performance and strive to perfect it.

BG Donald J. Ryder

When the Army developed the Active Defense strategy in 1976, the US was facing the Cold War scenario of central Europe. Military strategy and doctrine were related to a single, focused threat that revolved around the countries in the Warsaw Pact. We were an outnumbered and technically inferior force facing an armor-dominated European battlefield. The MP Corps supported the Active Defense strategy by tailoring its forces to meet the threat. In 1982, when the AirLand Battle strategy was developed, US forces were still outnumbered, but were no longer technically inferior. Still threat-based and focused on a central European conflict, the AirLand Battle strategy used a relatively fixed framework suited to the echeloned attack of soviet-style forces. It delineated and clarified the levels of war; emphasized closed, concerted operations of airpower and ground forces; balanced the offense and the defense; and highlighted the synchronization of close, deep, and rear operations. MP doctrine kept pace with the Army's AirLand Battle strategy by supporting the battlefield commander through four basic missions—battlefield circulation and control, area security (AS), enemy prisoner of war (EPW), and law and order (L&O).

OVERVIEW

1-1. In October 1983, MP capabilities in the AirLand Battle strategy were tested during operation Urgent Fury in Grenada. The MP performed missions that ranged from assisting the infantry in building-clearance operations to assisting Caribbean peacekeeping forces in restoring L&O. These actions secured the MP's place in the combat-support (CS) role, demonstrating the professional knowledge and flexibility necessary for rapid transition from combat to CS to peacetime missions. The changing battlefield conditions of operation Urgent Fury set the stage for the demand of MP units today.

1-2. Evolving simultaneously with the changing definition of the modern battlefield, MP performance in Operations Hawkeye, Just Cause, and Desert Shield/Storm galvanized their ability to perform at any point along the operational continuum. With the publication of FM 100-5 in 1993, the Army adopted the doctrine of full-dimensional operations, relying on the art of battle command to apply those principles and to shift the focus from AirLand Battle to force-projection doctrine. This new doctrine was based on recent combat experience in a multipolar world with new technological advances. Already trained and expected to perform in this new strategy, MP support was already in place and fully operational. The MP continued to perform their basic battlefield missions and to refine their capabilities while supporting the battlefield commander as he deployed to contingency operations throughout the world.

1-3. In 1996, the MP Corps went through a doctrinal review process to determine if it was properly articulating its multiple performance capabilities in support of US forces deployed worldwide (see Appendix B). The review process identified the need to restructure and expand the EPW mission to include handling US military prisoners and all dislocated civilians. This new emphasis transformed the EPW mission into the internment and resettlement (I/R) function. The review process also identified the need to shift from missions to functions. In the past, the four battlefield missions adequately described MP capabilities in a mature theater against a predictable, echeloned threat. However, that landscape is no longer valid. Accordingly, the four MP battlefield missions have become the following five MP functions:

- Maneuver and mobility support (MMS).
- AS.
- L&O.
- I/R.
- Police intelligence operations (PIO).

1-4. These new MP functions are shaped by the following factors:

- The application of stability and support operations where the integration of joint, multinational, and interagency capabilities are common occurrence.
- The lack of traditional linear battlefields, requiring theater commanders in chief (CINCs) to request forces that meet a specific function to accomplish operational requirements.
- The impact of asymmetric threats (such as drug traffickers and terrorist factions) and the effects of man-made and natural disasters.
- The impact of advances in information and communication technologies and specifically in understanding the increased vulnerabilities presented by these technologies.

1-5. Articulating MP capabilities along functional lines benefits the MP and the Army echelon commander as well as the combatant commander. Since there is a multinational, interagency, and sister-service overlap of security services, the importance of including MP leaders and staffs early in the operational planning process cannot be overemphasized. This means before units are designated, unit boundaries are drawn, and unit missions are assigned. Early involvement ensures the proper development of common security responsibilities, communication and connectivity, liaisons, processes, and the rules of interaction between all forces. The ultimate goal should be the optimal, phased employment of MP forces in support of a commander's operational plan. MP functions not only reflect and capture current capabilities, they define the MP Corps in the twenty-first century.

1-6. As the Army reshapes and focuses its resources on transformation, Force XXI, and other redesign efforts, the MP Corps stands proud and ready to support this progress and reiterate its commitment to assist, protect, and defend.

OPERATIONAL FRAMEWORK

1-7. The operational framework consists of the arrangement of friendly forces and resources in time, space, and purpose with respect to each other, the enemy, or the situation (see Figure 1-1). The operational framework for Army forces (ARFOR) rests within the combatant commander's theater organization. Each combatant commander has an assigned geographical area of responsibility (AOR), also called a theater, within which he has the authority to plan and conduct operations. Within the theater, joint-force commanders at all levels may establish subordinate operational areas such as areas of operation (AOs), joint-operations areas (JOAs) and joint rear areas

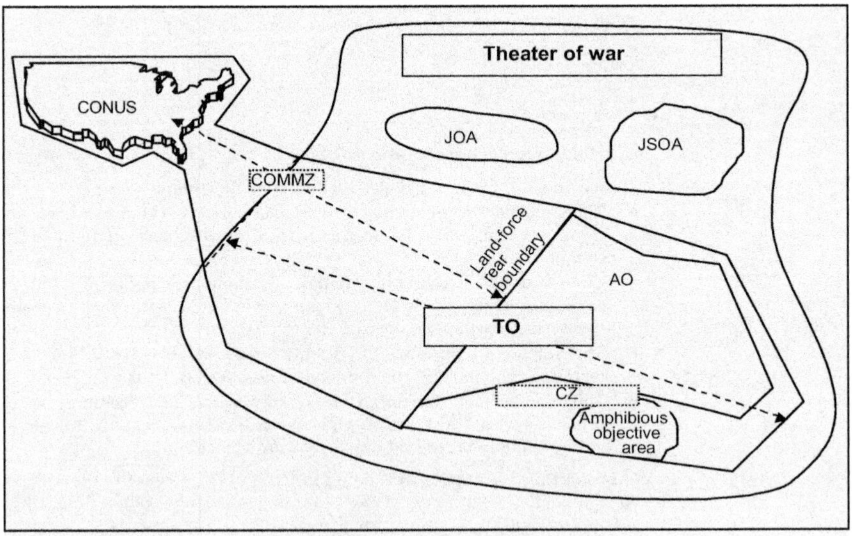

Figure 1-1. Operational Framework

(JRAs). The JRAs facilitate the protection and operation of bases, installations, and forces that support combat operations. When warranted, combatant commanders may designate theaters of war, theaters of operations (TOs), combat zones (CZs), and communications zones (COMMZs).

1-8. A theater of war is that area of air, land, or water that is, or may become, directly involved in the conduct of the war. A theater of war may contain more than one TO. It does not normally encompass the geographic combatant commander's entire AOR. A TO is a subarea (defined by a geographic combatant commander) within a theater of war in which specific combat operations are conducted or supported.

1-9. A CZ is the area required by combat forces for conducting operations. It normally extends forward from the land force's rear boundary. The COMMZ is the rear part of the TO (behind but contiguous to the CZ) that contains the lines of communication (LOC) and provides supply and evacuation support. Other agencies required for the immediate support and maintenance of field forces may also be located in the COMMZ. The COMMZ spans back to the continental US (CONUS) base, to a supporting combatant commander's AOR, or both.

1-10. An AO is an operational area defined by the joint-force commander for land and naval forces. An AO does not typically encompass the entire operational area of the joint-force command (JFC), but it should be large enough for component commanders to accomplish their mission and protect their forces. Army commanders use control measures to describe AOs and to design them to fit the situation and take advantage of the joint force's capabilities. Commanders typically subdivide the assigned AO by assigning subordinate-unit areas. These subordinate-unit areas may be contiguous or noncontiguous (see Figure 1-2). When friendly forces are contiguous, a boundary separates them. When friendly forces are noncontiguous, the concept of operations links the force's elements, but the AOs do not share a boundary. The intervening area between noncontiguous AOs remains the responsibility of the higher headquarters.

BATTLEFIELD ORGANIZATION

1-11. Battlefield organization is the arrangement of forces according to purpose, time, and space to accomplish a mission. Battlefield organization has both a purpose- and spatial-based framework. The purpose-based framework centers on decisive, shaping, and sustaining (DSS) operations. Purpose unifies all elements of the battlefield organization by providing the common focus for all actions. However, forces act in time and space to accomplish a purpose. The spatial-based framework includes close, deep, and rear areas. Despite the increasing nonlinear nature of operations, there may be situations where commanders describe DSS operations in spatial terms. Typically, linear operations involve conventional combat and concentrated maneuver forces. Ground forces share boundaries and orient against a similarly organized enemy force. In such situations, commanders direct and focus simultaneous DSS operations in deep, close, and rear areas, respectively (see FM 3-0).

1-12. The MP battlefield organization supports every Army echelon, from the Army service component command (ASCC) and the theater-support command

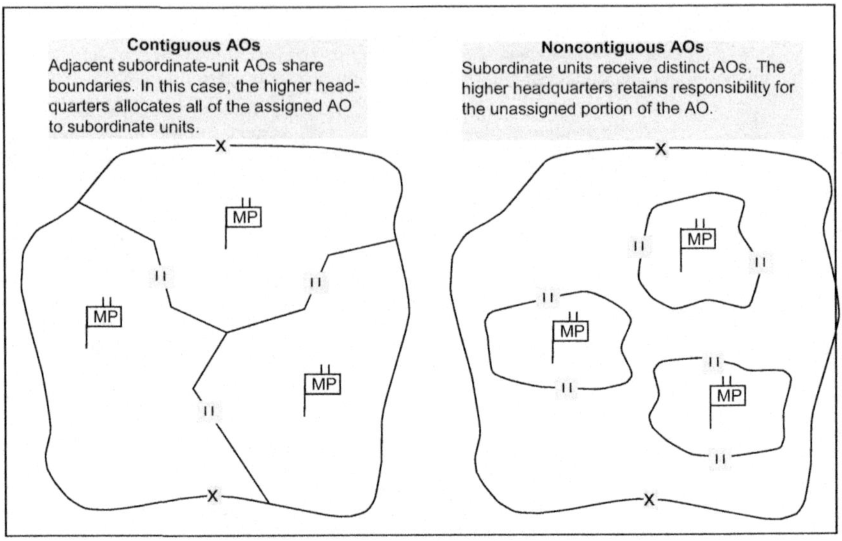

Figure 1-2. Contiguous and Noncontiguous AOs

(TSC) to the maneuver brigade. Regardless of the battlefield organization (purpose or spatial based), MP support to the Army commander is based on available resources and mission, enemy, terrain, troops, time available, and civilian considerations (METT-TC).

1-13. MP support throughout the theater of war may include MP units in the JOA and in the TO. If the combatant commander designates a COMMZ and a CZ within his TO, MP support will come from the established MP modified table of organization and equipment (MTOE) at the subordinate-command echelon. MP support to the JOA is also provided based on METT-TC and available MP assets. Typical MP support may include an I/R brigade liaison detachment (BLD), MP brigades and battalions, a division MP company, a military-working-dog (MWD) team, an L&O team, and a customs team. Figure 1-3, page 1-6, depicts a typical MP organization throughout the TO. In the COMMZ, Figure 1-3 depicts the different types of MP units that are assigned to echelons above corps (EAC) (the ASCC or the TSC). In the CZ, Figure 1-3 depicts the different types of MP units that are assigned to corps, division, and the separate brigades.

SUPPORT IN THE COMMZ

1-14. MP support in the COMMZ is provided by an array of multifunctional MP units. The following MP units provide MP support to EAC:

- The MP brigade (I/R). The MP brigade (I/R) may augment the ASCC or the TSC during wartime. Its mission is to provide command, staff

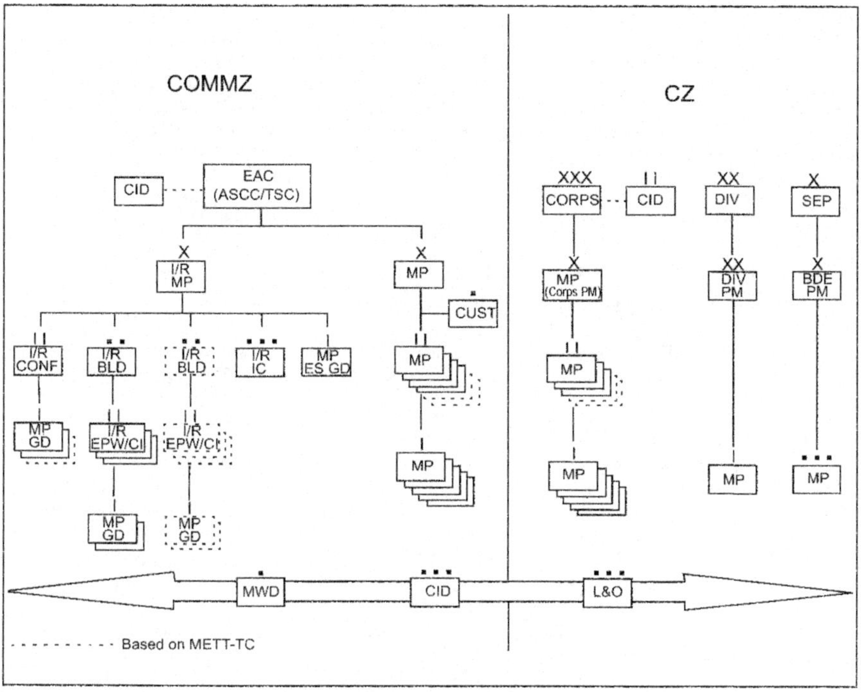

Figure 1-3. MP Structure in the TO

planning, and supervision of I/R operations. This includes coordination with joint and host-nation (HN) agencies, civilian police authority, nongovernmental organizations (NGOs), and US federal agencies.

- The MP brigade (CS). The MP brigade (CS) is assigned to the ASCC or the TSC during wartime (based on METT-TC). The MP brigade (CS) is capable of performing all five MP functions.
- The Criminal Investigation Division (CID) group. The CID group is a stovepipe organization that reports directly to the Commander, US Army Criminal Investigation Command (USACIDC). The CID group provides support to the ASCC and subordinate commands (TSC, corps, or division). See Chapter 9 for further discussion of CID support.

1-15. MP support to other EAC subordinate commands is performed only if MP resources are available. See Chapter 5 for further discussion of MP support to EAC.

SUPPORT IN THE CZ

1-16. MP support is provided in the CZ to each corps, division, and brigade (separate teams or initial/interim brigade combat teams [IBCTs]). An MP brigade (CS) is assigned to each corps, and the MP brigade commander is the corps's provost marshal (PM). A PM and his section, along with an organic division MP company, are assigned to each division. A PM cell and an MP platoon are organic to a separate brigade. A two-person PM cell is organic to the IBCT. The MP units assigned to corps, divisions, and separate brigades are capable of performing all five MP functions. They provide combat, CS, and combat-service-support (CSS) operations within their command's AO.

TYPES OF MILITARY POLICE UNITS

1-17. Most MP units supporting a TO and a JOA are capable of performing all five MP functions. However, the functions must be prioritized based on METT-TC and the availability of MP assets. Current MP structures are designed and tailored to better support the level of command deployed. For example, at the division level, division MP companies are organized as light, heavy, airborne, or air assault and are organic to their respective divisions. The EAC and corps MP brigades and battalions are equally designed to command and control a force mix of up to six battalions or companies. An MP escort-guard and guard company are designed to transport, guard, and provide security to EPWs, civilian internees (CIs), or dislocated civilians. The MP escort-guard company is assigned to the MP brigade (I/R), and the MP guard company is assigned to the MP battalion (I/R).

1-18. MP units can also be tailored and augmented to accomplish multiple, diverse, or specific missions. Customs, L&O, and MWD teams are examples of MP capabilities and flexible responses to a combatant commander's operational requirements. (See Table 1-1, pages 1-9 through 1-19, for a more complete description of MP units. See FM 19-10 and Army Regulation (AR) 190-12 for further information.) The battlefield workload analysis (BWA) is a tool used to determine the number of MP units required to perform some of these multiple missions (see Appendix C).

JOINT, MULTINATIONAL, AND INTERAGENCY OPERATIONS

1-19. In today's environment, the Army will rarely operate or fight alone. The high probability that the Army will operate in concert with its sister services, in an alliance with the forces of foreign nations, or in support of United Nations (UN) operations (when it is committed) is fully reflected in joint doctrine. In such operations, protecting LOC, key facilities, and command and control (C^2) centers will be a shared responsibility. Under this framework, MP units can expect to share the AO with joint, combined, multinational, or interagency resources. MP forces must be prepared to conduct a number of full-spectrum operations with a variety of government and nongovernmental agencies, other services, allied nations, and international agencies.

1-20. Corps and division commanders and staffs must plan (in advance) the transition from a single-service headquarters with joint representation to a joint headquarters capable of functioning as a joint task force (JTF) headquarters. When tasked to form a JTF headquarters, the corps or division must ensure that all of the staff sections and agencies have joint representation (see FMs 100-15 and 71-100). To this end, MP planners must ensure that the JTF is augmented with the appropriate MP forces and with the appropriate echeloned C^2.

1-21. Regardless of the force mix, the MP provide the force with unparalleled, multifunctional capabilities. Among these capabilities is the MP's ability to generate firepower or to handle populations such as EPWs/CIs, dislocated civilians, and refugees. Additionally, MP expertise in investigations and law enforcement enhances the capabilities of other joint, multinational, and interagency police and security forces.

1-22. MP security plans must reflect the joint synergy derived from combining the multiple and diverse capabilities of all participants. To capitalize upon that synergy, MP leaders must keep an open line of communication and coordination to offset the challenges presented by interoperability. Some of these interoperability challenges include—

- Differing political objectives.
- Differing capabilities.
- Cultural/language differences.
- Legal and policy constraints.
- Media impacts.
- Compromise of sensitive processes, procedures, and equipment.
- C^2.
- Communications (digital- and analog-equipment differences).

1-23. MP plans must also accommodate differences in planning capabilities, as well as differences in doctrine, training, and equipment. The intent is to match security missions with force capabilities. MP leaders must understand that operations will often involve multinational teams. While US forces routinely task-organize, this may be more difficult to accomplish with some multinational security forces. This kind of orchestration requires employing standardized procedures, communications, equipment, and liaison within the constraints of operations security (OPSEC).

1-24. Coordination is the key to mission accomplishment in multinational and interagency operations. A military coordination center or a civil-military operations center (CMOC) may meet this coordination requirement. The CMOC provides access for nonmilitary agencies desiring military (to include MP/CID) assistance and coordination. These nonmilitary agencies may include—

- Government organizations (GOs).
- NGOs.
- International organizations (IOs).
- International humanitarian organizations (IHOs).
- HN authorities and agencies.

1-25. The introduction of US Army MP forces in any joint, multinational, or interagency operation is based on METT-TC and the capabilities they bring to the operation. Effective integration of MP forces with other security forces reduces redundant functions, clarifies responsibilities, and conserves resources.

Table 1-1. Unit Descriptions

MP Unit	Mission	Action Elements	Support Role	Capabilities (Include But Are Not Limited to [See TOEs for Details])	Augmentation Needed for Mission Accomplishment
DIVISION MP UNITS					
MP Company (Airborne Division)	Perform the 5 MP functions.	A PM cell and 4 platoons. Each platoon has 3 squads and each squad has 2 three-man teams. Total: 24 three-man teams.	1 platoon deploys with each division brigade and provides DS during the assault phase. Then platoons revert to GS.	During the assault phase, 6 mounted, mobile or fixed-position teams deploy to provide DS for each brigade and 6 teams deploy to provide GS near the division main CP OR after the assault phase, 1 division EPW collection point (6 teams) and 18 mounted, mobile patrols or fixed-position teams provide GS OR 24 mounted, mobile or fixed-position teams are used in any combination.	The division band to provide close-in security for the division CP or to guard EPWs AND the corps CS MP to conduct combat operations within the division rear and provide sustained MMS, area security, special operations support, and escort for evacuating EPWs/CIs between division collection points and to the corps holding area AND the corps L&O MP and CID to conduct sustained L&O operations and criminal investigations AND division/corps transportation assets for evacuation of EPWs **NOTE: Both heavy division companies require corps CS MP augmentation for each forward support platoon (1 corps MP squad per forward support platoon).**
MP Company (Light Infantry Division)	Perform the 5 MP functions.	A PM cell and 3 platoons. Each platoon has 3 squads and each squad has 2 three-man teams. Total: 18 three-man teams	Provide GS for all units in the AO.	Provide screening security for the division main CP (6 teams), the division EPW collection point (6 teams), and 6 mounted, mobile or fixed-position teams.	
MP Company (Heavy Division)* *19333F000	Perform the 5 MP functions.	A PM cell and 2 division support platoons. Each platoon has 3 squads and each squad has a squad leader and 3 three-man teams AND 3 forward support platoons. Each platoon has 2 squads and each squad has a squad leader and 3 three-man teams. Total: 36 three-man teams	2 platoons provide GS and 3 platoons provide DS.	Provide screening security for the division main CP (6 teams), the division EPW collection point (6 teams), and 6 mounted, mobile or fixed-position teams AND EITHER 3 DS platoons providing their brigade with either 1 forward EPW collection point (3 teams) and 3 mounted, mobile or fixed-position teams or 6 teams in any combination OR 18 mounted, mobile or fixed-position teams are used in any combination.	
MP Company (Heavy Division)* *19333L000	Perform the 5 MP functions.	A PM cell and 5 platoons. Each platoon has 2 squads and each squad has 3 three-man teams. Total: 36 three-man teams	3 platoons provide GS and 3 platoons provide DS.	Provide screening security for the division main CP (6 teams), the division EPW collection point (6 teams), and 6 mounted, mobile or fixed-position teams AND EITHER 3 DS platoons providing their brigade with either 1 forward EPW collection point (3 teams) and 3 mounted, mobile or fixed-position teams or 6 teams in any combination OR 18 mounted, mobile or fixed-position teams are used in any combination.	

Table 1-1. Unit Descriptions (continued)

MP Unit	Mission	Action Elements	Support Role	Capabilities (Include But Are Not Limited to [See TOEs for Details])	Augmentation Needed for Mission Accomplishment
MP Company (Air Assault Division)	Perform the 5 MP functions.	A PM cell and 4 platoons. Each platoon has 3 squads and each squad has 2 three-man teams. Total: 24 three-man teams	Provide GS for all units in the AO.	24 mounted, mobile or fixed-position teams are used in any combination.	(See augmentation needs on previous page.)
MP COMBAT SUPPORT UNITS					
HHC MP Brigade	Provide C2 and coordinate the combat, CS, and CSS of all MP elements assigned or attached.	Command and staff elements and long-range planning section	Provide command, control, planning, and supervision for up to 6 MP . battalions and other assigned or attached subordinate HQ.	Provide continuous command, control, coordination, planning, and supervision for subordinate units.	None
HHD MP Battalion (Combat Support)	Provide C2 and coordinate the combat, CS, and CSS of all MP elements assigned or attached.	Command and staff elements plus a support element	Provide command, control, planning, and supervision for up to 6 MP companies and other assigned or attached subordinate HQ.	Provide continuous command, control, coordination, planning, and supervision for subordinate units.	None
MP Company (Combat Support)	Perform the 5 MP functions.	4 platoons with 3 squads. Each squad has a squad leader and 3 three-man teams. Total: 36 three-man teams	Provide GS for all units in the AO.	Provide security for a unified or combined HQ or higher HQ or 1 main CP and 1 tactical CP OR operate 1 EPW holding area (9 teams) and 27 mounted, mobile or fixed-position teams for use in any combination.	Corps band assets to guard EPWs AND corps transportation for evacuation of EPWs AND external MP to escort EPWs
MP Company (Combat Support)	Perform the 5 MP functions.	4 platoons with 3 squads. Each squad has a squad leader and 3 three-man teams. Total: 36 three-man teams	Provide GS for all units in the AO.	Provide security for a unified or combined HQ or higher HQ or 1 main CP and 1 tactical CP OR operate 1 EPW holding area (9 teams) and 27 mounted, mobile or fixed-position teams for use in any combination.	Corps band assets to guard EPWs AND corps transportation for evacuation of EPWs AND external MP to escort EPWs
MP Company Arctic Support Augmentation Detachment	*Augment MP* units when operating in an arctic environment.	*1 mechanic* (63B10)	Provide mobility support to MP units.	Provide support as designated by the commander.	None
MP Detachment (C2)	Provide command and administrative personnel for MP law enforcement teams.	19503LA has a platoon HQS; 3 individuals. 19503LD has a commander and support personnel; 8 individuals.	Provide L&O. 19503LA provides platoon-level C2. 19503LD provides company-level command, control, and support.	Commands and controls law enforcement teams, investigations teams, MWD teams, and physical-security teams.	None

Table 1-1. Unit Descriptions (continued)

MP Unit	Mission	Action Elements	Support Role	Capabilities (Include But Are Not Limited to [See TOEs for Details])	Augmentation Needed for Mission Accomplishment
MP Detachment (Old Guard)	Perform the 5 MP functions.	4 squads. Each squad has 3 three-man teams. Total: 12 teams	Support the HHC old guard battalion.	Provide support as directed by the commander.	None
MP LAW AND ORDER (L&O) UNITS					
MP HQ Team	Provide C2, personnel administration, and logistical support to attached L&O augmentation teams.	Commander and support personnel	Provide L&O.	Provide C2, personnel administration, and logistical support to attached L&O augmentation teams.	None
MP Operations Team	Provide mission supervision, staff planning, and technical supervision required to support the L&O mission.	L&O operations officer and operations staff	Provide L&O.	Provide overall mission supervision, planning, employment, and coordination of support. Provide technical supervision and administrative support of MPI, TAI, and force protection MP. Provide an evidence custodian/police intelligence NCO to be responsible for confiscated property used in criminal offenses and to receive and collect the initial analysis of criminal/operational information/intelligence.	None
MP Desk or Desk and Record Team	Serve as the primary control point for L&O-related incidents and operations initiated by MP patrols.	2 three-man teams	Provide L&O.	Receive/record complaints, dispatch investigative persons, maintain control of offenders/detained persons, maintain the status of investigations, prepare reports, and provide criminal data to the police intelligence NCO for analysis.	None
MP Traffic Accident Investigations Team	Provide the technical capability to investigate traffic accidents.	1 two- or three-man team: 19517AD00 has 1 two-man team. 19523LE00 has 1 three-man team.	Provide L&O.	Augment MP capabilities to enforce MSR regulations; man checkpoints, roadblocks, and dismount points; patrol traffic areas; and perform traffic escort duties.	The 19517AD00 accident investigation team is dependent on the supported unit for transportation.
MP Investigation Team	Provide the technical capability to investigate criminal incidents and conduct surveillance operations.	One, two, or three-man teams: 19517AE has 1 two-man team. 19533LA has 1 one-man team. 19533LB has 1 one-man team. 19533LC has 1 three-man team.	Provide L&O.	Investigate criminal incidents, conduct surveillance, work with HN military and civilian police, and collect criminal and operational intelligence. 19533LA00 provides supervision for the teams. 19533LB00 provides investigative support for populations of not less than 700 troops. 19533LC00 provides investigative support for populations of not less than 2,100 troops.	Dependent on the supported unit for transportation

Table 1-1. Unit Descriptions (continued)

MP Unit	Mission	Action Elements	Support Role	Capabilities (Include But Are Not Limited to [See TOEs for Details])	Augmentation Needed for Mission Accomplishment
MP Force Protection and Physical-Security Teams	Provide the expertise and technical capability to assist units in safeguarding personnel, equipment, and facilities.	One, two, or three-man teams: 19517AF has 1 two-man team. 19593LA has 1 three-man team. 19593LB has 1 one-man team.	Provide L&O and area security.	Conduct vulnerability assessments and identify and prioritize critical facilities and key terrain within the AO. Identify mission-essential activities that are vulnerable to criminal acts or disruptive activities. Ensure that activities are inspected to determine if safeguards are adequate. Assist base/base cluster commanders with the development of internal defense plans.	The 19517AF is dependent on the 19517AB operations team for transportation.
MP Detachment (Patrol Supervision Team)	Provide supervision for 3 MP teams.	1 MP NCO	Provide L&O.	Plan, direct, and supervise the employment of assigned or attached MP teams.	Dependent on the supported unit for transportation
MP Detachment (Motor and Dismounted Patrol Team)	Perform L&O operations.	1 three-man team	Provide L&O.	Protect designated personnel or facilities. Perform route and area reconnaissance, enforce MSR regulations, provide refugee and straggler control, and disseminate information. Perform peacekeeping operations to maintain surveillance over an area, observe activities, and report findings. Preserve or establish L&O. Apprehend absentees or deserters (US military personnel) in conjunction with civil law enforcement agencies.	The dismounted patrol team is dependent on the supported unit for transportation, when required.
MP Detachment (Registration Team)	Perform vehicle and firearm registration.	1 three-man team	Provide L&O.	Service up to 10,000 personnel for registration of individuals, vehicles, and firearms on a closed post or area. Issue credentials prescribing limits of circulation and privileges. Process up to 50 personnel daily for fingerprints and photography services.	Dependent on the supported unit for transportation
Senior Military Customs Inspector Supervisor	Perform technical supervision, staff planning, and coordination for customs inspection teams.	1 MP NCO	Provide L&O.	Supervise two senior military customs teams to ensure that personnel, equipment, and material meet customs, immigration, Department of Agriculture, and other federal agency requirements for units and personnel redeploying to the US.	None

Table 1-1. Unit Descriptions (continued)

MP Unit	Mission	Action Elements	Support Role	Capabilities (Include But Are Not Limited to [See TOEs for Details])	Augmentation Needed for Mission Accomplishment
Senior Military Customs Inspection Team	Conduct customs inspections and train and supervise redeploying unit personnel to augment US customs inspectors.	1 five-man team	Provide L&O.	Conduct inspections and advise redeploying units. Train selected individuals from the redeploying unit to augment US customs efforts. NOTE: One team supports one port of embarkation.	The redeploying unit undergoing customs inspection, to augment US Customs efforts. The number of personnel is proportional to the size of the unit.
MILITARY WORKING DOG (MWD) UNITS					
MWD Kennel Master Team	Supervise MWD teams.	One- or two-man team. 19537AA has 1 two-man team. 19583LG has 1 one-man team.	Provide L&O, area security, and I/R operations support.	Supervise, plan, and coordinate MWD operations and support requirements. Provide MWD handler/ canine proficiency certification.	None
MWD Explosive, Narcotics, and Patrol Team	Detect explosives and controlled substances and search for, detect, and control personnel in support of crime scene searches. Provide personal protection, MOUT, health and welfare inspections, and I/R and customs operations.	One-man/one-dog team. The 19537A000-series TOEs have 3 teams each. The 19583L000-series TOEs have 1 team each.	Provide L&O, area security, and I/R support.	The 19537A-series dog teams are capable of providing a 24-hour explosive, narcotics, and tracking handler/dog requirement or three concurrent short-duration missions each requiring 1 handler/dog team. The duration of the missions will vary based on the climate, the environment, and the individual dog's ability. The 19583L-series dog teams provide short- duration missions based on the dog's ability. Explosive and narcotics teams provide MWD patrol support when not employed with explosive and narcotics detection.	None
MILITARY POLICE INTERNMENT/RESETTLEMENT (I/R) UNITS					
MP Command (I/R)	Provide command, control, staff planning, and supervision of I/R operations performed by all assigned or attached elements.	Command and staff elements	Provide I/R operations support.	Provide command, control, and staff planning for 2 or more MP brigades.	MP IRIC AND MP EPW evacuation detachment

Table 1-1. Unit Descriptions (continued)

MP Unit	Mission	Action Elements	Support Role	Capabilities (Include But Are Not Limited to [See TOEs for Details])	Augmentation Needed for Mission Accomplishment
MP Brigade (I/R)	Provide command, control, and staff planning for I/R operations performed by all assigned or attached MP elements.	Command and staff elements	Provide I/R operations support.	Provide command, control, and staff planning for 2 to 7 MP I/R battalions or up to 21 MP I/R battalions when augmented by 7 brigade liaison detachments (BLDs). Plan and provide staff supervision of I/R collection and evacuation operations. Coordinate with HN military territorial organizations and civilian police authorities, nongovernment organizations, private volunteer organizations, and US federal agencies on I/R matters. Provide coordination and support for out-of-theater evacuation of EPWs, if required. Support posthostilities operations.	MP IRIC AND MP EPW evacuation detachment AND MP I/R BLD (TOE 19453LD00) **NOTE: The BLD expands the staff planning and coordination capabilities on a ratio of 1 BLD to 3 MP I/R battalions.**
MP Battalion (I/R)	Provide command, staff planning, administration, and logistical support for the operation of an I/R facility.	Command and staff elements	Provide I/R operations support.	Operate an internment facility for either EPWs/CIs or DCs or a confinement facility for US prisoners. (Never more that one category at the same time.)	MP detachment (I/R) (EPW/CI) for EPW/CI/DC internment missions AND MP detachment (I/R) (confinement) for US prisoner confinement missions AND MP company (EG) for security of the evacua-tion and/or movement of US prisoners or EPWs/CIs/DCs AND MP company (guard) for prisoner guard services AND Corps/EAC transportation assets for movement of EPWs/CIs/DCs

Table 1-1. Unit Descriptions (continued)

MP Unit	Mission	Action Elements	Support Role	Capabilities (Include But Are Not Limited to [See TOEs for Details])	Augmentation Needed for Mission Accomplishment
MP Detachment (I/R) (EPW/CI)	Augment the MP battalion (I/R) to provide supervision, administration, combat health support, and logistical support for operating an internment facility interning EPWs/CIs or housing DCs.	2 compound control sections (2 three-man teams), a work project section (1 two-man team), a personnel section, and a supply section.	Provide I/R operations support.	Provide command, control, and supervision of up to 1,000 EPWs/CIs or 2,000 DCs. Provide supply and subsistence support (to include supervision of food preparation) for up to 1,000 EPWs/CIs or 2,000 DCs. Direct the activities related to assigning and supervising work projects for EPWs/CIs. Augment the MP I/R battalion in processing and maintaining records for up to 1,000 EPWs/CIs or 2,000 DCs.	MP company (guard) for prisoner guard services
MP Company (Escort Guard)	Provide supervisory and security personnel for evacuating and/or moving EPWs/CIs.	4 platoons with 3 squads each. Each squad has a squad leader and 3 three-man teams. Total: 36 teams	Provide I/R operations support.	Provide security for the movement of the following numbers of POWs or CIs by the methods indicated: Marching - 1,000 to 1,500. Vehicle - 1,500 to 2,000. Rail - 2,000 to 3,000. Air - 2,125 to 2,406.	Corps/EAC transportation assets for movement of EPWs/CIs
MP Company (Guard)	Provide guards for EPWs/CIs or US prisoners, installations, and facilities.	3 platoons with 3 squads each. Each squad has a squad leader and 2 five-man teams Total: 18 teams.	Provide I/R operations support.	Provide security for a confinement facility containing up to 500 US military prisoners. Provide guards for securing 1 EPW/CI compound containing up to 2,000 EPWs/CIs or 4,000 DCs. Provide security guards for 3 railway terminals, each having up to 8 tracks. (Includes EPW/CI rail movement operations and protection of sensitive material within the terminal.) Provide guards for 1 military installation or facility up to 240,000 square yards in size, containing sensitive material. (This unit can provide guards for material transit.)	None
MP I/R Processing Squad	Process EPWs/CIs.	1 squad leader and 1 eight-man processing squad	Provide I/R operations support. Augment the I/R battalion processing capability to meet operational requirements.	Provide processing capabilities of about 8 EPWs/CIs per hour.	None

Table 1-1. Unit Descriptions (continued)

MP Unit	Mission	Action Elements	Support Role	Capabilities (Include But Are Not Limited to [See TOEs for Details])	Augmentation Needed for Mission Accomplishment
I/R Brigade Liaison	Expand the MP brigade (I/R) or ASCC C2 capabilities.	Liaison officer and support staff	Provide I/R operations support.	When assigned to the MP I/R brigade, provide staff augmentation that expands the brigade's staff planning, coordination, and C2 capabilities for 3 MP I/R battalions. When assigned to the ASCC, provide I/R staff augmentation and a liaison link with allied/HN forces to ensure that the care and handling of US captured EPWs/CIs is in compliance with the Geneva Conventions.	None
I/R Camp Liaison Team	Provide continuous accountability of EPWs/CIs captured by US forces and transferred to an allied/HN for internment.	Liaison officer and support staff	Provide I/R operations support.	Provide advice, as requested, to commanders and staffs of allied/HN-operated internment facilities. Verify arrival, forward records, and provide continuous accountability for US captured EPWs/CIs interned in allied/HN facilities. Monitor prisoner treatment to ensure compliance with the Geneva Conventions. Receive/certify allied/HN requests for reimbursement of expenses associated with interning EPWs/CIs captured by US forces.	None
I/R Processing Liaison Team	Process and verify the transfer of EPWs/CIs captured by US forces to a allied/HN facility.	Liaison officer and support staff	Provide I/R operations support.	Provide processing and transfer capability for US-captured prisoners to an allied/HN EPW/CI facility.	None
MP Detachment (EPW Evacuation)	Provide evacuation support of EPWs.	EPW coordination officer and staff plus 10 one-man POD/POE teams and 3 two-man POE teams	Provide I/R operations support.	Coordinate administrative, logistical, and transportation support for up to 7 PODs and 6 POEs used for out-of-theater EPW evacuation and the escort guard personnel that support the movements. Coordinate for theater security at PODs/POEs and intermediate refuel points.	Responsible POD/POE security forces

Table 1-1. Unit Descriptions (continued)

MP Unit	Mission	Action Elements	Support Role	Capabilities (Include But Are Not Limited to [See TOEs for Details])	Augmentation Needed for Mission Accomplishment
MP Detachment (Internment Resettlement *Information* Cener [IRIC])	Provide a US central agency in each theater of war to *receive,* process, maintain, disseminate, and transmit the required information and data relating to EPWs, American POWs, CIs, and DCs within the theater.	Command and staff personnel	Provide I/R operations support. Serve as the single source for collection *and storage of* EPW/CI information in theater. Forward information to the National Prisoner of War Information Center at DA.	Collect, process, and disseminate (to authorized agencies) information regarding interned personnel detained in theater and those released to the custody of allied/HN authorities within theater. *Receive, document, and disseminate to the theater* commander information received regarding American POWs, CIs, and foreign nationals who are captured, missing, or otherwise detained. Receive, store, and dispose of personal property belonging to interned personnel who have died, escaped, or been repatriated and any property belonging to enemy soldiers killed in action that is not disposed of through grave registration channels.	None
MP Detachment (I/R) (Confine-ment)	Augment an MP battalion (I/R) in operating a confinement facility. Provide supervision, administration, combat health support, and logistical support of US military prisoners.	Corrections officer and confine-ment facility staff	Provide I/R operations support.	Provides a control team for the command, control, and supervision of US military prisoners and a disci-plinary guard team to supervise custodial personnel for a maximum custody of 500 prisoners. Provide administration, health service, supply, and food ser-vice personnel to augment the battalion in providing these services for internees. (The food service per-sonnel supervise and train US military prisoners working in the internee dining facility.)	MP battalion (I/R) for command, staff, plan-ning, and operational support AND MP company (guard) for prisoner guard services

Table 1-1. Unit Descriptions (continued)

MP Unit	Mission	Action Elements	Support Role	Capabilities (Include But Are Not Limited to [See TOEs for Details])	Augmentation Needed for Mission Accomplishment
MP Confinement Facility Detachment	Provide C2, staff planning, administration, and logistical support for a confinement facility for US military prisoners.	Commander and facility staff	Provide I/R operations support.	Provide command, control, and administrative support for a confinement facility.	MP guard company to provide exterior perimeter security and required guard support to the confinement facility AND MP correctional team supervisor (TOE19553LE) to provide supervision of the guard force AND MP correctional team work supervisor (TOE 19553LF) to provide supervision of prisoners
MP Correctional Team Supervisor	Provide correctional supervision of prisoner work guard forces within a confinement facility.	2 two-man teams	Provide I/R operations support.	Provide correctional supervision to guard force personnel and US military prisoners within a modular-configured confinement facility.	None
MP Correctional Team Work Supervisor	Provide direct control and supervision of military prisoners within a confinement facility.	1 two-man team	Provide I/R operation support.	Provide correctional control, supervisory functions, and escort duties for military prisoners within a modular-configured confinement facility.	None
MILITARY POLICE CRIMINAL INVESTIGATION (CID) UNITS					
MP Group (CID)	Provide command, control, staff planning, and supervision for all CID elements within the theater.	Command and staff elements	Provide L&O.	Provide command, control, staff planning, and supervision over all CID elements within the theater. **NOTE: The Criminal Investigation Command is a stovepipe organization.**	None

Table 1-1. Unit Descriptions (continued)

MP Unit	Mission	Action Elements	Support Role	Capabilities (Include But Are Not Limited to [See TOEs for Details])	Augmentation Needed for Mission Accomplishment
MP Battalion (CID)	Provide command, control, staff planning, and supervision for all assigned or attached CID elements.	Command and staff elements	Provide L&O.	Provides command, control, staff planning, and supervision over all assigned and attached CID elements.	None
MP Detachment (CID) HQ Cell	Provide C2, evidence custody control, and investigative administration support. Coordinate for personnel administration and logistic support.	Commander and support staff	Provide L&O.	Provide C2 of assigned CID SA sections and/or teams performing all CID operations in their area of responsibility.	None
MP Detach-ment (CID)	Provide criminal investigative support to Army commanders at all echelons.	2-man teams. The number of teams varies by organization and is as follows: DSE heavy - 4 DSE light - 3 Section A - 4 Section B - 4 Supervisory team - 1 Senior team - 1 SA team - 1	Provide L&O.	Supervise and conduct criminal investigations.	When directed, MP CS, L&O, and MWD support

Chapter 2

Battle Command

Command is the authority a commander in military service lawfully exercises over subordinates by virtue of rank and assignment. Leaders possessing command authority strive to use it with firmness, care, and skill.

FM 101-5-1

Battle command is the exercise of command in an operation against a hostile, thinking opponent. Battle command includes visualizing the current state and the desired end state, then formulating concepts of operations to get from one state to the other at the least cost. In addition to visualizing and formulating concepts, battle command encompasses assigning missions; prioritizing and allocating resources; selecting the critical time and place to act; and knowing how and when to make adjustments in the fight. Battle command enables MP commanders to lead, prioritize, and allocate assets required in support of the Army commander. MP commanders must observe, orient, decide, and act on their decisions quickly. Information is the key element in the battle-command process; therefore, the commander must have accurate and timely information upon which to base his decisions.

OVERVIEW

2-1. The battle command of MP units is typically decentralized due to the nature of their CS functions, METT-TC, and the needs of the Army commander. This places the burden of sound, timely decision making to the lowest levels. MP leaders must develop a keen sense of situational awareness and visualization, and they must constantly track the actions of supported units.

BATTLEFIELD VISUALIZATION

2-2. The ability to visualize the battlefield is a critical element of battle command. Battlefield visualization is an essential leadership attribute and is critical to accomplishing the mission. It is learned and attained through training, practice, experience, technical and tactical knowledge, and available battle-command technologies. It results when the MP commander understands the higher commander's intent, his assigned mission, the enemy, and the friendly force's capabilities and limitations. See Appendix D for further information on command technologies.

2-3. Battlefield visualization includes the MP commander's view of what his forces will do and the resources needed to do the mission. He envisions a sequence of actions that will cause his MP forces to perform at the desired end state. Ultimately, the MP commander's battlefield vision evolves into his intent and helps him develop his concept of operations.

COMMANDER'S INTENT

2-4. The commander's intent is a key part of Army orders. It is a clear, concise statement of what the force must do to succeed with respect to the enemy, the terrain, and the desired end state. It provides the link between the mission and the concept of operations by stating key tasks. These tasks, along with the mission, are the basis for subordinates to exercise initiative when unanticipated opportunities arise or when the original concept of operations no longer applies. MP leaders at all echelons must ensure that the mission and the commander's intent are understood two echelons down (see FM 101-5).

CONCEPT OF OPERATIONS

2-5. The commander's intent does not include the method by which the MP units will accomplish the mission. This method is called the commander's concept of operations. It must—

- Convey the commander's vision of how to accomplish the mission in a manner that allows his subordinates maximum initiative.
- Build around intelligence gathering and the precise employment of MP resources.
- Provide the basis for task organization, scheme of maneuver, terrain organization, tasks to subordinates, and synchronization.

COMMAND AND CONTROL RELATIONSHIPS

2-6. MP units are assigned to, attached to, or placed under the operational control (OPCON) of MP or other units they support. OPCON is the authority to perform command functions over subordinate forces. This includes organizing and employing commands and forces, assigning tasks, designing objectives, and giving authoritative direction necessary to accomplish the mission. MP C^2 relationships may be changed briefly to provide better support for a specific operation or to meet the needs of the supported commander. MP units may be placed under the OPCON of another unit commander for short-term operations. The MP unit remains in this relationship only as long as it is needed for that operation.

> MP support to the Bosnian municipal elections consisted of one division and two corps MP companies. These MP assets, attached to Task Force (TF) Eagle, were task-organized from different sources. The division MP company and the PM cell were organic to TF Eagle's mechanized infantry division headquarters, but the two corps MP units were from US Army Forces Command (FORSCOM) MP battalions in CONUS.

SUPPORT RELATIONSHIPS

2-7. MP units on the battlefield provide two types of support—general support (GS) and direct support (DS). Corps and EAC MP units provide GS to their respective corps/EAC subordinate commands. Light, airborne, and air-assault MP companies provide GS to their respective divisions. Heavy-division MP companies provide GS to the division rear and DS to the division's subordinate brigades.

STAFF RELATIONSHIPS

2-8. The PM for each level of command is that command's advisor on MP combat, CS, and CSS operations. The PM—

- Advises the commander and staff about MP abilities/capabilities.
- Supervises the preparation of plans and dictates policies.
- Coordinates MP operations.
- Assists and supervises the interaction of supporting and supported units.
- Reviews current MP operations.
- Coordinates with allied forces and HN military and civil police.
- Ensures that MP plans and operations supporting the commander's tactical plan are carried out.
- Recommends when and where to concentrate the command's MP assets.
- Supervises or monitors MP support in the command's AO.

2-9. The PM works daily with the commander and staff officers who employ MP resources and whose AORs influence MP support. The PM works closely with the coordinating staff at the appropriate command level to coordinate MP support. He ensures that MP planning is practical and flexible, that plans are coordinated with staff sections and subordinate commands, and that plans reflect manpower and resources needed by MP. (This includes the need for C^2, fire support, equipment, and supplies. It also includes construction, communication, transportation, and aviation support.) As new information is received, the PM reviews, updates, and modifies the plans. He ensures that the echelon commander gets the necessary MP support.

2-10. In the absence of specific directions or orders, the PM plans the use of MP assets. He evaluates the current operations and projects the future courses of action (COAs). He bases his plans on assumptions consistent with the commander's intent and a thorough knowledge of the situation and mission. The PM considers—

- METT-TC.
- Current estimates developed by the intelligence preparation of the battlefield (IPB) and the police information assessment process (PIAP).
- The environment within the AO. This includes the climate, the terrain, and obstacles. It also includes the legal authority and status of the force; the width, depth, size, and location of built-up areas; and the attitudes and abilities of the local populace.

- The types of units operating in the area (to include joint, combined, multinational, and interagency units) and the missions and capabilities of these units. This knowledge is imperative to understand their capability to counter threats in their area.
- The specific missions of MP units in the area and the impact that rear-area security operations will have on the ability of these units to perform other functions.
- Personnel, vehicles, and equipment in the MP units.

2-11. Coordination and communication between the PM and Army commanders is essential. Such actions ensure timely and efficient MP support to all levels of command during any operation. The informal, technical chain of coordination is an open line of communication between PMs at different echelons. The informal chain of coordination fosters cooperation and help among the MP elements at each echelon. For instance, when the division PM needs more assets to accomplish added missions, he initiates coordination with the corps PM. If the corps PM can provide support, the division PM formalizes his request for assistance through the division Assistant Chief of Staff, G3 (Operations and Plans) (G3).

Chapter 3

The Threat

In the 40-odd years of the Cold War, in many locations around the world, the Army performed a deterrent role as part of the containment strategy. In other places, at other times, the Army fulfilled the Nation's expectation in operations too small to be called "wars," although no less dangerous. To the soldier on the ground, Operations Urgent Fury in Grenada and Just Cause in Panama were indistinguishable from combat operations of their forefathers. Operations Provide Comfort in Iraq and Restore Hope in Somalia, although peace operations, also proved to be dangerous.

FM 100-1

The end of the Cold War has reduced, but not eliminated, the most immediate threat to the security of the US and other western nations. However, the absence of a dominant, identifiable threat has produced a far more complex and confusing strategic environment than the one that was present during the Cold War. Forward-deployed and CONUS-based ARFOR and civilians are and will continue to be engaged in a range of military actions. These actions stem from deterring conflicts to conducting peacetime engagement operations to providing support to civil agencies at home and abroad.

OVERVIEW

3-1. During the past decade, the US has deployed forces in multiple operations that have included crisis response in combat situations as well as participation in noncombat activities. The Army's presence in South Korea, Saudi Arabia, and Kuwait and its deployments to Somalia, Haiti, Bosnia, and Kosovo are clear indicators that the military must be prepared to face not only the traditional threat, but also a nontraditional, nonecheloned enemy. To support Army commanders successfully, MP leaders must understand the nature and complexity of these threats and how they can potentially affect the desired strategic, operational, and tactical end states.

REAR-AREA AND SUSTAINMENT OPERATIONS

3-2. The rear area for any particular command is the area extending forward from its rear boundary to the rear of the area assigned to the next lower level of command. This area is provided primarily for the performance of support functions. Operations in the rear area assure freedom of action and continuity of operations, sustainment, and C^2. Sustainment operations are those that enable shaping and decisive operations by assuring freedom of action and continuity of operations, CSS, and C^2 (see FM 3-0). Sustainment operations include the following elements:

- CSS.
- Rear-area and base security.
- Movement control.
- Terrain management.
- Infrastructure development.

3-3. During the Cold War, the danger to rear areas included forces that would be deployed in support of major soviet-style operations. The adversaries using the soviet model could be expected to engage in intense combat activity in their enemy's rear area. Their forces were prepared to penetrate into the enemy's rear and to attack and destroy its reserve forces and rear-area installations. To protect the rear areas, the MP were among the first mobile fighting forces available to the battlefield commander and thus, a source of combat power. Today, the Army commander uses the MP's flexibility and their modular-force training, adaptability, and mobility to serve as a combat multiplier throughout his entire AO. During sustainment operations, the MP perform all functions to ensure freedom of maneuver in support of the overall operational effort.

3-4. Failure to protect our forces during sustainment operations normally results in failure of the entire operation. Sustainment operations determine how fast ARFOR reconstitute and how far they can exploit success. The likelihood of MP units encountering the enemy and engaging in direct combat (not only in the rear area, but also during sustainment operations) cannot be underestimated.

3-5. Threats to rear-area and sustainment operations exist throughout the full spectrum of military operations. These threats may be related or independently engaged, but their effects are frequently cumulative. Threats to rear-area and sustainment operations are usually theater-dependent and are not limited to those outlined in this manual. Joint Publication (JP) 3-10 further discusses the threat in the rear area. Although JP 3-10 defines the threat in the context of a JRA, MP leaders can expect the same level of activity anywhere that US forces are deployed.

RECEPTION, STAGING, ONWARD MOVEMENT, AND INTEGRATION OPERATIONS

3-6. Reception, staging, onward movement, and integration (RSOI) operations consist of essential and interrelated processes in the AO that transform arriving personnel and materiel into forces capable of meeting operational requirements. During RSOI operations, the threat encountered will depend mostly on the type of entry, the nature of the operation, and the enemy. During major contingencies, forces deploy from power-projection platforms within the US or forward bases. The PM must plan MP support during the initial stages of the deployment to ensure the protection of follow-on forces and the detection of potential threats (see FM 100-17-3).

3-7. MP support to RSOI operations includes, but is not limited to—

- Conducting AS operations to counter or prevent enemy actions against marshalling and staging areas.
- Conducting convoy, airport, and rail security operations.
- Conducting populace- and resource-control operations.
- Conducting other physical-security and force-protection measures.
- Conducting other MP functions (as determined by the PM).

THREAT LEVELS

3-8. The threat is divided into three levels. These levels provide a general description and categorization of threat activities, identify the defense requirements to counter them, and establish a common reference for planning guidelines. MP leaders must understand that this does not imply that threat activities will occur in a specific sequence or that there is a necessary interrelationship between each level.

LEVEL I

3-9. Level I threats include the following types of individuals or activities:

- **Enemy-controlled agents.** Enemy-controlled agents are a potential threat throughout the rear area. Their primary missions include espionage, sabotage, subversion, and criminal activities. Their activities span the range of military operations and may increase during both war and military operations other than war (MOOTW). These activities may include assassinating or kidnapping key military or civilian personnel or guiding special-purpose individuals or teams to targets in the rear area.
- **Enemy sympathizers.** Civilians sympathetic to the enemy may become significant threats to US and multinational operations. They may be the most difficult to neutralize because they are normally not part of an established enemy-agent network, and their actions will be random and unpredictable. During war and MOOTW, indigenous groups sympathetic to the enemy or those simply opposed to the US can be expected to provide assistance, information, and shelter to guerrilla and enemy unconventional or special-purpose forces operating in the rear area.
- **Terrorism.** Terrorists are among the most difficult threats to neutralize and destroy. Their actions span the full spectrum of military operations.
- **Civil disturbances.** Civil disturbances, such as demonstrations and riots, may pose a direct or indirect threat to military operations. Although this threat may not be of great impact during war, it may significantly change and affect MOOTW.

LEVEL II

3-10. Level II threats include the following types of forces:

- **Guerilla forces.** Irregular and predominantly indigenous forces conducting guerrilla warfare can pose a serious threat to military forces and civilians. They can cause significant disruptions to the orderly conduct of the local government and services.

- **Unconventional forces.** Special-operations forces (SOF) are highly trained in unconventional-warfare techniques. They are normally inserted surreptitiously into the rear area before the onset of an armed conflict. They establish and activate espionage networks, collect intelligence, carry out specific sabotage missions, develop target lists, and conduct damage assessments of targets struck.
- **Small tactical units.** Specially organized reconnaissance elements are capable of conducting raids and ambushes in addition to their primary reconnaissance and intelligence-gathering missions. Small (size or capability), bypassed conventional units, as well as other potential threat forces, are also capable of conducting raids and ambushes to disrupt operations.

LEVEL III

3-11. Level III threats are made up of conventional forces. Potential threat forces are capable of projecting combat power rapidly by land, air, or sea deep into the rear area. Specific examples include airborne, heliborne, and amphibious operations; large, combined-arms, ground-forces operations; and bypassed units and infiltration operations involving large numbers of individuals or small groups infiltrated into the rear area, regrouped at predetermined times and locations, and committed against priority targets. Level III forces may use a combination of the following tactics as a precursor to a full-scale offensive operation:

- **Air or missile attack.** Threat forces may be capable of launching an air or missile attack throughout the rear area. It is often difficult to distinguish quickly between a limited or full-scale attack before impact; therefore, protective measures will normally be based on the maximum threat capability.
- **Nuclear, biological, and chemical (NBC) attack.** Commanders must be aware that NBC munitions may be used in conjunction with air, missile, or other conventional-force attacks. The NBC weapons could also be used at Level I or II by terrorists or unconventional forces in order to accomplish their political or military objectives.

THREAT-LEVEL MATRIX

3-12. Table 3-1 lists the threat levels and their likely appropriate responses. The threat levels listed are based on the type of threat. The table should not be construed as restricting the response options to any particular threat.

THREAT PRIORITIES

3-13. The threat will attempt to perform the following operations against targets in the rear area:

- Detect and identify targets.
- Destroy or neutralize operational weapons-system capabilities.
- Delay or disrupt the timely movement of forces and supplies.

Table 3-1. Threat Levels

Threat Level	Example	Response
I	Agents, saboteurs, sympathizers, and terrorists	Unit, base, and base-cluster self-defense measures
II	Small tactical units, unconventional-warfare forces, guerrillas, and bypassed enemy forces	Self-defense measures and response forces with supporting fires
III	Large tactical-force operations (including airborne, heliborne, amphibious, infiltration, and bypassed enemy forces)	Timely commitment of a TCF

- Weaken the friendly force's C^2 network.
- Disrupt support to combat forces.
- Set the stage for future enemy operations.
- Create panic and confusion throughout the rear area.

3-14. Typical examples of enemy priority targets include the following:

- NBC-weapons storage sites and delivery systems.
- Key command, control, communications, computers, and intelligence (C4I) facilities.
- Air-defense artillery (ADA) sites.
- Airfields and air bases.
- Port facilities.
- Main supply routes (MSRs) and MSR checkpoints.
- Key LOC.
- Reserve assembly areas (AAs).
- Troop barracks.
- Critical civilian and logistics facilities.

THREAT LOCATION

3-15. The fact that the Cold War has ended does not imply that our traditional threat has ended. North Korea and Iraq are constant reminders of this fact. For the near future, Army commanders will fight units with Cold-War-era equipment and tactics. The Army trains and is prepared to fight an enemy capable of interfering with our freedom of maneuver throughout the battlefield. On an extended battlefield with asymmetric threats, the danger to high-value assets (HVAs) (including CSS, C^2, communication nodes, and MSRs) only increases. The idea that the danger to the rear area decreases as you travel farther away from the forward edge of the battle area (FEBA) is not true. Threat intensity does not depend on geographical location; it depends on what operations the enemy believes must be initiated (and to what degree) to achieve its objective in the rear area. Military commanders depend on the MP to delay and defeat threats in their AO with a mobile reaction force.

COMMUNICATIONS ZONE

3-16. The nature of the COMMZ will encourage Level I and II threats to concentrate along the LOC and other areas of military significance. MP units will encounter an enemy that is capable of disrupting operations throughout the COMMZ while employing terrorist activities, enemy-controlled agent activities, enemy sympathizers, and saboteurs. If the enemy is Level III capable, MP leaders must expect infiltrations and air, missile, or NBC attacks as a precursor to a major Level III operation.

CORPS REAR AREA

3-17. The activities in Levels I and II will be similar in composition and density as in the COMMZ, but they will target key corps units, key facilities, and corps sustainment capabilities. The threat activities, especially at smaller unit levels, may even precede hostilities. MP leaders must be alert and prepared to encounter unconventional forces conducting diversionary or sabotage operations and small combat units conducting raids, ambushes, or reconnaissance operations or collecting special warfare intelligence. With the fast tempo of offensive operations, MP leaders must also be alert and prepared to encounter bypassed forces that can disrupt operations in the corps rear area.

DIVISION REAR AREA

3-18. The division rear area (DRA) contains many types of CS and CSS units and conducts many complex operations. As in the COMMZ and the corps rear area, the full spectrum of Level I, II, and III activities may occur in the DRA. The main target will be the division's HVA (including key C^2 facilities; airfields; artillery, aviation, and air-defense assets; LOC; and essential CSS units). The threat may conduct diversionary attacks, sabotages, raids, ambushes, and reconnaissance operations to affect the commander's freedom of maneuver and the continuity of operations. Unlike corps MP, the likelihood of division MP encountering bypassed enemy forces is expected. Failure to delay or defeat these forces will impact division operations.

OTHER TYPES OF THREATS

3-19. As US forces are deployed throughout the world, they will have to face nontraditional, asymmetric threats (other than those listed in Table 3-1, page 3-5) that may be geographically specific. As part of situational awareness, and in coordination with military intelligence (MI) and CID personnel, MP leaders must evaluate and assess the impact of these threats in their AO. A TO is vulnerable to any or a combination of the following threats:

- National or international organized crime.
- Narcotics traffickers.
- Narcotics terrorists.
- Extremist groups.
- Paramilitary groups.
- Ethnic or religious disputes.
- Trade in illegal weapons or strategic materials.

3-20. MP leaders must be aware that other threats exist and that they have the same potential as the Level I and II threats to disrupt operations in rear-area or sustainment operations. In some instances, the above threats' capabilities or the massing of personnel may have the same potential threat as a Level III threat.

COUNTERING THE THREAT

3-21. The disruption of rear-area and sustainment operations directly affects military efforts. Three types of forces may be used to counter the threat in these areas—a base/base-cluster self-defense force, a response force, or a tactical combat force (TCF).

BASE/BASE-CLUSTER SELF-DEFENSE FORCE

3-22. A base cluster is established when the appropriate echelon rear-operations cell or command post (CP) places geographically contiguous or noncontiguous bases under the control of a headquarters. The base cluster becomes the next higher tactical C^2 headquarters of those bases. The rear-operations cell or the rear CP may also establish a base cluster for a corps support group (CSG), an area support group (ASG), or other CSS units operating in the corps or division rear areas.

3-23. US ARFOR have the inherent responsibility to contribute as many forces as possible for base defense and local security for themselves and their facilities, installations, and activities. Each base and base-cluster commander must develop a defense plan to detect, defeat, and minimize the effects of Level I and limited Level II threat attacks on his base or base cluster (including NBC attacks). To maximize the unit's mission accomplishment, defense plans must be flexible and allow for differing degrees of security based on the probability of threat activity. Defense plans are given to MP units operating near the base or base cluster. The base commander most often employs a series of defense measures providing internal and perimeter security. His internal reaction forces use organic weapons to neutralize and defeat most low-level threat activity. Although not fully equipped to engage major conventional or unconventional enemy forces that may confront him, a base commander must deploy his personnel to defend themselves until MP, HN, local police, or combat forces (if available) can respond.

RESPONSE FORCE

3-24. A response force is summoned when the base or base cluster is faced with threat forces that are beyond their self-defense capability. If the MP are the designated response force, they must—

- Coordinate with the supported bases or base-cluster commanders to conduct a joint IPB.
- Review base and base-cluster self-defense plans.
- Exchange signal-operating-instructions (SOI) information.
- Identify MP contingency plans to counter likely enemy activities.
- Integrate ADA, engineer, chemical, field-artillery (FA), Army-aviation, and close-air-support (CAS) fire support into their plans (if available).

3-25. MP units help the base or base cluster return to its primary mission by defeating Level II threats. MP units closely watch likely avenues of enemy approach, possible landing zones (LZs), drop zones (DZs), C^2 facilities, and other key installations. They accomplish this through the MMS, AS, and PIO functions.

3-26. If MP units are not the designated response force, they may become the initial response force for units within their AO. When this occurs, they block, delay or, if possible, destroy enemy elements within their capability. If the attack is by a larger or more capable force, they will maintain contact and continue to develop the situation or delay until the appropriate response force appears or the battlefield commander commits the TCF.

TACTICAL COMBAT FORCE

3-27. When the MP response force encounters or engages threats beyond its ability to defeat, it immediately notifies the higher headquarters. The battlefield commander will then evaluate the situation and commit the TCF to defeat the Level III threat. The TCF is normally a combined-arms organization tailored by the corps or division G3, based on METT-TC. The TCF normally receives fire, aviation, or other support needed to fight and defeat the threat. Once the TCF is identified and before it is committed to battle, it will conduct direct coordination with the MP or other response forces regarding the exchange of reconnaissance information, battle-handoff procedures, and contingency plans for TCF operations. Once the TCF is committed, the MP unit performing as the response force becomes OPCON to the TCF commander.

MP AS THE TCF

3-28. The MP brigade or battalion is capable of conducting TCF missions only when properly augmented. MP augmentation may be in the form of fires, small combat units, aviation assets, or CAS. The specific type of augmentation is METT-TC dependent. Additionally, the MP C^2 headquarters must receive the respective liaison officers to ensure that augmentation forces are synchronized and employed according to their capabilities. The MP commander's situational awareness and battlefield visualization are key elements to TCF operations. Once designated as the TCF, the MP unit commander establishes liaison with the appropriate rear CP to obtain—

- The current rear-area IPB.
- The friendly unit disposition.
- Defense plans.
- Priorities for protection.
- The fire-support plan.

3-29. Based on the above information, the MP commander conducts his own IPB and develops a concept of operations. He then forwards it to the appropriate higher echelon for coordination and approval.

3-30. The MP's ability to employ organic MP assets as part of the TCF is limited by the following factors:

- **MP availability.** Normally, all MP assets available are committed at all times. The specific function and scope that MP units perform during the operation are determined by the Army commander's needs, the intensity of the conflict, and the availability of MP resources. The commander, with advice from the MP leader, must decide which MP operations must be scaled back, delayed, or shifted before the MP unit can be designated as part of the TCF.
- **MP dispersion.** MP units are normally displaced over a large geographical area. Technological capabilities and mobility allow them to operate over great distances. In today's battlefield, a typical MP company employment covers between 1,000 and 1,200 square kilometers and performs numerous missions in support of all five functions. The distance between elements, the reprioritization and movement of other MP units, the difficult terrain, poor roads, and bad weather may slow down the MP's commitment as a TCF.

Chapter 4

Military Police Functions

Military police support the Army commander's mission to win the battle. They help the commander shape the battlefield so that he can conduct decisive operations to destroy enemy forces, large or small, wherever and whenever the Army is sent to war.

MAJ(P) Anthony Cruz

The MP Corps supports shaping and sustainment operations while performing its five basic functions as a flexible, economy-of-force organization. Through these functions, MP units are able to provide the commander with an array of CS operations across the full spectrum of military operations (see Table 4-1).

Table 4-1. MP Functions

	Functions				
	MMS	**AS**	**PIO**	**L&O**	**I/R**
Subtasks	Support to river-crossing and breaching operations and passage of lines Straggler and dislocated-civilian control Route R&S MSR regulation enforcement	Recon operations ADC Base/air-base defense Response-force/ TCF operations Critical site, asset, and HRP security Force protection/ physical security Antiterrorism	IPB support PIAP Active and passive roles Information collection and dissemination Joint, interagency, and multinational coordination	Law enforcement Criminal investigations US Customs operations Related L&O training	EPW/CI handling US military prisoner handling Populace and resource control Dislocated civilians
NOTE: Subtasks not all-inclusive.					

OVERVIEW

4-1. MP assets are limited. Specific functions are performed at any given time and are determined by the supported commander's need, the intensity of the conflict, and the availability of MP resources. The supported commander, through the command's PM, sets the priorities for MP operations.

4-2. The PM (based on METT-TC and the available assets) continuously evaluates the trade-off between the MP support that the commander requires and the MP support that can be provided. To meet the priorities set by the commander's tactical plan, the PM recommends the allocation and employment of MP assets for MP combat, CS, and CSS operations.

MANEUVER AND MOBILITY SUPPORT

4-3. The MMS function involves numerous measures and actions necessary to support the commander's freedom of movement in his AOR. The MP expedite the forward and lateral movement of combat resources and ensure that commanders get forces, supplies, and equipment when and where they are needed. This is particularly important in the modern battlefield where there is a greater geographical dispersal of forces and lengthened LOC.

4-4. The MP maintain the security and viability of the strategic and tactical LOC to ensure that the commander can deploy and employ his forces. The MP support the commander and help expedite military traffic by operating traffic-control posts (TCPs), defilades, or mobile patrols; erecting route signs on MSRs or alternate supply routes (ASRs); or conducting a reconnaissance for bypassed or additional routes. The MP move all units quickly and smoothly with the least amount of interference possible.

4-5. As part of the MMS function, the MP support river-crossing operations, breaching operations, and a passage of lines. They also provide straggler control, dislocated-civilian control, route reconnaissance and surveillance (R&S), and MSR regulation enforcement.

RIVER-CROSSING OPERATIONS

4-6. US forces conduct river-crossing operations to move a large force across a river obstacle with a minimum loss of momentum. The MP play a vital role by assisting the force commander in crossing the river as quickly and efficiently as possible. The crossing is usually planned and conducted by the headquarters directing the crossing. As such, a division crossing operation is conducted by a corps. Whether a brigade or division is crossing, the division MP company may also cross to provide uninterrupted support to the division. In these instances, there is a total reliance on the corps MP to support the crossing. The same is true for breaching operations and a passage of lines.

4-7. MP support for river-crossing operations reduces the crossing time and promotes the efficient movement of vehicles. It reduces congestion, speeds the crossing, and enables the maneuver commander to continue his momentum toward his primary objective. The MP establishes staging and holding areas and TCPs to control movement to and from these areas (according to the traffic-control plan). The MP may be called on to provide security for crossing forces at the crossing sites. In most cases, the MP TCPs and engineer regulation points (ERPs) are located on both sides of the river to improve communications and coordination between the units.

4-8. MP employment for river crossing is influenced by METT-TC. The number and placement of MP assets supporting a river-crossing operation varies with the size of the crossing force, the direction of the crossing (forward or retrograde), and the degree of enemy resistance expected or encountered.

4-9. The MP operating inside the crossing areas are OPCON to the crossing-area commander for the duration of the operation. The MP operating outside of the crossing area are under the command of their appropriate echelon commander.

4-10. The main thrust of MP support to river-crossing operations is within the immediate river-crossing site. The MP direct units to their proper locations (such as holding areas and staging areas) and ensure that units move through the area within the time listed on the movement schedule. This is a highly critical aspect of river crossing because the number of crossing sites is limited. MP assets are placed where they can stress MMS operations on MSRs leading into the crossing area.

4-11. The MP also provide AS to allow crossing forces to cross the river without losing momentum or forces. On both near and far sides, the MP are used to recon the crossing unit's flanks and rear to enhance security (see FM 19-4).

BREACHING OPERATIONS

4-12. Breaching operations are conducted to allow forces to maneuver through obstacles. Obstacle breaching is the employment of a combination of tactics and techniques to advance an attacking force to the farside of an obstacle that may be covered by fire. It is perhaps the most difficult combat task a force can encounter. Breaching operations begin when friendly forces detect an obstacle, and they end when the battle handover has occurred between the follow-on forces and a unit conducting the breaching operation (see FM 90-13-1).

4-13. The MP support breaching operations in numerous ways. MP assets are employed based on METT-TC, the available resources, and the commander's priorities. As a minimum, MP support may include, but is not limited to—

- Establishing TCPs along routes leading to or departing from the breaching site.
- Establishing holding areas.
- Establishing TCPs at the breaching site.
- Assisting engineers with temporary route signs.
- Establishing straggler-control operations.
- Conducting AS operations.

4-14. The most critical MP support is provided at the breaching site. The MP provide the commander with a means to control traffic flow to appropriate lanes. When multiple lanes branch off of a single far-recognition marker, the MP assist in directing the formation through various lanes. They also assist in modifying the traffic flow when lanes have been closed for maintenance or expansion. The MP conduct close coordination with the crossing-force commander and the TF commander executing the breaching operation. The MP enable the commander to make last-minute changes in traffic flow, thereby giving him increased flexibility to react to the enemy situation.

PASSAGE OF LINES

4-15. A passage of lines is an operation in which a force moves forward or rearward through another force's combat positions with the intention of moving into or out of contact with the enemy. The passage of lines is a high-risk military operation that requires close coordination between the passing unit, the stationary unit, and the MP providing the support.

4-16. The MP help reduce confusion and congestion during a passage of lines. They provide security in areas surrounding passage points and passage lanes to ensure that the passing unit has priority for using routes to and through the areas. The headquarters directing the operations sets the route's priority. The MP can support a forward, rearward, or lateral passage of lines. Before the actual operation, the MP in the AO conduct an area or zone reconnaissance to become familiar with the routes to, through, and beyond the area of passage. This enables the MP to extend the commander's C^2 by providing directions at passage points and by guiding the units through the passing lanes. Maintaining unit integrity and reducing incidents of stragglers is vital to maintaining the passing unit's momentum in a forward passage of lines. The MP perform aggressive straggler- and dislocated-civilian-control operations to prevent possible infiltration of the enemy.

4-17. A passage of lines is usually planned and coordinated by the headquarters directing the passage. A division's passage of lines is planned and coordinated by the corps headquarters. The detailed plans are made and coordinated between the divisions involved. Close coordination between the division and corps PMs is essential. An MP unit may be the unit involved in passing through the lines of another unit. When conducting a delay of a Level II threat, the MP are likely to conduct a passage of lines with the TCF. To avoid fratricide, close coordination between the MP response-force commander and the TCF is imperative (see FM 19-4).

STRAGGLER CONTROL

4-18. Mobile patrols, TCPs, and checkpoint teams return stragglers to military control as part of their operations. Most stragglers are simply persons who become separated from their command by events in the CZ or while moving through the COMMZ. If a straggler is ill, wounded, or in shock, an MP must give him first aid and, if needed, call for medical evacuation (MEDEVAC). If a straggler is uninjured, an MP directs him to his parent unit or to a replacement unit (as command policies dictate). The MP ensure that stragglers attempting to avoid return to their units are escorted back to their command (as a minimum).

4-19. The MP set up special posts for straggler control following NBC attacks or major enemy breakthroughs that result in large numbers of lost, dazed, and confused military personnel. Mobile MP teams operate between posts, and they also direct or collect stragglers. Straggler collection points may be needed if many stragglers are present in a combat theater. If allied forces are present in the theater, each nation establishes a collection point for its own personnel. MP teams are aware of each allied location and are prepared to assist allied soldiers in returning to their respective command. The MP use available transportation assets to transfer stragglers from TCPs and checkpoints to a straggler collection point. At the collection point, they are screened and sorted for removal to a medical facility or returned to their units to reconstitute the tactical commander's combat force.

4-20. The MP report information about stragglers with whom they come in contact. This information is compiled by the MP headquarters and forwarded through appropriate channels to the higher command. Information given by stragglers that is of immediate tactical value is reported without delay.

DISLOCATED-CIVILIAN CONTROL

4-21. The MP expediting traffic on MSRs may encounter dislocated civilians that could hinder military traffic. The MP assist and divert dislocated civilians from MSRs and other areas to I/R facilities. They may also deny the movement of civilians whose location, direction of movement, or actions may be a threat to themselves or to the combat operation. The HN government is responsible for identifying routes for the safe movement of dislocated civilians out of an AO. If needed, the MP assist the civil-affairs unit and the HN in redirecting dislocated civilians to alternate routes established by the HN government.

4-22. The US forces do not assume control of dislocated civilians unless requested to do so by the HN or unless operating in an environment with a hostile government. When the senior US commander assumes responsibility, the MP coordinate with civil affairs to set up TCPs at critical points along the route to direct dislocated civilians to secondary roadways and areas not used by military forces. As required, MWD teams may be used as a show of force or as a deterrent to assist with uncooperative personnel.

ROUTE R&S

4-23. The MP conduct hasty and deliberate route reconnaissances to obtain information on a route and nearby terrain from which the enemy can influence troop movement. A route reconnaissance focuses on continually monitoring the condition of MSRs, ASRs, and other areas. MP patrols look for restricting terrain, effects of weather on the route, damage to the route, NBC contamination, and enemy presence or absence. When enemy activity is spotted, the MP report it, maintain surveillance, and develop the situation. To gather information for proposed traffic plans, they look at the type and number of available routes; and they check load classifications, route widths, obstructions, and restrictions.

MSR REGULATION ENFORCEMENT

4-24. The MP undertake MSR regulation enforcement to keep the routes free for DSS operations. MP units support the command's MSR regulation measures as stated in the traffic-regulation plan (TRP). The TRP contains specific measures to ensure the smooth and efficient use of the road network. It assigns military route numbers and names, the direction of travel, highway regulation points, and preplanned MP TCPs. Most importantly to the MP, it gives the route's control classification. The MP ensure that classified routes are used only by authorized traffic. Vehicles traveling on roads too narrow for their passage or on roads unable to support their weight can obstruct the route.

4-25. To expedite traffic on MSRs, the MP operate special circulation control measures such as—

- Temporary route signing.
- Static posts such as TCPs, roadblocks, checkpoints, holding areas, and defilades at critical points.
- Mobile teams patrolling between static posts and monitoring traffic and road conditions.

4-26. They also gather information on friendly and enemy activities and help stranded vehicles and crews. The MP also place temporary route signs to warn of hazards or to guide drivers unfamiliar with the route. Using these measures, the MP exercise jurisdiction over the road network in the AO and coordinate with the HN (whenever possible) to expedite movement on MSRs.

AREA SECURITY

4-27. The MP perform the AS function to protect the force and to enhance the freedom of units to conduct their assigned missions. The MP who provide AS play a key role in supporting forces in rear-area and sustainment operations. The MP are a response force that delays and defeats enemy attempts to disrupt or demoralize military operations in the AO. The MP's mobility makes it possible for them to detect the threat as they aggressively patrol the AO, MSRs, key terrain, and critical assets. The MP's organic communications enable them to advise the appropriate headquarters, bases, base clusters, and moving units of impending enemy activities. With organic firepower, the MP are capable of engaging in decisive operations against a Level II threat and delaying (shaping) a Level III threat until commitment of the TCF.

4-28. Throughout all aspects of the AS function, the MP perform counteractions to protect the force and to prevent or defeat a Level II threat operating within the MP's AO. MP countermeasures may include implementing vulnerability assessments, developing procedures to detect terrorist actions before they occur, hardening likely targets, and conducting offensive operations to destroy the enemy. The MP use checkpoints and roadblocks to control the movement of vehicles, personnel, and materiel and to prevent illegal actions that may aid the enemy. The use of these control measures serves as a deterrence to terrorist activities, saboteurs, and other threats. However, at the same time, checkpoints and roadblocks expose the MP to these potential threats. To counter this fact, the MP may upgrade or harden vehicles and defensive positions.

4-29. The MP provide combat power to protect the C^2 headquarters, equipment, and services essential for mission success. The MP provide the battlefield commander with a light, mobile fighting force that can move, shoot, and communicate against any threat. Major subtasks associated with the AS function include reconnaissance operations; area damage control (ADC); base/air-base defense; response-force operations; and critical site, asset, and high-risk personnel (HRP) security.

RECONNAISSANCE OPERATIONS

4-30. As part of their AS mission, the MP serve as the eyes and ears of the battlefield commander by seeking out the enemy and reporting information obtained by recon patrols. The MP conduct area and zone reconnaissances, screening, surveillance, and countersurveillance to gain information to help guard against unexpected enemy attacks in the AO. The MP monitor likely avenues of approach and potential LZs and DZs. They become familiar with towns and other populated areas, ridgelines, woods, and other terrain features from which the enemy can influence movements along road networks. The MP pay close attention to areas near facilities designated critical by the commander. These areas include key MSR bridges and tunnels, depots, terminals, logistics-support bases, ammunition supply points (ASPs), communications centers/nodes, and C^2 headquarters. The MWD teams provide explosive detection and personnel detection/tracking capabilities that enhance reconnaissance operations (especially in urban terrain).

AREA DAMAGE CONTROL

4-31. MP units take measures to support ADC before, during, and after hostile actions or natural and man-made disasters. The ADC actions integrate CS and CSS functions for many units. Engineers, medical personnel, and Army aviators work closely to ensure quick relief operations. The MP provide MSR regulation enforcement, refugee control, and some local security when required. As with reconnaissance operations, the MP may use MWD explosive-and personnel-detection capabilities to augment all MP missions in rear-area and sustainment operations.

BASE/AIR-BASE DEFENSE

4-32. The MP are the base and base-cluster commanders' links for detection, early warning, and employment against enemy attacks. The information gathered is dispersed throughout the rear area to help apprise the commander of enemy activities near bases. Base defense is the cornerstone of rear-area security. When the threat exceeds the base/base-cluster capability, the base/ base-cluster commander requests MP assistance through the appropriate C^2 element.

4-33. Air-base defense requires special MP coordination with the US Air Force (USAF). The MP treat air bases like any other base or base cluster. A USAF air base may house the base-cluster commander, or it may be a cluster by itself. The MP are responsible for the air base's external defense. Its internal defense is primarily the responsibility of the Air Force's security forces. The security force provides in-depth defense for weapons, weapons systems, command centers, personnel, and other priority resources established by the base commander.

4-34. The security force is trained and equipped to detect, delay, and deny Level I and II threats. If a Level III threat is present, the security force is tasked with delaying actions; however, the HN, a sister service, or other support must be employed to defeat this threat. If the security force requires assistance to defeat a Level II threat, it may rely on MP response forces or another response force to assist in the defense. If available, the MP response force will react to the air-base defense, just as it would for any other base or base cluster within the MP's AO. However, the key to successful MP employment depends on the critical exchange of information before and during the MP employment. Good communications, an understanding of the defense plan, and liaison operations are vital in preventing responding forces from entering a situation that could result in fratricide.

RESPONSE-FORCE OPERATIONS

4-35. The MP are the base and base-cluster commanders' response force against enemy attacks in rear-area or sustainment operations. The MP gather information about the enemy while performing missions throughout the AO. This information provides commanders with enemy activity near bases. When needed, the MP provide a mobile response force to respond to bases under attack and to destroy the enemy. A base commander's defense plan is the cornerstone for protecting rear-area and sustainment operations. The base commander is responsible for defeating all Level I threats. When this threat exceeds his capabilities, he requests MP support. The MP located near bases or patrolling or conducting AS operations consolidate their forces, respond as quickly as possible, and conduct combat operations to destroy the enemy. If needed, the MP conduct a battle handover to the TCF.

4-36. MP forces performing as a response force are capable of conducting the following offensive operations:

- A movement to contact.
- A hasty ambush.
- A hasty attack.
- A delay.
- A call for fire.
- A repel attack against critical sites.
- A defense of critical sites.

4-37. To conduct these missions, the MP consolidate into squads or platoons to delay, defeat, or defend against the threat. See FMs 71-3 and 71-100 for more information on battle-handover operations.

CRITICAL SITE, ASSET, AND HRP SECURITY

4-38. The MP perform their AS function across the entire designated AO. When the MP provide security around a critical site or asset, they usually provide a mobile security screen, taking advantage of its weapons and communications platforms. This standoff protection detects and defeats the threat before it can move within direct-fire range of the facilities. The MP may be tasked to provide detail security to key facilities, assets, and personnel.

4-39. The MP provide security to major CPs and other facilities within the AO. Their employment maximizes mobility, lethality, and communications capabilities as a security screen. They may be required to establish local AS measures (such as checkpoints and listening/observation posts) to further protect these facilities. The MP provide internal access-control points to critical facilities, and they act as a response force. When the critical CP relocates, the MP provide in-transit security. Other types of critical site security include ASPs; deep-water ports; petroleum, oil, and lubricants (POL) terminals and pipelines; trains and railways; and air bases.

4-40. The MP may provide convoy security for top-priority units transporting especially critical supplies to combat forces. MP assets should be employed primarily on aggressive patrolling, route, area, and zone reconnaissance measures that would create a safe and secure environment for all types of vehicular and unit movement.

4-41. The MP and the CID provide protective services to designated key personnel by providing access control to restricted areas within CPs, providing in-transit security, or providing static security measures around the clock. The MP coordinate with the CID when close-in protection of key personnel is needed. The MP and the CID also provide training for personal-protection countermeasures. The MWD teams may be employed to enhance MP and CID detection and protection capabilities.

INTERNMENT AND RESETTLEMENT

4-42. The Army is the Department of Defense's (DOD's) executive agent for all EPW/CI operations. Additionally, the Army is DOD's executive agent for long-term confinement of US military prisoners. Within the Army and through the combatant commander, the MP are tasked with coordinating shelter, protection, accountability, and sustainment for EPWs/CIs. The I/R function addresses MP roles when dealing with EPWs/CIs, dislocated civilians, and US military prisoners.

4-43. The I/R function is of humane as well as tactical importance. In any conflict involving US forces, safe and humane treatment of EPWs/CIs is required by international law. Military actions on the modern battlefield will result in many EPWs/CIs. Entire units of enemy forces, separated and disorganized by the shock of intensive combat, may be captured. This can place a tremendous challenge on tactical forces and can significantly reduce the capturing unit's combat effectiveness. The MP support the battlefield commander by relieving him of the problem of handling EPWs/CIs with combat forces. The MP perform their I/R function of collecting, evacuating, and securing EPWs throughout the AO. In this process, the MP coordinate with MI to collect information that may be used in current or future operations.

4-44. Although the CS MP unit initially handles EPWs/CIs, modular MP (I/R) battalions with assigned MP guard companies and supporting MWD teams are equipped and trained to handle this mission for the long term. A properly configured modular MP (I/R) battalion can support, safeguard, account for, guard, and provide humane treatment for up to 4,000 EPWs/CIs; 8,000 dislocated civilians; or 1,500 US military prisoners.

EPW/CI HANDLING

4-45. The MP are tasked with collecting EPWs/CIs from combat units as far forward as possible. The MP operate collection points and holding areas to temporarily secure EPWs/CIs until they can be evacuated to the next higher echelon's holding area. The MP escort-guard company assigned to the MP brigade (I/R) evacuate the EPWs/CIs from the corps's holding area to the COMMZ's internment facilities. The MP safeguard and maintain accountability, protect, and provide humane treatment for all personnel under their care.

4-46. In a mature TO, MP (I/R) units process EPWs/CIs collected by MP teams and other units in the CZ. MP guard companies assigned to the MP (I/R) units guard EPWs/CIs at designated camps (see FM 19-40).

POPULACE AND RESOURCE CONTROL

4-47. Populace and resource control (PRC) denies adversaries or insurgents access to the general population and resources. The MP supports civil-affairs personnel and the tactical commander in planning and conducting PRC programs employed during all military operations. These programs may consist of curfews, movement restrictions, resettling dislocated civilians, licensing, ration control, regulation enforcement, amnesty programs, inspecting facilities, and guarding humanitarian-assistance distributions. The MP also direct dislocated civilians to resettlement camps where they are cared for while NGOs work to coordinate their relocation.

4-48. The MP's security capability, acceptability, and interface with the populace make them suitable as the primary forces of choice in these operations. The MP I/R units are specifically designed to fill this need (see FM 41-10).

US MILITARY PRISONERS CONFINEMENT

4-49. The MP detain, sustain, protect, and evacuate US military prisoners. Whenever possible, soldiers awaiting trial remain in their units, unless reasonable grounds exist to believe that they will not appear at the trial, the pretrial hearing, or the investigation or that they will engage in serious criminal misconduct. Under either of these two pretrial confinement instances, the commander must also reasonably believe that a less severe form of restraint (such as conditions of liberty, restriction in lieu of apprehension, or apprehension) are inadequate. When these circumstances exist and other legal requirements are met, US military personnel may be placed in pretrial confinement under the MP's direct control. Convicted military prisoners are moved as soon as possible to confinement facilities outside of the AO.

4-50. MP confinement operations parallel (but are separate from) the MP's EPW/CI operations. No member of the US armed forces may be placed in confinement in immediate association with an EPW or other foreign nationals who are not members of the US armed forces. A confinement facility is maintained within the TO only if distance or the lack of transportation requires such a facility. When military prisoners are retained in theater, temporary field detention facilities may be established in the CZ and a field confinement facility in the COMMZ (see FM 19-40 and AR 190-47).

LAW AND ORDER

4-51. The L&O function consists of those measures necessary to enforce laws, directives, and punitive regulations. The MP's L&O function extends the battlefield commander's C^2. The MP, in close coordination with the CID, work to suppress the chance for criminal behavior throughout the AO. By coordinating and maintaining liaison with other DOD, HN, joint, and multinational agencies, the MP at all levels coordinate actions to remove conditions that may promote crime or that have the potential to affect the combat force. Crime-prevention measures and selective enforcement measures are also performed as part of other functions. For example, the MP investigate traffic accidents and regulate traffic as part of the MMS function. The L&O function includes major areas such as law enforcement, criminal investigations, and support to US Customs operations. The primary units conducting L&O are the L&O detachments, customs teams, and CID units. Both the MWD team and the MP company (CS) also support the L&O function.

LAW ENFORCEMENT

4-52. Law-enforcement operations assist the battlefield commander in preserving his force. The MP dedicate assets to conduct law-enforcement operations based on the commander's needs. Since MP L&O and CID assets may be limited during the initial stages of any operation, the PM depends on the supported commander's development of an effective crime-prevention program and uses established investigative tools (such as inquiries and AR 15-6 investigations) to enforce rules. This will allow the PM to employ limited MP assets to perform other functions. Law-enforcement operations include responding to civil disturbances, conducting raids, investigating traffic accidents, conducting vehicle searches, supporting the commander's force-protection program, and providing support to HN and civil-enforcement agencies. Law enforcement also includes employing special-reaction teams (SRTs), marksman/observer (M/O) teams, and MWD teams.

4-53. The Army conducts counterdrug-support operations that generally fall within several DOD counterdrug-mission categories. The MP support the Army's role rather than directly participating in civil law-enforcement activities (such as searches, seizures, and arrests). When tasked, the MP provide training to law-enforcement agencies in common soldier skills, physical security, and tactical planning and operations. US Code (USC) Title 18, Posse Comitatus Act, Section 1385 prohibits the use of DOD assets to enforce civilian law (federal and state) except in cases and under circumstances expressly authorized by the Constitution or by an act of Congress.

4-54. In multinational operations, the MP may assist with the creation of multinational police units. Circumstances that may support the establishment of these police forces include existing or negotiated terms of international agreements or security-assistance programs, a multinational operational agreement, or appropriate military directives. The MP provide the capability to train foreign MP and/or reconstitute indigenous constabulary forces as part of stability and support operations. The MP can provide the initial mentoring to these forces and provide temporary, emergency law-enforcement capabilities until the foreign military or civilian police forces are functional (see FM 100-8).

CRIMINAL INVESTIGATIONS

4-55. The MP investigate offenses against US forces or property committed by persons subject to military law. Investigations against minor crimes (such as low-value, personal-property thefts or simple assaults) are normally investigated by the MP's L&O detachment. Investigations against major incidents involving wrongful death, serious bodily injury, and war crimes are referred to special agents of the USACIDC. The USACIDC conducts death investigations in the absence of HN agreements or in conjunction with the HN. The USACIDC special agents are authorized to investigate any alleged criminal conduct in which there is an Army interest unless prohibited by law or higher authority.

4-56. The USACIDC's investigative authority and investigative responsibility outside of the US are determined by international treaty or agreement (including status of forces agreements [SOFAs]), the policies of the HN government, the US ambassador, and AR 195-2. In the absence of such provisions, the following guidelines apply:

- On Army-controlled installations, the USACIDC has the authority to investigate alleged crimes.
- Outside of an Army-controlled installation, the USACIDC may investigate after coordinating with HN authorities.

4-57. In all environments, the USACIDC has the responsibility to investigate all felony crimes involving Army personnel, DA civilians and agencies, and companies working for the Army. The USACIDC investigates war crimes and crimes involving personal and government property affecting the Army's mission (logistics security [LOGSEC]). Other investigations (such as those based on international treaties, SOFAs, and joint investigations with the HN) may be undertaken if requested by the supported commander in support of the overall Army mission. See Chapter 9 for a complete discussion of the USACIDC.

US CUSTOMS OPERATIONS

4-58. The MP support the US Customs Service (USCS), the US Department of Agriculture (USDA), other federal agencies, joint staffs, and commanders who enforce the laws and regulations of the US concerning customs, agriculture, and immigration border clearances. Support to the USCS also includes assistance to federal agencies to eliminate the illegal traffic of controlled substances and other contraband through Army channels. MP support to customs operations are normally performed by specially trained MP customs teams. Although other MP units are not trained in all facets of customs operations, they may assist MP customs teams, the USCS or the USDA, and other federal agencies in the enforcement of applicable laws and regulations. When tasked, the MP/CID supports the investigation of violations of US Codes, DOD or DA regulations, and applicable provisions of SOFAs.

4-59. The MP report violations of customs laws, regulations, inspections, and investigative results to the installation's PM, the supported commander, and affected units. During redeployment from outside the continental US (OCONUS) to CONUS installations, the MP support the USCS or USDA efforts to ensure that personnel, equipment, and materiel meet customs, immigration, and agriculture requirements as stated by all applicable laws and regulations. As with other functions, MWD teams may be employed in support of customs operations for the detection of explosives or narcotics.

> During operation Just Cause, an MP platoon temporarily assumed the customs mission at the main terminal of the Torrijos International Airport located just outside of Panama City. Their mission supported the air evacuation of foreign-national civilians and the redeployment of some of the initial-entry US forces.

POLICE INTELLIGENCE OPERATIONS

4-60. The PIO function supports, enhances, and contributes to the commander's protection program, situational awareness, and battlefield visualization by portraying relevant threat information that may affect his operational and tactical environments. This threat information—whether it is police, criminal, or combat information—is gathered while conducting MP functions. The PIO function—

- Demonstrates the MP's/CID's capability to collect relevant threat information actively or passively.
- Ensures that all information collected while conducting MMS, AS, I/R, and L&O functions continues to be reported through the proper channels so that it can be analyzed by the Intelligence Officer (US Army) (S2) or the Assistant Chief of Staff, G2 (Intelligence) (G2) with support from the appropriate MP echelon.
- Coordinates with USACIDC elements to employ data developed by the USACIDC's programs. These programs include—
  - The Combating Terrorism Program as outlined in AR 525-13 and CIDR 195-1.
  - The Criminal Intelligence Program (CIP).
  - Personal-security vulnerability assessments (PSVAs).
  - A crime threat analysis.
  - Logistics-security threat assessments (LSTAs).

NOTE: The MP/CID must ensure that criminal information is released according to existing controls and restraints.

- Maintains constant liaison and communication with the higher echelon S2/G2; psychological-operations (PSYOP) units; HN police and other law-enforcement agencies; joint, combined, interagency, and multinational forces; the staff judge advocate (SJA); the CMOC; civil-affairs teams; and the force-protection officer.

4-61. The MP brigade commander, the battalion commander, and the PM are responsible for the PIO function. As such, each one must determine the best way to employ the available staff resources to monitor the execution of the PIO function within his command.

IPB PROCESS AND PIO FUNCTION

4-62. The IPB is a systematic, continuous process for analyzing the threat and environment in a specific geographic area. It is designed to support staff estimates and military decision making. Applying the IPB process helps the commander selectively apply and maximize his combat power at critical points in time and space on the battlefield by—

- Determining the threat's likely COA.
- Describing the environment the unit is operating within and the environmental effects on the unit.

4-63. The IPB process consists of the following four steps:

- **Define the battlefield environment.** The S2/G2 identifies the battlefield characteristics that will influence friendly and threat operations, establishes the limits of the area of interest (AOI), and identifies gaps in current intelligence holdings.
- **Describe the battlefield effects.** The S2/G2 evaluates the environmental effects with which both sides must contend. The S2/G2 identifies the limitations and opportunities that the environment offers on the potential operations of friendly and threat forces. This evaluation process focuses on the general capabilities of each force until COAs are developed later in the IPB process. This environmental assessment always includes an examination of terrain and weather, but it may also include discussions of characteristics of geography and infrastructure and their effects on friendly and threat operations.
- **Evaluate the threat.** The S2/G2 and his staff analyze the command's intelligence holdings to determine how the threat normally organizes for combat and conducts operations under similar circumstances. When facing a well-known threat, the S2/G2 can rely on historical databases and threat models. When operating against a new or less known threat, he may need to develop his intelligence databases and threat models concurrently.
- **Determine the threat's COA.** Given what the threat normally prefers to do and the effects of the specific environment in which he is operating, his likely objectives and the COAs available to him are determined. The S2/G2 develops enemy COA models that depict the threat's available COAs. He also prepares event templates and matrices that focus intelligence and identify which COA the enemy will execute (see FM 34-130).

4-64. Although the S2/G2 has the staff responsibility for the command's IPB, he is not the only one who conducts or needs to understand and use the IPB. Every Army commander and staff member must understand and apply the IPB process during the staff planning process. The MP on the battlefield are no exception. The MP employ the IPB process as their first step in developing and implementing the PIO function within their commands.

4-65. During the IPB process, the S2/G2 uses all available databases, intelligence sources/products, and related MI disciplines to analyze the threat and the environment. The PIO function supports this process by providing the S2/G2 with collected police, criminal, and combat information that can directly and significantly contribute to the success of the MI effort. In addition to the combat information, the PIO function provides additional information on possible criminal threats and COAs that may support the S2's/G2's IPB process and that can be used by the commander to upgrade the force-protection posture.

POLICE INFORMATION ASSESSMENT PROCESS

4-66. Like the S2/G2 uses the IPB process to continuously analyze the threat and the environment in a specific geographical area, MP leaders use the PIAP as a tool to continuously collect, organize, interpret, and gain access to police/criminal information in support of the IPB process. Criminal trends may have an impact on the tactical scenario, and the PIAP is a method used to consider this threat and its impact on friendly forces. MP leaders cannot use the PIAP as a substitute for the IPB process—the PIAP complements the IPB (see Appendix E).

PASSIVE MODE PIO

4-67. Every MP conducts the PIO function in a passive mode during their normal day-to-day operations and across the full spectrum of military operations. In the passive mode, PIO are not a stand-alone function; as such, they cannot be separated from other MP functions.

4-68. During the performance of MMS, AS, I/R, and L&O functions, the MP develop and exchange information with other organizations in the AO. The MP obtain information through contact with civilians, NGOs, IHOs, local and HN police, multinational police, and other security forces. If the MP receive, observe, or encounter information (police, criminal, or combat) while performing these functions, they will immediately submit a report to relay information up the chain of command. This report may be in the form of a size, activity, location, unit, time, and equipment (SALUTE) report; a spot report (SPOTREP); or another appropriate report. When the higher echelon (brigade, battalion, or PM) receives this information, it is simultaneously integrated into the ongoing IPB/PIAP and forwarded to the higher echelon S2/G2 (see Figure 4-1, page 4-16).

4-69. If police/criminal information is obtained, the MP—

- Update the previous police/criminal estimates provided to the S2/G2.
- Identify new or potential criminal threats or trends in the AO.
- Consider recommending that the supported commander upgrade the force-protection level.
- Notify adjacent units of the potential criminal threat that may affect their forces.
- Consider reprioritizing MP support to the identified threat area.
- Share information with HN/local police and other agencies.

4-70. If combat information is obtained, the MP—

- Forward the information to the higher headquarters S2/G2.
- Forward the information to the MP chain of command, integrate it into the MP's IPB process and, if necessary, take appropriate action.
- Notify the adjacent unit of the potential threat that may affect their forces.

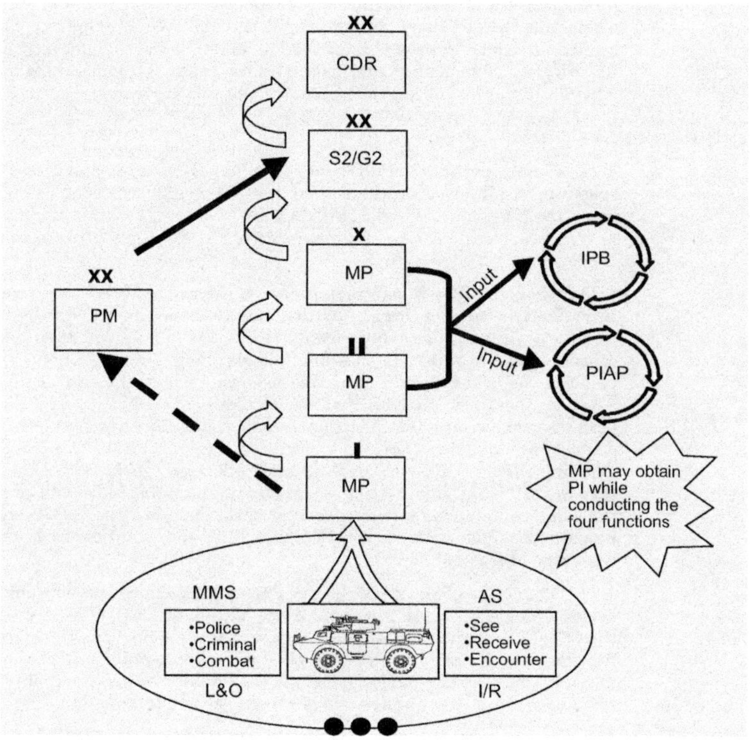

Figure 4-1. PIO—Passive Mode

During Operation Uphold Democracy, an MP team was conducting a TCP as part of a cordon and search operation in Port-au-Prince, Haiti. While performing the task, two civilians approached the MP team informing them of criminal activity in the neighborhood. During the interview, the MP team prepared a sketch of a house and surrounding areas. The team also obtained information describing the criminals and their weapons. Recognizing that the criminal activity was in fact the actions of a political/mercenary group named FRAP, the MP team radioed the platoon leader and forwarded the field interview to higher headquarters. Two days later, a unit from the 10th Mountain Division raided the house, capturing weapons, ammunition, and equipment.

4-71. The preceding vignette demonstrates the MP performing the PIO function in the passive mode. The MP team received the information while conducting a TCP and submitted it through the appropriate chain of command, which resulted in an action taken. This example stresses the importance of submitting information up the chain of command regardless of whether it may be police, criminal, or combat information.

ACTIVE MODE PIO

4-72. The MP perform the PIO function in the active mode and across the full spectrum of military operations when directed by higher headquarters. In this mode, the MP conduct specific MMS, AS, I/R, and L&O missions with the intent to collect information actively in support of the S2's/G2's IPB process or the PIAP.

4-73. When the S2/G2 identifies a gap in the command's knowledge of the threat and the current threat situation, it may be recommended to be included as priority intelligence requirements (PIR). The S2/G2 will then develop a collection plan to assist him in filling this gap. Part of his collection strategy is to select the best collectors available to cover each intelligence requirement. After a thorough analysis (which includes availability, capability, and performance history), the collection manager identifies which collection asset can best be used in response to a given requirement, and the Operations and Training Officer (US Army) (S3)/G3 tasks the asset. If the S2/G2 determines that the MP is the right force to serve as collectors, the S3/G3 will go through the appropriate request channels and task the PM. The S2/G2 will then provide the PM with a specific guideline and a prioritized collection requirement (see FM 34-2).

4-74. On receipt of the mission, the PM will conduct a mission analysis to decide which specific MP function (MMS, AS, I/R, or L&O) is needed to satisfy the S2's/G2's requirement. Once the analysis is completed and the appropriate function selected, the PM will then task subordinate units with the collection mission. Once the mission is completed, the PM may receive another collection tasking or continue with previous MP tasks.

4-75. When the MP commander or the PM conducts the PIAP, he may also encounter a police/criminal information gap. This gap may become the MP commander's police/criminal information requirements (PCIR). If the gap cannot be filled with available data from the S2/G2, the CID, the HN, and other agencies, the MP commander/PM may task subordinate MP units or request CID assistance to support the collection effort. On receipt of the mission, the MP will then conduct a mission analysis and decide which specific MP function (MMS, AS, I/R, or L&O) is needed to satisfy the PM requirement. Once the mission is completed, the MP may receive another collection tasking or continue with previous MP tasks (see Figure 4-2, page 4-18).

4-76. Since any soldier on the battlefield can report police or criminal information, the MP commander or the PM must constantly coordinate and communicate with the S2/G2, PSYOP, and other agencies to obtain information that could be of MP/CID interest. This constant coordination is a key factor for ensuring that the MP/CID has visibility over the police/criminal information that is reported through non-MP channels.

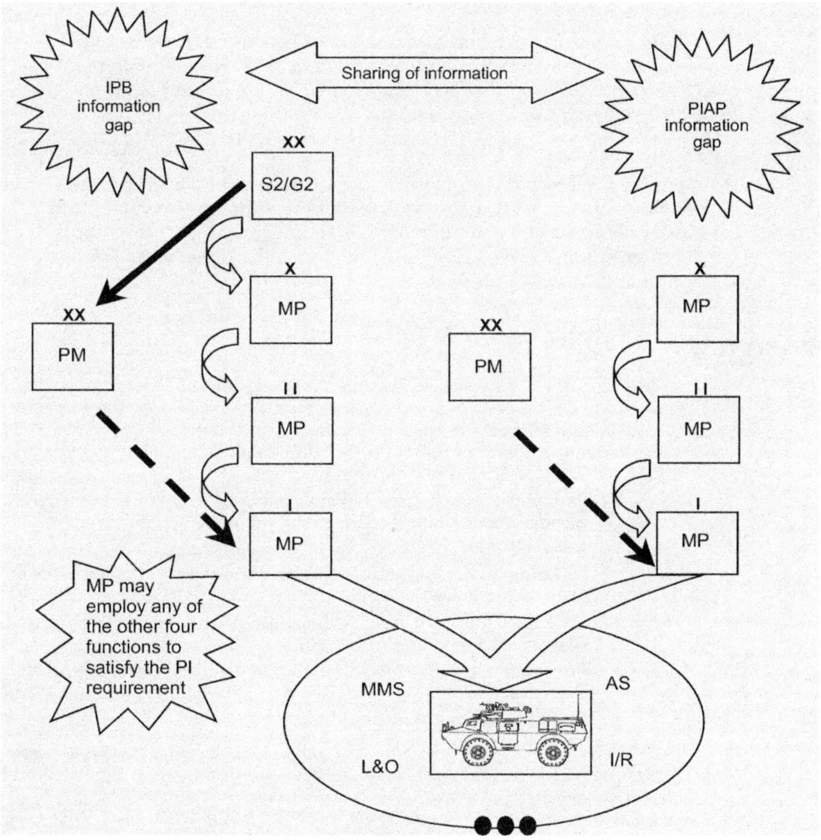

Figure 4-2. PIO—Active Mode

Chapter 5

Military Police Support to Echelons Above Corps

MP units supporting EAC perform combat, CS, and CSS operations. Like the MP supporting corps and divisions, MP units supporting EAC units perform the five MP functions based on available assets and the supported commander's needs.

USAMPS

MP support to EAC includes support to the ASCC and the TSC. The ASCC is responsible for Army Title 10 requirements in support of a combatant commander. This support includes recruitment, organization, supply, equipment, training, servicing, mobilizing, demobilizing, administration, and maintenance functions.

OVERVIEW

5-1. The ASCC may also be responsible for significant DOD- and combatant-commander-designated Army support to other services. As the senior Army commander in the AOR, the ASCC commander tailors and designates ARFOR to accomplish operational-level tasks while conducting major land operations. The ASCC's operational responsibilities include—

- Recommending the proper employment of Army-component forces to the joint-force commander or to the subunified commander.
- Accomplishing operational missions as assigned.
- Selecting and nominating specific Army units for assignment to subordinate theater forces.
- Informing the combatant commander of the Army's CSS effects on operational capabilities.
- Providing data to the supporting operations plans (OPLANs) as requested.
- Ensuring signal interoperability.

5-2. The ASCC provides administrative and logistics (A/L) services to assigned ARFOR and to those of subordinate JFCs. When appropriate, the ASCC delegates the authority for support tasks to a single subordinate Army headquarters. In major operations, the TSC (along with other EAC support commands) would be the ASCC's lead organization for planning, coordinating, executing, or providing required support functions (see FM 100-10).

5-3. The TSC is the senior Army support organization in a theater. Its commander reports to the ASCC or ARFOR commander. The TSC normally operates at the operational level of CSS with links to the strategic and tactical levels. Unity of command is the critical element that the TSC brings to the fight. The TSC is a multifunctional organization that centralizes the command, control, and supervision of support functions at EAC as directed by the ASCC or ARFOR commander. The TSC's mission is to maximize throughput and follow-on sustainment, including all CSS functions, of ARFOR and other designated supported elements. The TSC is capable of synchronizing logistics and other support operations for the ASCC. It provides area support to EAC units in the COMMZ and sustainment support to tactical forces. This support may include supply, procurement, property disposal, maintenance, transportation, field services, health services, civil-military affairs, MP support, engineer support, religious support, finance support, and personnel and administrative services.

5-4. Units and commands requiring support coordinate with the TSC support-operations staff to secure their initial support, to reestablish support, or to resolve support problems. In a fully developed theater, the TSC coordinates with a corps support command (COSCOM) for support of combat forces, although direct coordination with a division support command (DISCOM) is sometimes necessary. The TSC, augmented by a rear operations center, is also responsible for security operations as directed by the ASCC/ARFOR commander (see FM 63-4).

MILITARY POLICE SUPPORT

5-5. MP support to EAC units is provided through an array of multifunctional MP units. The nature of the operation, METT-TC, and the requirements of the supported commander will determine which type of MP unit is appropriate to augment, assign, attach, or place under OPCON to an EAC unit. The types of MP units that support EAC include CS, I/R, CID, and L&O teams (such as MWD or customs teams).

THE MP BRIGADE (CS)

5-6. The MP brigade (CS) provides MMS and AS to extended LOC within the COMMZ. These supply corridors include ports, inland waterways, railways, pipelines, airfields, MSRs, and MSR critical points. The MP support the users of the COMMZ's LOC by aggressively patrolling the area along the LOC. They play an important role in securing rear areas by performing combat operations against the threat. When properly augmented, the MP brigade headquarters may serve as the TSC's/ASCC's TCF headquarters. The MP provide MMS on the COMMZ MSRs leading into the corps's rear area. The MP implement the plans of HN and US staff elements to control the forward movement of combat resources along the LOC.

5-7. If resources are available, the MP brigade (CS) provides escorts to move US noncombatants (if present) from AA points to theater embarkation terminals. Until the MP brigade (I/R) arrives in theater, the CS MP units also perform EPW, confinement, and other operations normally performed by the MP I/R units.

5-8. The organization of an MP brigade (CS) supporting EAC includes the following:

- A brigade headquarters and headquarters company (HHC).
- Up to six MP battalions (each with up to six companies).

- Numerous L&O detachments and MWD teams.

5-9. Additionally, the ASCC's PM or commander may attach or direct OPCON of customs teams to the MP brigade (CS). Battalion and company organization in the MP brigade is the same as that in the corps MP brigade; however, METT-TC determines the number of battalions and companies. The MP brigade (CS) has additional MP companies to provide security for EAC-specific units/activities such as—

- A unified command.
- An ASCC and TSC headquarters.
- LOC seaports, airfields, and railways.
- EAC ammunition storage areas.
- EAC petroleum terminals and pipelines.

5-10. While the corps MP brigade (CS) MWD teams are employed in a GS role, the MP brigade (CS) MWD teams are employed to augment seaport security and to conduct inspections of postal items to detect explosive materiel and narcotics.

5-11. The C^2 within the MP brigade (CS) is consistent with that in any Army brigade. The MP brigade commander works directly for the EAC commander, the battalion commander works for the brigade commander, and the company commander works for the battalion commander. However, this usual C^2 relationship may be altered briefly (based on METT-TC) to enhance the overall EAC combat capability for responding to a Level II threat. For example, MP units operating within an ASG's AO may be under the OPCON of the ASG's rear-area operations center (RAOC), which directly tasks MP units responding to Level II threats. The same is true for placing MP units under the OPCON of the EAC's TCF headquarters for responding to Level III threats. Any conflict in mission priorities is resolved through MP C^2 channels.

5-12. The MP brigade commander is both the MP brigade commander and the EAC's PM. He employs his assets according to METT-TC and the commander's concept of operations. Factors affecting his employment of MP assets include the—

- Nature of the operation (joint, combined, or multinational).
- HN's ability to provide MP-related support (such as port security).
- Custody and location of EPWs/CIs during internment operations until I/R units arrive in theater.
- Number of kilometers of the MSR in relationship to movement-control requirements.
- Number and kinds of critical facilities.
- Number of HRP requiring close-in security.
- HN's ability to control the civilian populace, refugees, and dislocated civilians.
- Supply distribution strategy.
- Risk acceptance and threat in the AO.
- Communications requirements (such as using teams as relays).

5-13. Whenever possible, the MP brigade's AO coincides with the territorial responsibility of the supported command. The MP brigade commander assigns the MP battalion's AO by the above factors as well as by METT-TC. For example, the AOR for an MP battalion may be a large population center of larger geographical areas in which CSS complexes and MSRs are located. But as employment factors and the commander's needs change, so will the MP's AOR. The MP brigade commander must move and tailor his forces to meet the current and projected mission requirements. Unlike many other EAC assets, MP units require 100 percent mobility to shift AOs frequently and rapidly. The following vignette depicts the required MP flexibility to support EAC operations:

> During Operation Desert Shield/Storm, MP units were supporting and moving units throughout their AOs at a such an extraordinary rate that many of them had to relocate their headquarters multiple times just to keep pace with their changing AO.

5-14. Most EAC MP assets are employed along LOC and around areas of high troop concentration. Few EAC MP are dedicated to support fixed commitments (such as ports, air bases, and headquarters [discussed in paragraph 5-8]). When supporting fixed commitments, MP units provide a mobile security screen, and they man static positions when directed or when based on METT-TC. The MP brigade (CS) may have to plan for and actually perform the evacuation and internment of EPWs/CIs and the confinement of US military prisoners until the MP brigade (I/R) arrives.

THE MP BRIGADE (I/R)

5-15. US policy requires that all persons held in Army custody be accorded humane care and treatment from the moment of custody to their final release. The policy applies to detained or interned civilians as well as to EPWs and confined US military personnel. This policy is equally binding on all US troops (see FM 19-40).

5-16. The ASCC or the TSC supports US laws, regulations, policies, and international agreements by providing personnel, administrative, morale, internment, resettlement, and confinement services for the TO. The TSC's MP brigade (I/R) in the TO provides this support. However, since most I/R units are in the reserve components, the initial I/R operations (as mentioned above) may have to be conducted by the MP brigade (CS). Once the I/R unit arrives in the AO, it is responsible for—

- Providing firm but humane control of EPWs/CIs and dislocated civilians.
- Coordinating with HN personnel, military territorial organizations, civilian police authorities, NGOs, private volunteer organizations, and US federal agencies on matters pertaining to I/R operations.
- Performing C^2 operations for all I/R units.
- Controlling, employing, and releasing EPWs/CIs as set forth by the Geneva convention and other international laws and by the UN and other governmental bodies.
- Handling US military prisoners.

5-17. In a mature theater in which large numbers of EPWs are captured, the EPWs' requirements may exceed the capacity of the MP brigade (I/R). In this instance, an I/R command is established. An MP command (I/R) has two or more MP brigades (I/R) and will normally be assigned to the ASCC. When the MP command (I/R) assumes OPCON of the MP brigades (I/R) from the lower echelon, it assumes that echelon's I/R mission.

5-18. If the US decides to transfer captured EPWs/CIs to the HN or to another nation, the US must ensure that the nation is a party to the Geneva convention and is willing and able to comply with the convention. In this case, the number and type of I/R MP units required for processing and retaining EPWs/CIs before the transfer is based on agreements and on METT-TC. Additionally, the MP brigade (I/R) is assigned I/R teams that are located at the processing and transfer points and at the HN or third-country EPW camps. The MP brigade (I/R) liaison team will supervise these dispersed teams to ensure that the HN or the third country provides adequate care and security of US-captured EPWs/CIs and that accountability is maintained according to the Geneva convention.

5-19. The MP brigade (I/R) HHC is the C^2 element for the brigade's assets. It consists of the following elements:

- A brigade command section.
- A company headquarters.
- An Adjutant (US Army) (S1).
- An S2.
- An S3.
- A Supply Officer (US Army) (S4).
- A Civil-Affairs Officer (US Army) (S5).
- A Communications Officer (US Army) (S6).
- Finance and accounting.
- Medical operations.
- Public affairs.
- A unit ministry team.
- An SJA.
- An inspector-general (IG) section.

5-20. Other brigade assets may include the following:

- An MP I/R BLD.
- An I/R information center.
- An MP escort-guard company.
- An MP I/R battalion headquarters.
- MP I/R (EPW/CI) detachments.
- MP I/R (confinement) detachments.
- MP guard companies.
- MWD teams.
- Processing squads, processing liaison teams, camp liaison teams, and evacuation teams (all as required to support EPW transfer or to conduct an out-of-theater evacuation).

5-21. The MP brigade (I/R) subordinate units are employed most often in the COMMZ near CSS facilities and are placed near sea, air, and rail terminals. They receive, process, and intern EPWs/CIs; confine US military prisoners; or assist in the resettlement of refugees or dislocated civilians.

5-22. The MP brigade (I/R) escort-guard company supports the evacuation of EPWs/CIs from the CZ. The company has a company headquarters and four platoons. The escort-guard company requires sufficient mobility to transport MP personnel to the CZ and to operate mobile teams while escorting the EPWs/CIs. The MP go forward to the corps's holding area to take custody of the EPWs/CIs. They may go forward to division collecting points, if distances and resources permit. Using any means of available transportation, the MP ensure that the EPWs/CIs are quickly evacuated to MP battalions (I/R) in the COMMZ. Close coordination with the EAC and corps movement-control centers and the corps MP brigade is required to ensure that transportation assets returning to the COMMZ are employed to evacuate EPWs/CIs from the corps's holding area. Walking wounded EPWs/CIs are evacuated by the same means as other EPWs/CIs, while litter patients are evacuated through medical channels. Guarding EPWs/CIs while in the MEDEVAC channels and during their hospitalization is not an MP mission; therefore, there is not an MP force structure to support this mission. In most instances, the impact of having the MP perform this mission causes trade-offs in missions for which they are responsible.

5-23. The theater MP brigade (I/R) and out-of-theater MP brigade (I/R) subordinate units will evacuate EPWs to internment sites within CONUS (if directed). The theater brigade structure is based on the projected capture rate over time and available out-of-theater transportation assets (frequency and capacity). The out-of-theater brigade structure is based on the total EPW/CI population, the number of internment sites, transportation nodes, and escort requirements. Theater escort-guard MP move the EPWs/CIs to the seaport and aerial port of embarkation (SPOE/APOE). The escort-guard MP assigned to the out-of-theater brigade escort the EPWs/CIs from the theater ports of embarkation (POEs) to the out-of-theater internment sites. The out-of-theater brigade is assigned an I/R evacuation detachment, which is employed at and coordinates the evacuation from the theater POEs, through the out-of-theater ports of debarkation (PODs), to the out-of-theater facilities.

THE MP BATTALION (I/R)

5-24. The MP battalion (I/R) is a modular organization and can be configured to operate internment facilities for EPWs/CIs, confine US military prisoners, or resettle dislocated civilians. When performing EPW/CI internment operations, the MP brigade (I/R) has up to 7 MP battalions (I/R); when augmented with the appropriate number of BLDs, it has up to 21 MP battalions (I/R). The ASCC, the TSC, and the MP brigade (I/R) must consider that the requirement to establish an MP battalion (I/R) internment facility is resource intensive. Therefore, MP I/R units, other supporting units, supplies, and equipment for the EPWs/CIs should arrive in theater ahead of the projected EPW/CI arrival at the internment facilities. Early arrival should be based on the time required to establish fully operational facilities (construct and man) and resupply operations before the EPWs/CIs arrive.

5-25. The MP battalion (I/R) has a command section, a company headquarters, and various staff sections. The staff sections provide the core battalion-level capabilities to conduct internment operations. The modulated design expands as the EPW/CI population increases. The battalion is assigned up to four detachments, two guard companies and, if needed, a processing squad. When fully operational, an MP battalion (I/R) operates an enclosure capable of interning 4,000 EPWs/CIs. The battalion mission centers on eight 500-man compounds. The battalion operates the compounds in close proximity to maximize its resources for the security and internment of the EPWs/CIs. This includes the resources needed to employ EPWs/CIs as a labor force according to the provisions of the Geneva convention.

5-26. Each MP detachment (I/R) (EPW/CI) operates two 500-man compounds and provides augmentation to the battalion staff sections to support 1,000 EPWs/CIs. Each guard company is capable of providing security for 2,000 EPWs/CIs. The guard company has a company headquarters and three platoons. The guard company requires sufficient mobility and communications to support routine battalion missions. While minimum mobility and communications is required to support EPW/CI internment operations, on-site guard personnel must often move considerable distances guarding labor groups performing work projects throughout the COMMZ.

5-27. The MP (EPW/CI) processing squad is capable of processing eight EPWs/CIs per hour and includes interpreters to support the processing. If processing squads are required to augment MP battalions (I/R), the operational requirements will be based on METT-TC.

MP-Battalion Resettlement Operations

5-28. The basic organization used for EPW/CI internment is used for resettlement operations. The primary mission-focus change is from guarding EPWs/CIs to protecting and controlling dislocated civilians. As such, an MP battalion (I/R) with four detachments and two guard companies is capable of supporting 8,000 dislocated civilians. However, the MP battalion (I/R) may require augmentation to conduct L&O operations associated with the resettlement. Augmentation may include the full scope of PM functions (operations, investigations, physical security, MP-station operations, and patrols) and civil affairs.

MP-Battalion Confinement Operations

5-29. When configured with the MP detachment (I/R) (confinement), the MP battalion (I/R) is capable of confining US military prisoners. The MP detachment (I/R) (confinement) provides trained corrections and support personnel required for confinement operations. As with the EPW/CI configuration, the modular confinement structure expands as the US prisoner population increases. With three confinement detachments and three guard companies, the MP battalion (I/R) has a maximum capacity of handling 1,500 US prisoners. Generally, only one MP brigade battalion (I/R) is configured for confinement, but the actual number will depend on the number of US prisoners requiring confinement within a theater. While theater policy for confinement operations remains with the ASCC commander, it is the MP brigade (I/R) that executes the mission.

5-30. When possible, soldiers awaiting trial remain in their units unless reasonable grounds exist to believe that they will not appear at the trial, the pretrial hearing, or the investigation or that they will engage in serious criminal misconduct. Under either of these two pretrial confinement instances, the commander must also reasonably believe that a less severe form of restraint (such as conditions of liberty, restriction in lieu of apprehension, or apprehension) is inadequate. When these circumstances exist and other legal requirements are met, US military personnel may be placed in pretrial confinement under the MP's direct control. Commanders may choose to establish field confinement facilities within their AO. However, corps and division MP companies have the expertise to operate only a field detention facility for a limited period of time. These units cannot operate a confinement facility and have neither the resources nor the capability to operate such a facility on an extended basis. Therefore, all confinement is consolidated in the COMMZ whenever possible.

5-31. All assets of the MP battalion (I/R) (confinement) are employed to detain, confine, sustain, and protect US prisoners. As with the battalions conducting EPW/CI and resettlement operations, the battalion conducting confinement operations is generally located in the rear of the COMMZ, near logistics and transportation support. This allows US prisoners to be moved as quickly as possible from the corps's area to the COMMZ's confinement facility. Movements of US prisoners from the COMMZ to CONUS will be according to DA policy.

Chapter 6

Military Police Support to Corps

Corps MP support their command by performing the MP functions critical to the success of their tactical commander's concept of operations.

USAMPS

Corps are the largest tactical units in the US Army. They are the instruments by which higher echelons conduct operations at the operational level. Higher headquarters tailor corps for the theater and the mission for which they are deployed. They contain organic combat, CS, and CSS capabilities to sustain operations for a considerable period of time. Corps are capable of operating in a joint and multinational environment, providing C^2 for up to five divisions and covering up to 35,000 square kilometers.

OVERVIEW

6-1. MP support to a corps is provided by an MP brigade (CS) assigned to each corps. The MP brigade provides combat, CS, and CSS throughout the corps's AO. However, subordinate MP units are not assigned to subordinate corps units. Instead, the MP brigade commander gives them an AO based on the corps commander's concept of operations. When possible, MP battalion AOs coincide with those of the CSG RAOCs.

6-2. The corps MP provide combat power within the command's rear area. They perform combat operations to counter Level II forces and to support the defeat of Level III forces. When properly augmented, the MP brigade may serve as the corps's TCF. The corps MP also provide a critical link between MP operations in the division and in the COMMZ. The corps MP support division commanders by helping the division MP conduct sustainment operations. The corps MP coordinate with the division MP for mutual support.

THE MILITARY POLICE BRIGADE (COMBAT SUPPORT)

6-3. The MP brigade (CS) supporting a corps contains a brigade headquarters, up to six MP battalions (CS), numerous L&O detachments, and MWD teams. Each MP battalion (CS) has up to six MP companies (CS). As with the EAC's MP brigade (CS), the number of battalions and companies is determined by METT-TC. The corps MP brigade (CS) has additional companies to support each division and to provide security for the corps and COSCOM headquarters and corps ammunition storage areas (see Figure 6-1, page 6-2).

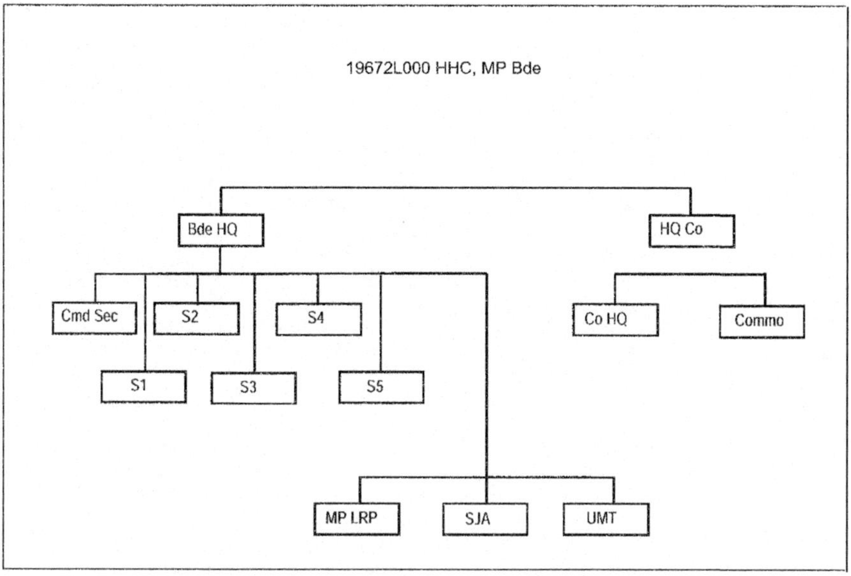

Figure 6-1. Military Police and USACIDC Organizations

6-4. The MP brigade HHC provides C^2 and A/L support to the brigade. The brigade HHC consists of a company headquarters and a brigade headquarters that contains the commander's immediate staff. The staff officers supervise the brigade's major organizational elements, including the—

- S1, S2, S3, S4, and S5.
- MP long-range plans (LRP) section.
- Communications section.
- SJA.
- IG.
- Public affairs.
- Unit ministry team (UMT).

6-5. The MP brigade (CS) command sergeant major (CSM) requires mobility and communications capabilities to execute his duties and responsibilities. The MP LRP section works with the corps G3 plans element, operating out of the corps main CP.

6-6. The MP battalion headquarters and headquarters detachment (HHD) provides C^2 for MP CS companies and any other assigned or attached MP elements. The battalion HHD consists of a detachment headquarters and a battalion headquarters that contains the battalion commander's staff. The staff officers supervise the major organizational elements, including the—

- S1, S2, S3, and S4.
- Communications and support sections.
- UMT.

6-7. As with the MP brigade (CS), the battalion CSM requires mobility and communications. The support section is vital to an MP battalion commander's ability to sustain his widely dispersed assigned or attached units during the performance of all five MP functions, primarily the MMS and AS functions.

6-8. The MP company (CS) provides support to an assigned AO. The company has a company headquarters, an MP operations center, a combat-medic section, and four platoons. The company headquarters provides maintenance, supply, communications, mess, and medical support to the unit. An MP operations center supports the unit's operation, conducting and planning for all five MP functions. The MP operations center includes three MP teams as the company-level response force.

6-9. The MP L&O detachments provide support to an assigned AO. The headquarters provides A/L support. The operation team plans and supervises desk operations, the traffic-accident and MP-investigation teams, and the force-protection teams. As with the EAC MP brigade (CS) detachments, requirements are based on the population supported and on METT-TC.

6-10. The number of assigned MWD teams is significantly less than those assigned to the EAC MP (CS) or (I/R) brigades. A kennel master, five explosive/patrol teams, and four narcotics/patrol teams are normally assigned to the corps MP brigade (CS). They are employed based on METT-TC.

COMMAND AND CONTROL

6-11. The C^2 in a corps MP brigade (CS) is consistent with that of any Army brigade. The MP brigade commander commands the brigade and all attached personnel. Battalion commanders work for the brigade commander. The MP company commanders receive their orders and work for their respective battalion commanders and direct their platoon leaders according to mission requirements.

6-12. When corps MP assets are sent to augment the division PM or the MP company, they are placed under the OPCON of the division PM for as long as the unit is needed in the division area. This command relationship is applicable to company-size organizations and smaller. If an MP battalion (or larger MP unit) augments the division, then the division PM will not exercise OPCON of that MP unit. The MP battalion will be placed OPCON as designated by the division commander but not under the division PM. One L&O detachment is normally attached to the supported division and placed OPCON to the division PM. However, MP leaders at each level establish an MP C^2 relationship based on METT-TC and the supported commander's needs.

EMPLOYMENT

6-13. Each corps MP brigade commander employs his assets according to METT-TC, the needs of the forces operating in his AO, and the priorities of the corps commander. Few MP assets in the corps area are employed to support fixed commitments. Instead, based on a broad consideration of the enemy and friendly situations, the corps MP are employed to support friendly forces engaged in combat, CS, and CSS operations. Since MP forces are dispersed throughout the corps area, the concentration of US forces, the location and vulnerability of critical sites, and the number of kilometers of the MSR to be controlled influence the designation of MP assets in the AO.

6-14. The MP brigade commander establishes his MP concept of operations based on the corps commander's concept of operations. His successful employment of MP assets depends on his foreseeing where the battle will be rather than where the battle is. Based on the rear-area IPB and PIAP, the MP brigade commander allocates and shifts resources to ensure the accomplishment of priority missions. This ensures the continuous support and forward sustainment of combat units and the safety of CSS units operating in the corps rear and sustainment areas.

6-15. To support the MP brigade commander's planning, the MP at brigade headquarters operate from several locations. Most of the staff locates in the vicinity of the corps rear CP where they can interface with the corps staff responsible for planning and executing rear operations. The control element of the brigade headquarters must be located where it can command and control its subordinate units. The brigade commander and his staff decide the best place to locate this element. The planning element of the headquarters locates near the corps main CP where it can interface with the corps commander's coordinating, special, and personal staffs. From there they monitor MP operations, integrate MP support with the corps plans cell for future operations, and learn the enemy situation through the G2's threat analysis almost immediately. The detailed information on rear-area activities and operations provided by the corps staff enhances the accuracy of the MP LRP.

6-16. The brigade S3 section provides the day-to-day planning and execution of an MP mission. The section provides a responsive CP that can relocate frequently throughout the AO. The S3 will normally provide liaison personnel to the corps rear CP, selected COSCOM units, interagencies, or other headquarters (based on mission requirements). When possible, battalion AOs coincide with the AOs of the CSG RAOCs. The MP brigade commander usually tailors battalions' AO boundaries to ensure responsive and flexible support across the corps's AO. He pays particular attention to the LOC behind the most heavily committed division and the critical bases and facilities in that area. He also ensures that the MP are available to respond quickly to combat operations throughout the entire corps rear area or during sustainment operations.

6-17. The MP brigade commander, coordinating with his battalion commanders, locates the MP companies where they can provide combat and CS power throughout the corps's AO. He bases his decision on the—

- Number and composition of urban areas.

- Location of CS and CSS units.
- Location of critical facilities (such as the headquarters, ammunition storage areas, and airfields).
- Expected threat.
- Level and frequency of support needed by the commander.
- Current and projected tactical situation.
- MSR network, including choke points and critical bridges and tunnels.
- Number of supported divisions and requirements.

6-18. The prioritization of MP missions is especially crucial during the early stages of the deployment when it is unlikely that an MP brigade commander will have a full complement of CS companies. Augmenting the division MP company with corps MP assets may not be possible at that time. Until it is, the MP brigade commander must locate corps MP assets to help meet the needs of the division while fulfilling the needs of the corps.

6-19. Like the brigade commander, battalion commanders plan the employment of their companies and platoons using METT-TC. Certain corps needs are constant. One MP company is allocated to provide security for the COSCOM, and one MP company is allocated to provide security for the corps's main CP. One platoon from that company may be used to secure the corps tactical CP or the jump CP. One or more platoons will also help secure the corps's rear CP. The number of MP assets allocated for a corps-level EPW/CI holding area and escort is based on METT-TC. However, a minimum of one platoon is dedicated to operate the corps's EPW/CI holding area and a minimum of one platoon per division is allocated for evacuating EPWs/CIs from division collection points. Additional MP assets may be allocated to provide security for the corps's ammunition storage area and ASPs supporting the divisions. The MP battalion commander places his companies where—

- MP assets support the brigade commander's concept of operations.
- The MP can support troop concentration, bases and base clusters, road networks, and critical areas.
- The MP can aggressively patrol critical terrain and monitor LZs and DZs to detect or deny enemy interference.
- The MP can respond to Level II threats.
- The MP can support the movement of combat resources throughout the AO.
- The MP can remove EPWs/CIs from division collection points.
- The MP can influence stragglers, refugees, and dislocated civilians.

6-20. Battalion commanders may choose to place a company behind the division rear boundary. This can help to ease the coordination between the corps MP and the division MP.

Chapter 7

Military Police Support to Division

Division MP provide essential support to the forward tactical commander. The MP provide a flexible, mobile, and lethal force ready to be employed in combat or CS operations.

USAMPS

The division is a large Army organization that trains and fights as a tactical team. Largely self-sustaining, it is capable of independent operations. The division is a unit of maneuver organized with varying numbers and types of combat, CS, and CSS units. Divisions fall mainly into two categories—heavy or light. The first category includes heavy divisions, their variants, and the Division XXI digitized heavy division. The second category includes the light infantry, airborne, and air-assault divisions. Each division can conduct operations over a wide range of environments. The success of Army operations depends on the success of its divisions.

OVERVIEW

7-1. In corps operations, divisions are normally comprised of three maneuver brigades, each with up to nine maneuver battalions, artillery battalions, aviation battalions, and supporting CS and CSS units. Divisions perform a wide range of tactical missions and are self-sustaining for limited periods of time. The corps augments divisions as the mission requires. All divisions must be able to deploy and conduct offensive, defensive, stability, and support operations. Airborne and air-assault divisions must be able to conduct forced-entry operations. Divisions may be part of a JTF or serve as the ARFOR headquarters.

7-2. MP support to a division is provided by an organic MP company that includes the PM and his staff (assigned to each division) and is augmented by the corps MP brigade. The division MP company provides support in the DRA and in the maneuver brigade rear area. It is fully mobile in order to relocate frequently under short notice. While division MP companies are capable of performing all five MP functions, they must be augmented by corps MP/CID to fully perform all five MP functions throughout the division's battle space. The division PM must receive at least one corps MP company for GS missions in the DRA, one L&O detachment for L&O missions, and one CID detachment for criminal investigations and LOGSEC. In a mature theater (or based on METT-TC), the division PM can expect two corps MP companies. In addition, the division PM may request a corps MP battalion headquarters to provide C^2 to corps MP units operating in the DRA or to perform missions as a TCF. Because the need for MP support exceeds division organic assets (and many times exceeds augmenting corps MP assets), careful planning of MP employment is essential.

DIVISION MILITARY POLICE COMPANY

7-3. The organization of the division MP company is set by specific tables of organization and equipment (TOE) and is designed to support a specific type of division. Regardless of the type of division, the assets in a division MP company enable a PM and his supporting staff to supervise the performance of all five MP functions. The PM coordinates the employment of the MP assets in the division's AO through his PM section. The division MP company contains a headquarters, a PM section, a medic section, and three to six platoons.

COMMAND AND CONTROL

7-4. The C^2 of the division MP company extends downward from the tactical commander to the division PM. The division PM is the division commander's special staff officer. Under normal circumstances, he works for the chief of staff. The division PM has OPCON of the division MP company and MP assets that have been provided by the corps. The division PM plans and employs all of the organic, assigned, or attached MP according to METT-TC. He anticipates the support needs of the division commander and plans the employment to meet his needs. After learning the division commander's concept of operations and estimating the situation, the division PM decides which MP functions are critical to accomplish the commander's mission. He then recommends the allocation of resources, tasks his assets, and establishes the priority of support.

7-5. When the division MP company is augmented by corps MP platoons, the platoons are placed under the direction of the division MP company commander, who assigns an AO to each platoon. When corps MP assets augmenting the division make up a company, the company is attached to the division. The company then comes under the OPCON of the PM, who assigns it an AO. The augmenting corps MP company is usually given an AO from the division rear boundary forward. However, this assignment is flexible and based on the division commander's needs as set forth by the division PM. The four platoons of that corps MP company perform all five MP functions in that AO. Additionally, the corps MP brigade (CS) L&O detachment supporting a division is attached to the division and OPCON to the division PM.

7-6. The division MP company commander directs the employment of company assets. Through his company headquarters, he provides administrative, maintenance, and logistical support to the PM section and to the company's platoons that are dispersed throughout the division's AO. These sections are mobile to support the platoons without delay despite the extended distances that may be separating them. The company commander is also mobile. He travels the extended distances from one platoon area to another when checking on the status of his soldiers and resolving the problems affecting mission accomplishment. The division HHC provides mess support. Personnel administration is handled by the company headquarters and the DISCOM. The number and kind of assets in a division MP company and the configuration of its platoons and squads are determined by the type of division to which the company is assigned.

NOTE: See the consolidated TOE update that is published yearly for complete details (base and objective) for all MP TOEs.

7-7. At the direction of the division commander, the division band may be available to perform its secondary mission of supporting MP operations. When the intensity of the conflict prevents the band from performing its musical mission, the division commander may direct it to augment the MP company. The band, normally employed as a unit under the direction of the bandmaster, is under the OPCON of the MP company commander while augmenting the MP. When tasked to augment MP assets, band members may be employed to augment MP security at the division main CP or to augment perimeter security at the division EPW/CI collection point. The band is released to perform its primary mission as soon as the tactical situation permits.

MILITARY POLICE EMPLOYMENT

7-8. The employment of division MP companies differs somewhat with the type of division to which they are assigned. In heavy divisions, the division MP companies provide GS to the DRA and DS to the division's maneuver brigades. When the platoon is providing DS to a maneuver brigade, the MP platoon leader coordinates all logistical support with the supported brigade. A platoon's DS relationship is seldom interrupted. However, METT-TC may require weighing MP support to the maneuver brigade tasked as the main effort or employing all platoons in a division MP company to function as a unit (such as during division-size river-crossing operations or deliberate breaching operations). In light divisions, the MP provide only GS to the DRA or during sustainment operations. Support exceptions are addressed under MP support to airborne and air-assault divisions, discussed later in this chapter.

7-9. In the division (where flexible support of an austere force is crucial), the division PM must have a clear understanding of situational awareness. To obtain current information for projecting MP needs in the division area, he must be mobile and be able to conduct split-cell operations. The assets available to the PM include the division MP company and at least one corps MP company. Corps augmentation is required for sustained operations and for special operations such as river crossings, dealing with dislocated civilians, and refugee internment or resettlement. The division PM coordinates with the corps PM and the MP brigade or CID commanders for—

- Evacuating and guarding EPWs/CIs from division to corps.
- Providing law-enforcement assistance to HN forces in the division's AO.
- Providing corps augmentation for the division's AO, convoy security, LOC security, AS, R&S, L&O, and other missions. These other missions may include augmentation for security of the division main CP, ASPs, and other critical facilities.
- Integrating, sharing, and exchanging police intelligence between corps and division MP elements.
- Providing CID support.

SUPPORT TO THE HEAVY DIVISION

7-10. The Army's armored and mechanized infantry divisions (normally referred to as heavy divisions) provide mobile, armor-protected firepower that is normally employed for their mobility, survivability, lethality, and psychological effect (shock) on the enemy. These divisions destroy enemy armored and mechanized forces, and they can seize land areas and secure key terrain. Because of their strategic lift requirements, heavy divisions are slow to deploy from home staging bases into an AO. They have high consumption rates of supplies and have limited use in restrictive terrain. These capabilities and limitations are key factors in planning effective and efficient MP support throughout the division's AO.

7-11. In the heavy divisions, the PM section is organized to support split-cell operations at the main and rear CPs. The PM must be mobile to ensure that he is fully aware of the current status of critical MP operations. Therefore, the deputy PM (DPM) locates in one of the PM cells and—

- Handles the section's routine operations.
- Monitors ongoing division operations and MP support.
- Helps provide long-range planning and interface with the primary division staff.
- Forwards PM taskings to the division MP company commander.

7-12. The DPM and the operations sergeant normally set up operations at the division main CP where they can coordinate requirements with the division staff personnel. The company headquarters is initially located near the division rear CP in the division support area (DSA). Once augmentation arrives from the corps, the headquarters relocates with one of its GS platoons operating behind the brigade rear boundary. However, this location could be changed based on METT-TC, the supported commander's needs, and the company commander's idea of where he can exercise better C^2 for his unit.

7-13. The Army of Excellence (AOE) heavy division MP company has six platoons. Three platoons provide support to each maneuver brigade and are designated as DS. The other three platoons are designated as GS platoons. One MP platoon provides security for the division main CP; one provides security for the division's EPW central collection point; and one performs other MP operations within the division rear.

7-14. The GS MP platoons' AOs are configured based on METT-TC and the availability of MP augmentation from the corps. The DS MP platoons' AOs coincide with the supported maneuver brigade's boundary. Each platoon headquarters locates within its brigade's support area or any other area where it can best provide and receive support. To accomplish its mission, each DS platoon requires a minimum of two squads, each with three teams. One squad operates the EPW/CI collection point. The other squads perform MMS and AS operations. All MP platoons are capable of performing all five MP functions. However, performance of these functions is prioritized based on METT-TC and the division commander's concept of operations. The division PM, the company commander, and METT-TC dictate how these platoons should be tasked-organized to accomplish the mission.

SUPPORT TO THE LIGHT INFANTRY DIVISION

7-15. The AOE light infantry division (LID) is one of the most rapidly and strategically deployable divisions. It fights as part of a larger force in war or conducts missions as part of a joint force in MOOTW. Its C^2 structure readily accepts any augmentation forces, permitting task-organizing for any situation. The augmentation required for the division is largely determined by METT-TC. The division's capabilities allow it to exploit the advantages of restricted terrain and limited visibility. It achieves mass through the combined effects of synchronized small-unit operations and fires rather than through the physical concentration of forces on the battlefield. These characteristics are key factors in planning and employing MP assets in support of the LID.

7-16. The LID MP companies are capable of performing all five MP functions. However, their performance of these functions is prioritized based on METT-TC and the division commander's concept of operations. Contrary to the heavy division MP company, the LID MP companies are much smaller. The constrained size of the LID MP companies makes corps MP augmentation crucial to the sustainment of MP operations. Additionally, the LID MP company is the only MP unit with the capability of antiaircraft support through the use of shoulder-fired air-defense weapons.

7-17. The company has three GS platoons to support the division. No platoons are provided to the maneuver brigade. One platoon is normally located in the vicinity of the division main CP so that its resources can help support CP security. Another platoon locates in the DSA and operates the division EPW/CI collection point. The last platoon has an AO configured according to METT-TC and the commander's priority of MP missions. Each GS MP platoon has a headquarters and three squads, each with two teams. The PM section is located in the vicinity of the division main CP. The exact location is based on the current operational status and on METT-TC.

SUPPORT TO THE AIRBORNE DIVISION

7-18. The AOE airborne division can rapidly deploy anywhere in the world to seize and secure vital objectives. It conducts parachute assaults to capture initial lodgments, executes large-scale tactical raids, secures intermediate staging bases (ISBs) or forward operating bases (FOBs) for ground and air operations, and rescues US nationals besieged overseas. It can also serve as a strategic or theater reserve as well as a reinforcement for forward-presence forces. The airborne division can assault deep into the enemy's rear area to secure terrain or to interdict enemy supply and withdrawal routes. It can seize and repair airfields to provide a FOB and airheads for follow-on air-landed forces. It is capable of all other missions assigned to LIDs. The airborne division does not have sufficient armored protection to defeat heavier armored formations at close range. Therefore, engagements with enemy armored formations require special considerations. Antitank weapons in the division compensate for, but do not completely offset, this deficit.

7-19. Airborne divisions conduct operations in two phases—assault and defense. The division accomplishes the assault phase in three stages. First, they deploy; second, they establish the airhead; and third, they extend the logistics base and build their force.

7-20. Like other division MP companies, those assigned to the airborne division are employed to support their division commander's concept of operations. The airborne MP company has four platoons, each having three squads with two teams. Normally, the company headquarters and one of the platoon headquarters colocates with the PM section. The division PM's exact location depends on METT-TC. During the assault phase, the airborne division MP company provides DS to the assault brigade. The remaining platoons provide MMS and AS in the vicinity of the division main CP.

7-21. The nature of airborne operations makes the capture of EPWs likely. Therefore, during the first stage of the assault phase, the priority of MP support is given to EPW operations. After assembling the DZ or LZ, the MP collect EPWs captured during the assault. Combat elements are relieved of EPWs as far forward as possible. In airborne operations, EPWs are held for later movement to a central collection point. During the first stage of the assault, the MP perform limited straggler and refugee control and undertake AS operations, when possible.

7-22. MP support is reevaluated after the airhead is established in the second stage of the assault. The PM takes in consideration personnel and equipment flow, roadways, and security requirements to shift MP support priorities. As the entire operation matures, MP support may expand to include all five MP functions.

SUPPORT TO THE AIR-ASSAULT DIVISION

7-23. The air-assault division executes tactical missions at operational depths to achieve strategic results. It is capable of launching brigade-size air assaults of nearly 4,000 soldiers from either an ISB or a tactical assembly area (TAA). Within 6 hours, this air-assault task force (AATF) can attack an opposing force, occupy and defend key terrain, or establish a FOB (out to 150 kilometers) from which even deeper operations can be executed. Air-assault operations are terrain independent, but they rely on suitable weather and a detailed attrition/assessment of enemy capability—particularly air-defense assets along air corridors and in the objective area. The air assault is preceded by detailed, lethal and nonlethal condition settings, culminating in a comprehensive condition check before execution.

7-24. As the corps AATF unit, the brigade combat team (BCT) task-organizes and habitually trains with both aviation lift and attack (Apache) battalions. An air-assault division MP company has four platoons, each having three squads with two teams. To facilitate operations, air-assault division MP platoons are habitually aligned with each of the three AOE BCTs; however, as with aviation assets, the MP are task-organized to support the main effort's BCT/AATF. When conducting these operations, two MP platoons may be tasked to provide support; one is available to posture and participate in the air assault, and the other conducts AS and MMS operations at the pickup zone (PZ). This second MP platoon may further provide MMS and accompany a ground-assault convoy (GAC) to the objective, or it may remain at the PZ to receive EPWs returning on aircraft from the objective area. The mission of conducting MMS along this vulnerable ground LOC is particularly critical if the AATF objective is a FOB from which further division operations will expand.

7-25. MP participation in the actual air assault competes for heavy lift with FA and air-defense systems, as well as with vital Class V resupply. However, the lethality and versatility of the MP bode well for their employment, and two MP platoons are available to support the brigade air assault as the division's center of gravity.

7-26. When possible, habitually aligned platoons remain with their brigades, and corps assets perform GS missions. However, when no corps assets are available and two division platoons are employed as stated above, the two remaining platoons conduct division EPW collection-point operations and other MP functions based on METT-TC. Normally, the EPW platoon and the MP company headquarters colocate in the DSA. As required (and based on METT-TC), airflow planning includes EPW/CI evacuation from the AATF/ FOB collection point back to the DSA. The PM section operates from the division rear CP to facilitate I/R operations and to coordinate MMS and AS with key logistical staff. Due to potentially extreme distances on the air-assault battlefield, the DPM normally locates with the division main CP to serve as a key G3 battle-staff member and to coordinate PIO with the G2.

SUPPORT TO DIVISION XXI (HEAVY DIVISION)

7-27. Division XXI represents a significant change in the manner in which division operations are conducted. These changes are brought on by information-age capabilities; an increased integration of service components into an effective battle team; more lethal, survivable, and agile systems; and more capable soldiers and leaders. The Division XXI operates in a larger battle space and at a higher tempo than the AOE division. The division is evolutionary in design, but revolutionary in its use of information technology. It improves the Army's deployability while enhancing its ability to dominate in decisive fights. The following are characteristics of the Division XXI operational environment:

- **Multidimensional.** The division will operate in an extended battle space that goes beyond the traditional physical dimension of width, depth, and height. It includes portions of the electromagnetic spectrum; and it extends beyond the physical boundaries of the division through its communication and digital connectivity to other Army, joint, and coalition elements, even reaching back to CONUS.
- **Precise.** Precision operations go beyond a precision strike to include every aspect of military operations from deployment through combat and through redeployment or transition to other operations. Precision in decisive operations is enabled by three emerging capabilities. First, digitization provides soldiers and leaders at each echelon the information required for making decisions. Second, a full suite of strategic, operational, and tactical sensors linked to analytical teams fuses combat information into situational awareness across the battle space. Third, simulations enable Army elements to be tailored and operations to be planned, war-gamed, and rehearsed—yielding precision execution.

- **Nonlinear.** Nonlinear operations do not seek a battle-space grid of close, deep, and rear operations. Instead, the battle space is fluid, changing as METT-TC changes through the duration of mission preparation and execution. Another dimension of this characteristic is the synchronization of near-simultaneous operations to achieve nonlinear effects across the battle space.
- **Distributed.** Division operations are distributed or executed where and when required to achieve decisive effects concentrated at a decisive point. Dispersion empowers subordinates to operate independently within the commander's intent, leading to synergistic effects that exceed the effects of a centralized headquarters.
- **Simultaneous.** The concept of decentralization operations that are multidimensional, precise, distributed, and nonlinear yields the capability to conduct simultaneous operations across the battle space. Simultaneous operations seize the initiative and present the enemy leadership with multiple crises and no effective responses. Rather than a single, concentrated attack, the division executes a series of attacks (lethal and nonlethal) as simultaneously as possible.
- **Integrated.** Division operations are fully integrated with joint, multinational, and nongovernmental partners. Integrated operations enable the Army to leverage the full suite of capabilities that the services bring to the battle space.

7-28. The Division XXI heavy division is one of total integration—not rounding up or down. Reserve-component soldiers assigned to the division are an integral part of the division mission and task accomplishment. The division cannot operate without them. The division is focused on massing effects on the enemy—enabled by a new foundation of information technology and distributed logistics. The division is built around the pattern of "conduct decisive operations," accomplished by either fires or maneuver. The design performs complementary battle-space shaping in concert with the overall corps plan. Its long-range assets are not typically held in reserve nor just employed in the close fight; they are used to best support the corps and division commanders' plans of operations.

7-29. Although the basic structure of the division has not changed—three maneuver brigades, a division artillery, and a DISCOM—the internal structure and operational construction have changed in a near revolutionary way. The C^2 structure and the basic organizational structure are very similar to the divisions discussed above. However, by integrating the capabilities of digital technology and by using enhanced situational awareness, it is better able to execute combat operations.

7-30. The maneuver brigades have their own scouts—a brigade reconnaissance troop (BRT); therefore, they no longer rely exclusively on the battalion scouts. The maneuver battalions are reduced to three companies with a total number of 45 combat platforms. The division artillery has three Multiple-Launch Rocket System (MLRS) batteries per battalion. The cavalry squadron is under the aviation brigade's C^2 but retains and performs the division's traditional cavalry mission. Figure 7-1 depicts the organizational structure of Division XXI.

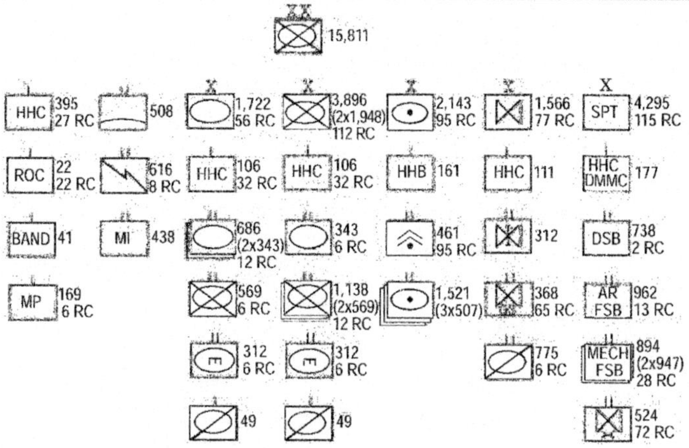

NOTE: Personnel figures are estimates; they may change.

Figure 7-1. Division XXI Structure

7-31. The division's CSS is centralized. Instead of each unit having its own organic CSS personnel, they are now centralized in CSS units. The CSS assets are reorganized and reassigned to the DISCOM. Forward-support companies (FSCs) are habitually associated with each maneuver battalion and forward-support battalions (FSBs) will support each maneuver brigade. This operational structure enables distribution-based rather than supply-based logistics concepts to support the division. The overall logistics footprint for the division is now reduced and the division has organic self-sufficiency to operate unsupported for up to 30 hours.

7-32. A major difference between the AOE fix-forward doctrine and the Division XXI maintenance support is the change in the actual repair site. Division XXI maintenance units directly supporting a given customer will diagnose faults and will replace components forward in the battlefield and will repair rearward. Units will be manned with maintenance personnel capable of performing DS and limited GS maintenance.

7-33. The division operates in a battle space two and a half times larger than an AOE division, and it will usually fight as part of a corps or a JTF. The division conducts offensive, defensive, stability, and support operations in cooperation with other elements of the corps/JTF to gain tactical or operational results significant to the joint-force commander's campaign goals.

7-34. Although the division is a heavy division, MP support is somewhat different from that of an AOE heavy division. The PM (based on METT-TC) establishes operations where MP assets can best be commanded. At the same time, overall MP operations can be synchronized to support operations within the division rear and the division commander's intent can be effectively supported. To this end, the PM conducts split-cell operations within the division main CP. Normally, the PM section is distributed between the security-operations cell (SOC) and the mobility/survivability (M/S) cell. The DPM (who works out of the M/S cell) is also part of the division planning team. He ensures that MP operations are integrated into division plans according to the PM's intent. The DPM attends coordination meetings and assumes coordination responsibility with the rest of the division staff when the PM relocates away from the division main CP.

7-35. The PM, as a staff officer, is involved with every aspect of planning and with the military decision-making process (MDMP). He uses the Force XXI Battle Command—Brigade and Below (FBCB2) C^2 system to understand the battlefield. He understands the mission and the commander's intent and priorities. The PM has complete situational awareness, establishes an effective staff relationship with the G3, and recommends the best way to employ MP resources.

7-36. The Division XXI heavy division MP company has five platoons. Although tasked-organized based on METT-TC and the commander's needs, three MP platoons provide DS for each maneuver brigade. The DS platoon requires a minimum of three squads (each having three teams) to accomplish the mission. The other MP platoons are normally employed forward in sector, providing AS and force protection for the division main and tactical (TAC) CPs and for other HVAs in the vicinity of the maneuver brigade rear boundary. As with the AOE heavy division, the Division XXI heavy division requires augmentation with at least one corps MP company for CS missions in the DRA and can expect two companies in a mature theater (based on METT-TC) as the main effort. In addition, the division may request an MP battalion headquarters to command and control CS MP operations in the rear or to conduct TCF operations.

7-37. While the division MP company is capable of performing all five MP functions, it must be augmented by the corps MP to fully perform all of these functions throughout a mature division's battle space. The PM prioritizes MP functions (based on the supported commander's needs), conducts a mission analysis, and requests needed corps assets through the division G3. Corps MP companies, L&O detachments, and CID assets are needed for sustainment operations. When corps MP companies and L&O detachments are attached to the division, they are placed under OPCON of the PM. The PM will then assign an AO and missions based on METT-TC.

7-38. The PM supports the division's distribution-based logistics concept by conducting aggressive R&S operations that secure the LOC. Since the main effort is to secure the LOC, traditional MMS missions (such as convoy security and escorts) are performed by exception. Through the allocation of increased MP to conduct LOC R&S operations aggressively, the division is able to detect and destroy threat forces before they interrupt operations in rear-area or sustainment operations.

7-39. Digitization gives the division the situational awareness that allows it to cover much larger distances and to move much faster. The systems that provide this digitization are generally unprotected HVAs on the battlefield, which require aggressive AS and force-protection operations.

7-40. The EPW operations are much more fluid, sometimes requiring the division GS MP to move forward and assume brigade collection points, thus allowing DS platoons to move forward with their brigades. Similarly, corps MP may be required to move forward and assume the division's central collection points as the division's rear boundary is pushed forward to shorten the LOC.

7-41. As with all AOE divisions, until corps L&O and CID detachments augment the division, L&O operations are on an exception basis. The MP collect intelligence during every facet of mission execution. The PM develops police intelligence in the division main CP through coordination with the SOC, the G2, the Assistant Chief of Staff, G5 (Civil Affairs) (G5), and the Assistant Chief of Staff, G6 (Communications) (G6).

Chapter 8

Separate Brigades and Initial/Interim Brigade Combat Teams

The MP support separate brigades and IBCTs designed to provide the brigade commander with MP assets that can serve as a force multiplier and that can be employed as an economy of force. The brigade MP platoon is as lethal, flexible, and capable as any other platoon in the MP Corps.

USAMPS

The Army uses separate brigades to inject a small but powerful force where it is needed. It must be able to fight and win while operating on its own for a sustained period of time. It must be able to defend itself on a 360-degree front in war or MOOTW. While there are some variations of separate brigades, it is the heavy separate brigade that is most commonly found within the force.

OVERVIEW

8-1. The Army's IBCT is a full-spectrum, wheeled combat force. It is employed in all operational environments against all projected future threats. However, it is designed and optimized primarily for employment in small-scale contingency operations in complex and urban terrain, confronting low-end and midrange threats that may employ both conventional and asymmetric capabilities. The IBCT deploys very rapidly, executes early entry, and conducts effective combat operations immediately on arrival to prevent, contain, stabilize, or resolve a conflict through shaping and decisive operations. The IBCT participates in war (with augmentation) as a subordinate maneuver component within a division or a corps and in a variety of possible roles. The IBCT also participates (with appropriate augmentation) in stability and support operations as an initial-entry force or as a guarantor to provide security for stability forces by means of its extensive capabilities.

SUPPORT TO SEPARATE BRIGADES

8-2. MP support to separate brigades is provided by a four-squad MP platoon organic to the brigade HHC. A separate PM cell within the brigade HHC serves as the C^2 element for the platoon (see Figure 8-1, page 8-2). Support to the platoon and the PM section for maintenance, supply, mess, and communications is provided by the brigade HHC. Since the platoon and PM section have no organic support, the MP leadership must perform close coordination for this support. However, the MP platoon must compete with other brigade HHC assets for priority of repair for weapons, vehicles, and communications equipment.

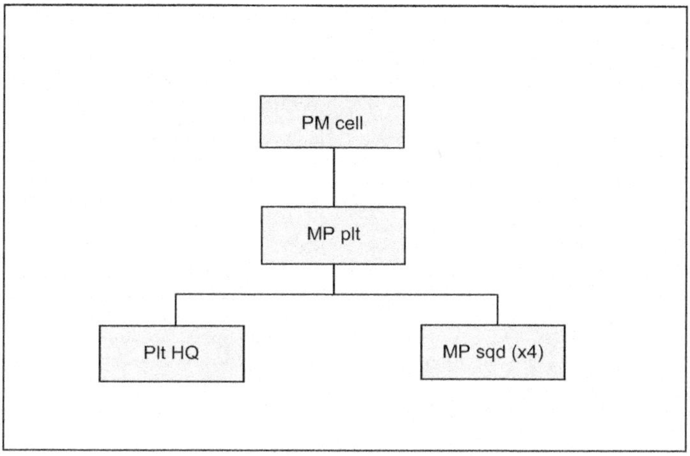

Figure 8-1. MP Support to a Separate Brigade

8-3. The C^2 of an MP unit supporting separate brigades also extends down from the tactical commander. The separate brigade PM has OPCON of brigade MP assets the same way the division PM has OPCON of division MP assets. This includes any MP assets that may have been provided from the corps. The PM advises the commander of a separate brigade on matters pertaining to MP operations. The platoon leader directs the execution of his platoon's mission based on the priorities set forth by the PM and the supported commander. Since the separate brigade's organic MP platoon is more robust than an MP platoon supporting a division maneuver brigade, METT-TC will determine the requirement for augmentation. However, corps L&O and CID augmentation is required.

8-4. The separate brigade's MP platoon is capable of performing all five MP functions. However, its resources are quite limited. Although the MP squads are employed according to METT-TC, the platoon supporting the separate brigade may have—

- One squad operating in the EPW collection point.
- One squad providing a mobile security screen and providing AS around the brigade's main CP.
- Two squads conducting MMS and AS throughout the brigade's rear area.

SUPPORT TO THE INITIAL/INTERIM BRIGADE COMBAT TEAM

8-5. The IBCT is a divisional brigade. It is designed to optimize its organizational effectiveness and to balance the traditional domains of lethality, mobility, and survivability with the domains required for responsiveness, deployability, sustainability, and a reduced in-theater footprint. Its two core qualities are its high mobility (strategic, operational, and tactical) and its ability to achieve decisive action through a dismounted infantry assault. The major fighting components are its motorized infantry battalions. The IBCT has a unique reconnaissance, surveillance, and target acquisition (RSTA) squadron to enhance situational understanding.

8-6. To achieve a rapid deployment threshold, the brigade's design capitalizes on the widespread use of common vehicular platforms—particularly a highly mobile, medium-weight, combat/CS platform coupled with the minimization of the personnel and logistical footprint on theater. Encompassing a personnel strength of about 3,500 and preconfigured in ready-to-fight combined-arms packages, the entire IBCT can deploy within 96 hours of "first aircraft wheels up" and begin operations immediately upon arrival at the APOD. In essence, the APOD is the TAA. The IBCT cannot conduct forced-entry operations, but it provides the JFC with an improved capability to arrive immediately behind forced-entry forces and begin operations to shape the battle space and to execute decisive action to expedite decisions. Once committed, the IBCT can sustain operations for up to 180 days without relief.

8-7. The IBCT's organization is expandable either through augmentation or scalability (according to METT-TC) in any given contingency. The IBCT is scalable in terms of its ability to accept like-type additional forces to expand core tasks and functions already resident within the IBCT (for example, adding additional infantry or RSTA organizations). The IBCT is also capable of accepting augmentation consisting of units or elements that execute tasks or functions not resident within the IBCT (for example, adding armor, air-defense, MP, or aviation assets). In both cases, added units execute their normal mission-essential task list (METL) tasks and, therefore, will not require extensive training to deploy and operate with the IBCT. In both cases, the IBCT includes the command, control, and communications (C^3) capabilities necessary to permit the rapid integration of additional enabling capabilities, particularly for operations outside the scope of small-scale contingency operations (such as stability operations, support operations, and war).

8-8. The IBCT organization excludes other unit-based capabilities often provided in a division slice. However, for each missing capability, the brigade headquarters includes staff cells capable of conducting rudimentary planning and analyses to ensure that all functional-area considerations are incorporated in route planning and preparation for operations. The first MP elements in support of the IBCT are at the brigade headquarters level.

8-9. The MP planning cell is composed of a two-person (a major [MAJ] and a sergeant first class [SFC]) planning cell located within the HHC's maneuver-support cell and under the direct staff supervision of the brigade S3. The role of the MP planning cell is significantly different from that of a traditional division PM or a separate brigade PM. The main difference is in the IBCT's lack of organic or habitual MP assets. However, the absence of organic MP assets makes the job of the MP planning cell that much more critical. The MP planning cell must—

- Understand the organization, capabilities, and limitations of the IBCT.

- Conduct effective liaison with higher headquarters PM elements.
- Become an effective planner and anticipator of MP requirements.
- Prepare MP staff estimates and employment recommendations.
- Plan for MP deployment via air, sea, rail, or land.
- Task-organize MP units effectively and efficiently.
- Assume C^2 of incoming MP forces or, if operating under a division or corps, relinquish C^2 to the division/corps PM or battalion commander (if appropriate for effective employment of MP forces).

8-10. Depending on METT-TC, the brigade may be augmented by MP elements ranging from a platoon to a battalion. In any case, the MP planning cell then becomes a staff planner and coordinator of functional matters pertaining to—

- Force protection, physical security, and vulnerability countermeasures.
- Operations for collecting, processing, and evacuating (possibly) EPWs, CIs, and detainees.
- US military prisoner operations.
- Operations for processing and controlling dislocated civilians and refugees.
- Customs and counterdrug operations.
- Criminal investigations and CID-/MP-investigator support.
- PIO and establishing links with HN police agencies and other international/interagency law-enforcement agencies.
- MMS, AS, and L&O operations.
- Coordination of MP or CID division/echelons above division (EAD) augmentation forces, MWD teams (explosives, narcotics, and patrol), L&O detachments, and MP I/R units through the division, corps, and major Army command (MACOM) PMs.
- Training guidance to other US forces required to perform MP functions in the absence of MP forces.

8-11. Since the IBCT is a divisional brigade, the division PM and the IBCT MP planning cell play an important role in developing an optimum MP force package to support the brigade commander's concept of operations. Despite the brigade's early-entry timelines, the MP planning cell must consider and plan for MP augmentation forces as early as possible to free valuable combat resources. Small-scale contingency operations that result in numerous EPWs, civilian detainees, and refugees will hamper momentum and freedom of maneuver.

8-12. Once the initial brigade receives MP augmentation (see Figure 8-2), the MP priority of effort during the offense will focus on ensuring that routes remain unencumbered and secure for movement of ground combat, CS, and CSS forces. The MP's priority of effort during the defense will focus on conducting AS and counterreconnaissance along the LOC, C^2 centers, and CSS bases. The MP may conduct response-force operations or become part of the TCF.

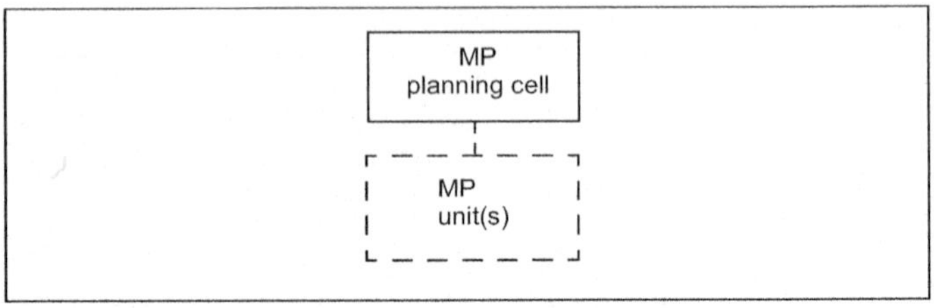

Figure 8-2. MP Support to the IBCT

8-13. Stability and support operations present some unique challenges. The IBCT may be deployed to a geographical area that is politically unstable, that lacks civil control, or that is in complete turmoil. The MP planning cell plans for and coordinates MP support according to available resources and the supported commander's needs. In this scenario (and based on METT-TC), an MP battalion TF could be required to deal with the challenges of—

- A significant number of refugees or dislocated civilians.
- AS or force-protection issues.
- MMS operations.
- Black-market and criminal investigations.
- Restoration of order.
- Civil unrest.
- Intervention of private and nongovernmental organizations.

Chapter 9

The United States Army Criminal Investigation Command

The history of the USACIDC goes back to World War I, when General John J. Pershing organized the CID in France. Today, the USACIDC is the Army's sole agent responsible for investigating felony crimes on and off the battlefield. The USACIDC provides investigative support to commanders at all echelons.

USACIDC

The USACIDC investigates offenses committed against US forces or property, or those committed by military personnel or civilians serving with US forces or where there is a military interest. The USACIDC agents investigate violations of international agreements on land warfare. They conduct special investigations at the direction of the USACIDC's commanding general (CG) or a higher authority. In general, the USACIDC's missions include—

- Investigating and deterring serious crimes.
- Conducting sensitive/serious investigations.
- Collecting, analyzing, processing, and disseminating criminal intelligence (CRIMINTEL).
- Conducting protective-service operations for designated personnel.
- Providing forensic-laboratory support.
- Maintaining Army criminal records.
- Enhancing the commander's crime-prevention and force-protection programs.
- Performing LOGSEC operations.

OVERVIEW

9-1. The USACIDC's operations help the commander maintain discipline and order by preventing or investigating crimes that reduce a unit's ability to fight. During the investigation of serious crimes, the USACIDC concentrates its efforts on investigating serious crimes such as wrongful deaths, controlled-substance offenses, theft, fraud, sex crimes, and assaults. The USACIDC also conducts sensitive and special investigations involving matters pertaining to senior Army officials and those associated with classified programs.

9-2. The USACIDC agents collect, analyze, process, and disseminate criminal intelligence/information relating to crime within or directed toward the Army. Specific information relating to modus operandi, crime techniques, investigative leads, gang violence, and terrorism is shared with the appropriate intelligence and law-enforcement agencies. Conversely, USACIDC agents solicit and receive crime-related information from the MP and from local, national, and foreign law-enforcement and intelligence agencies. Special agents identify and evaluate crime-conducive conditions and indicators of potential attacks against Army property, facilities, or personnel. They then provide reports to the appropriate commander.

COMMAND AND CONTROL

9-3. The USACIDC is a centralized (stovepipe) MACOM whose special agents in the field report through the USACIDC's chain of command (detachment to battalion to group) to the CG, who in turn reports directly to the Army Chief of Staff and the Secretary of the Army (see Figure 9-1).

9-4. As an independent MACOM, the USACIDC's tactical units are not under the C^2 of supported organizations. The USACIDC's elements may be temporarily attached to a supported commander when required to accomplish a nonstandard mission. Reassignment, promotion, accreditation, and disciplinary actions are retained by the USACIDC. Attachments will be coordinated with the appropriate USACIDC headquarters planning the specific mission and approved by the CG, USACIDC. Additionally, although there is no formal staff relationship, USACIDC commanders advise their supported commanders on criminal-investigation matters. This enhances the quality, reliability of information, support, and trust between USACIDC elements and their supported commanders.

WARTIME SUPPORT

9-5. The USACIDC supports each echelon of command from the division to the ASCC. The theater USACIDC structure is comprised of a C^2 headquarters

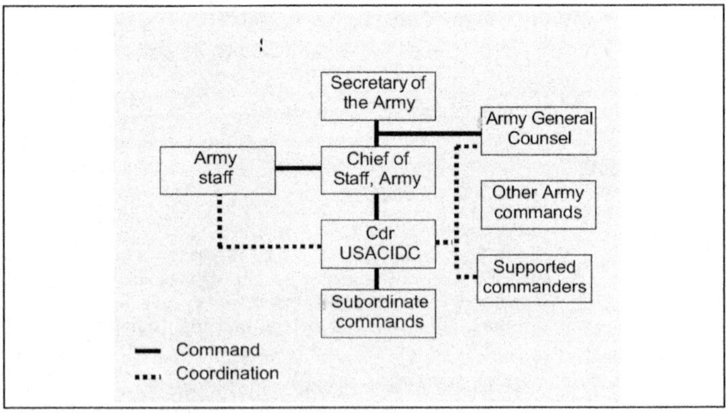

Figure 9-1. USACIDC Chain of Command

and mobile, modular, and tailorable investigative detachments. The USACIDC supports combatant commanders with the following functions:

- **LOGSEC.** Tracking and protecting materials and equipment from the manufacturer to the soldier on the battlefield.
- **CRIMINTEL.** Collecting, consolidating, analyzing, and disseminating intelligence associated with criminal and terrorists activities targeted at Army interests.
- **Criminal investigations.** Investigating war crimes and, in some cases, crimes against coalition forces and HN personnel.
- **Protective-service operations.** As in peacetime, protecting key personnel anywhere on the battlefield.

9-6. The USACIDC's LOGSEC function protects the Army's supply pipeline against criminal activities from the manufacturer, through logistics channels, all the way to the frontline soldier. It involves preventing, detecting, and investigating criminal and terrorist activities such as supply diversion, destruction, and sabotage or product substitution. The USACIDC's LOGSEC operations assess LOGSEC, identify weaknesses, and provide a prioritization of threats so that commanders can implement preventive measures to reduce the vulnerability of the logistics pipeline. Whenever possible, the USACIDC will initiate actions to recover logistical losses and return them to Army control.

> The Army let a $22 million contract for the manufacture of its mobile kitchen trailer (MKT)-90 mobile field kitchens in 1990. These kitchens were distributed to Army units worldwide, including those units serving in Saudi Arabia during Operation Desert Shield/Desert Storm. After distribution of the MKT was well underway, a US government quality-assurance representative received information from the manufacturer's employees indicating that the kitchens were not built to contract specifications. This information led to an investigation by the USACIDC's major fraud procurement unit (MFPU). The investigation revealed that inferior material and manufacturing processes were substituted, which allowed the contractor to realize an additional $228,000 in profit. The US Army Environmental Hygiene Agency determined that the MKT-90 had serious sanitation defects and posed a safety hazard to troops using and being fed from these kitchens. The problem with the MKT was identified and expeditiously resolved to ensure the safety of soldiers in the field. The primary subject was indicted by a grand jury and subsequently pled guilty to one count of major fraud. He was found guilty, sentenced to 10 months confinement, and ordered to pay restitution.

9-7. The USACIDC's CRIMINTEL collection efforts focus on the identification and prevention of terrorist and nonterrorist crimes against US and allied military personnel, facilities, and other interests. Upon collecting this information, agents recommend countermeasures to combat subversive activities through coordination with MI, the PM, rear-area operations officers, and HN military and civil intelligence agencies. Additionally, when directed, the USACIDC becomes the lead US military investigative agency at theater level. When this happens, it is tasked with leading the prevention-of-terrorism effort from all services, not just the Army component.

The MI analysts, the MP, and US Army criminal investigators deployed in Kosovo formed a combined all-source information center. This center provided threat assessments to units assigned to TF Falcon. Fifty-five percent of these assessments referenced some form of criminal activity. These intelligence products provided commanders with reliable information concerning potential incidents or criminal activity. With nearly 7,000 US soldiers deployed within the region, these assessments were absolutely crucial for the force-protection planning effort.

The fusion cell within the information center developed intelligence products from national, theater, and operational sources. Due to the significant threat posed by criminal elements in the region, two CID military agents and two CID civilian analysts were attached to the fusion cell to facilitate the police-intelligence function. The CID personnel, in cooperation with MP soldiers, played a key role by linking CRIMINTEL to specific groups and events. The CRIMINTEL collection effort was specifically targeted on weapons, drugs, organized crime, and the identification of smuggling routes. The identification of smuggling routes resulted in a significant increase in the number of weapons being confiscated. The timely transfer of CRIMINTEL products to tactical units enabled a rapid response to serious confrontations, increased confiscation of arms and ammunition, and improved stability in TF Falcon's AO.

9-8. During war and MOOTW, the criminal-investigation effort includes—

- War crimes and, when directed, crimes against coalition forces and HN personnel.
- Major felony crimes committed by US personnel or EPWs/CIs and, when directed, crimes committed by foreign nationals in which there is an Army interest.
- Criminal acts by indigenous personnel, factions, and ad hoc groups.
- Special/sensitive investigations, hostage negotiations, polygraph operations, and force-protection operations.
- CRIMINTEL operations.

9-9. The USACIDC detachments provide enhanced coverage and protection for designated key and essential leaders during war and MOOTW. The protection requirement for senior JTFs and Army commanders may be significantly greater during MOOTW than during peacetime or war as the propensity of asymmetrical threats (such as criminal and terrorist groups) operating in the AO also increases.

ORGANIZATION

9-10. The USACIDC is organized with an array of multifunctional units. As with many other Army organizations, its force is in a transition process. During peacetime, the USACIDC groups provide brigade-level area support for worldwide Army operations. The group provides C^2, staff planning, and administrative support for USACIDC elements assigned to an AOR or deployed to a contingency operation. During war, a group provides the same capabilities in support of an ASCC and provides C^2 for all USACIDC units in the theater. Regardless of the operational environment, the group ensures the connectivity between all USACIDC units. It establishes and maintains links with supported units and interagency, joint, multinational, allied, and HN authorities on matters pertaining to Army and USACIDC operations. The group headquarters has a command section, a detachment headquarters, an S1, an S2, an S3, an S4, an SJA, and a communications section. It provides C^2 for up to six subordinate battalions. Major differences in the AOE and Force XXI groups are in an increase of support and mission personnel and the movement of all polygraph support down to battalion level.

9-11. During peacetime, the USACIDC battalions provide area support for worldwide Army operations. They provide C^2, staff planning, forensic-science support, CRIMINTEL, polygraph support, technical guidance and supervision of USACIDC operations, and administrative support for USACIDC elements assigned to an AOR or deployed to a contingency operation. During war, the USACIDC battalion provides the same capabilities in support of each corps and TSC. Under AOE designs, one battalion supports each corps and TSC. Under the Force XXI design, up to two battalions support each TSC. Like the groups, each USACIDC battalion ensures connectivity and establishes and maintains links with other units/agencies within its AOR. The battalion headquarters consists of a command section, a detachment headquarters, an S1, an S2, an S3, an S4, and a communications section. It provides C^2 to five USACIDC detachments on a sustained basis. Its span of control can accommodate up to seven detachments for a short period of time. Major differences in the new battalion are the addition of a CSM, an S6 officer, a detachment commander and support personnel, and CRIMINTEL and forensic personnel.

9-12. The USACIDC investigative team is the smallest operational investigative element. Each team consists of two special agents (a warrant officer and a noncommissioned officer [NCO]). METT-TC requirements dictate that these teams have the capability to operate independently from the detachment headquarters. The investigative teams are the building block for both AOE and Force XXI USACIDC detachments. Both designs provide levels of flexibility to task-organize without splitting units apart. The detachment (with its complement of teams) performs the full range of criminal-investigative functions in all operational environments, but it has limited CRIMINTEL and LOGSEC management capabilities and no forensic or polygraph capability. Each detachment includes a headquarters section and up to eight investigative teams. METT-TC requirements dictate that these teams have the capability to operate independently from the detachment headquarters. The major differences between the AOE and Force XXI detachments' design begin with replacing the heavy and light division support elements (DSEs), the corps-area support element (CASE), the TSC-area support element, (TASE), and the port-area support element (PASE) designs with a modular design consisting of a standardized headquarters element and two special-agent sections. Each section consists of four investigative teams. Personnel administration, supply, and the bulk of administrative support found in the AOE organizations have been consolidated at battalion.

Additionally, the Force XXI detachment is commanded by a special-agent warrant officer.

9-13. Both the AOE and the Force XXI units retain the Army's standard dependencies upon other units for support requirements (religious, health services, finance, photographic processing, and so forth). The new USACIDC TOEs have enhanced A/L support capabilities. However, the new units still retain the AOE support requirement for food services and emergency Class III/V resupply from CS MP brigades, battalions, and companies.

THE FIELD INVESTIGATIVE UNIT

9-14. Sensitive investigations are normally conducted by the field investigative unit (FIU). The FIU is a one-of-a-kind organization within the DOD that enables the Secretary of the Army to conduct sensitive investigations requiring access to special information or programs that are highly classified. The FIU works closely with the Army IG, the Judge Advocate General of the Army, and the Army General Counsel to support commanders in the special-operations and intelligence communities and in the area of classified acquisition programs. The FIU may also be assigned investigations involving senior Army personnel or those of special interest to the Army leadership.

THE COMPUTER-CRIME INVESTIGATION UNIT

9-15. The computer-crime investigative unit (CCIU) deals with intrusions involving classified networks or multijurisdictional offenses. The CCIU works closely with MI and federal law-enforcement agencies to coordinate military actions, nonmilitary-affiliated offenders, and foreign-intelligence services. In addition, it provides technical assistance to CID elements that are conducting computer-related investigations.

THE PROTECTIVE-SERVICE UNIT

9-16. The protective-service unit (PSU) provides worldwide protective services to designated personnel to protect them from assassination, kidnapping, injury, or embarrassment. The PSU plans, coordinates, and executes executive protection for—

- The Secretary of Defense.
- The Deputy Secretary of Defense.
- The Chairman of the Joint Chiefs of Staff.
- The Vice Chairman of the Joint Chiefs of Staff.
- The Secretary of the Army.
- The Army Chief of Staff.
- The Army Vice Chief of Staff.
- Others (as directed).

9-17. During war and MOOTW, the USACIDC routinely provides protective-service support to senior Army ground commanders and, in some cases, to JTF commanders. The USACIDC agents also provide training to the MP details providing protective services to senior field commanders.

> The President of the US hosted the North Atlantic Treaty Organization's (NATO's) 50th summit anniversary celebration in Washington, DC during the period of 22 to 26 April 1999. The President invited 28 heads of state or prime ministers, 48 foreign ministers, 88 ministers and chiefs of defense, and the chairman of the military committee to the summit. During the conference, the principal guests resided in 23 hotels in Washington, DC. The responsibility for providing protection for the ministers, chiefs of defense, and the chairman for the military committee was given to the USACIDC's PSU. In preparation for this mission, additional training on protective-service operations and evasive driving techniques was given to over 200 reserve agents. These individual mobilization augmentees, troop program units, and other USACIDC agents were critical to the successful execution of this mission.

US ARMY CRIMINAL-INVESTIGATIONS LABORATORY

9-18. Forensic-laboratory support is provided by the US Army Criminal-Investigations Laboratory (USACIL) that is currently located at Fort Gillem, Georgia. The USACIL is a nationally accredited forensic laboratory that provides all three services with dedicated, highly responsive, deployable, state-of-the-art scientific and forensic investigative support. The six forensic divisions of the USACIL include the following:

- Chemistry (serology, drugs, trace, and deoxyribonucleic-acid [DNA] testing).
- Questioned documents (alterations and comparisons of handwriting, typing, and printing).
- Firearms and tool marks (weapons and obliterated numbers).
- Photography (evidence, charts and photos for court).
- Latent prints.
- Computer forensics.

9-19. The USACIL collects, receives, and examines evidence and prepares reports of findings. It also provides expert-witness testimony at court-martials.

US CRIME RECORDS CENTER

9-20. The US Crime Records Center (CRC) maintains the Army's criminal records. It maintains more than two million MP reports and USACIDC reports of investigations. These records are retained for at least 40 years before being destroyed. During their retention, the records are actively maintained and frequently researched.

> During the course of an investigative search, civilian police authorities seized 19 hand grenades. A routine search of stolen property and criminal index files failed to identify the grenades as having been stolen from a nearby military installation. However, a search of the CRC files identified a USACIDC investigation relating to the theft of a case of grenades that occurred 20 years earlier. Further investigation revealed that the lot number of the recovered grenades was identical to that of the stolen grenades. The investigative leads resulting from the CRC file search contributed to the resolution of both cases.

9-21. The CRC conducts more than 10,000 criminal-history name checks each month to identify victims and perpetrators of criminal offenses. The checks are requested not only by USACIDC agents, but also by other military and civilian law-enforcement officials. The CRC is also the Army's agent for Freedom of Information Act (FOIA) and Privacy Act requests relating to criminal investigations and MP reports. The CRC also manages the Army's polygraph programs and their support to Army installations around the world.

EMPLOYMENT

9-22. Whenever possible, USACIDC units are colocated with the MP. This provides unified MP support across the battlefield, and it facilitates logistical support (primarily mess) provided by MP units. However, the USACIDC maintains its vertical C^2. Based on METT-TC, the MP group (CID) will colocate with the supported ASCC and the MP battalions (CID) supporting the corps and EAC will colocate with the MP brigade (CS). When feasible, MP detachments (CID) will colocate with MP battalions/companies and division MP companies. Figure 9-2 illustrates a typical battlefield array for AOE organizations, and Figure 9-3 illustrates a Force XXI array.

9-23. Since USACIDC detachments are austere organizations with little organic A/L capability, they rely on the battalion for unit-level maintenance, supply, and personnel support. However, detachments must often operate independently when performing criminal-investigation functions throughout the AOR. During these times of long-distance separation, detachments must seek A/L support from the supported unit.

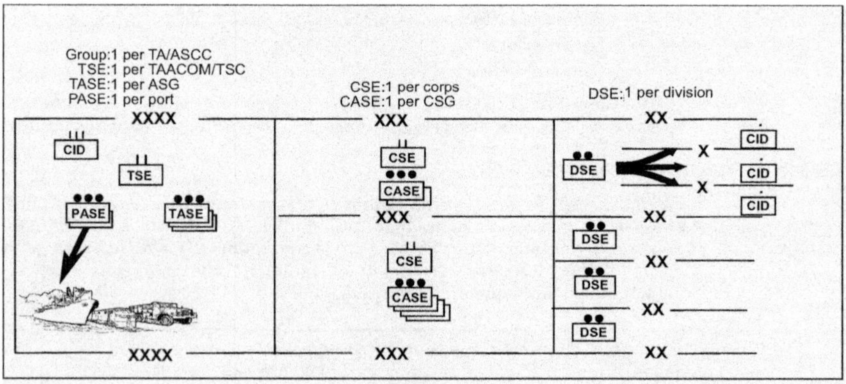

Figure 9-2. AOE Battlefield Array

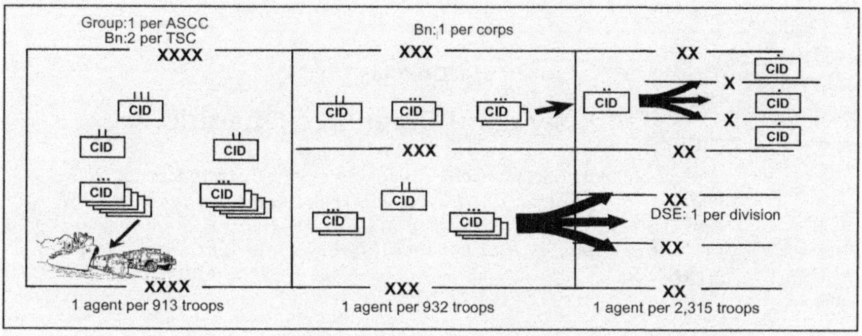

Figure 9-3. Force XXI Battlefield Array

Chapter 10

Offensive and Defensive Operations

The whole of military activity must relate directly or indirectly to the engagement. The end for which a soldier is recruited, clothed, armed, and trained; the whole object of his sleeping, eating, drinking, and marching is simply that he should fight at the right place and the right time.

Clausewitz

Army doctrine considers the full range of operations from war to MOOTW. An operation is a military action or the carrying out of a military mission needed to gain the objectives of any battle or campaign. Offensive and defensive operations normally dominate military operations in war and in some smaller-scale contingencies. An offensive operation is aimed at destroying or defeating the enemy and imposing our will on him for a decisive victory. Defensive operations defeat an enemy attack, buy time, economize forces, or develop conditions favorable for a counteroffensive that regains the initiative and defeats the enemy.

OVERVIEW

10-1. The offense is the decisive form of battle. Offensive campaigns and major operations are designed to achieve operational and strategic objectives quickly and decisively at the least cost. Offensive operations combine the terrain and the force to achieve the objective. The main purpose of the offense is to defeat, destroy, or neutralize the enemy force. Offensive operations are characterized by a rapid shift in the main effort to take advantage of opportunities by momentum and simultaneous destruction of enemy defenses.

10-2. The immediate purpose of defensive operations is to defeat an attack. Military forces defend only until they can gain sufficient strength to attack. Commanders choose to defend when they need to buy time, to hold a piece of terrain, to facilitate other operations, or to erode enemy resources at a rapid rate while reinforcing friendly operations. The ARFOR conduct defensive operations as part of major operations and campaigns and in combination with offensive operations (see FM 3-0).

SUPPORT TO OFFENSIVE OPERATIONS

10-3. MP support to offensive operations varies according to the type of operation being conducted. The MP functions support each echelon commander based on METT-TC. In the offense, MP priorities are often placed on MMS, I/R, and AS operations. Regardless of the offensive effort, MP leaders supporting an offensive operation must—

- Understand the commander's intent and his concept of operations.
- Anticipate that selective elements of the offensive force may need to pause, defend, resupply, or reconstitute while other forces attack.
- Anticipate changes in the operational tempo and prepare to weight the MP effort to meet the maneuver commander's intent.
- Provide MP support within the context of linear and nonlinear operations. They must understand how both types of operations affect the security of rear areas and LOC, the protection of C4I, the means of sustainment, and the land force's mobility.
- Support the maneuver commander's intelligence efforts by conducting the PIO function. The MP must observe, identify, and report information.
- Assist maneuver forces to mass. The MP protect C^2 nodes such as the CP that orchestrates the mass efforts and resources that sustain the offensive move. The MP must deny incursions by enemy reconnaissance forces into the AO. They also maintain surveillance, provide early warning, and impede and harass the enemy with supporting and organic fires.
- Speed the commitment of follow-on forces and reserves. They must also conduct MMS operations to assist the commander in a smooth and quick shift of forces to support the main effort.
- Anticipate the pursuit and exploitation by positioning MP forces that can support continuous operations. The MP support as far forward as possible while protecting resources needed to seize the opportunity. The EPWs/CIs are collected as far forward as possible. The MP coordinate the treatment of dislocated civilians and refugees with the HN or foreign forces.
- Tailor the MP support based on the unit being supported (corps, divisions, and brigades).
- Consider the location and composition of response forces and the TCF and the strength of rear-area threats.
- Integrate a force into the offensive plan that is capable of conducting aggressive combat operations against designated targets.

> During Operation Just Cause, an MP platoon received the mission to disable an armed Panamanian Defense Force (PDF) patrol boat docked near a key objective. After receiving the mission, the platoon leader maneuvered his force into position and attacked after receiving enemy fire. The platoon successfully disabled the patrol boat and captured the objective while suffering minimum casualties.

10-4. During offensive operations, MP support provided by teams, squads, and platoons provide the commander with an agile, flexible, versatile, and capable force ready to contribute to the overall mission success.

SUPPORT TO DEFENSIVE OPERATIONS

10-5. The MP support defensive operations consistent with the commander's intent and priorities. They support a defending force by assisting subordinate elements in the execution of different, yet complementary actions. For example, the MP may support the maneuver of an attacking unit by conducting MMS operations while simultaneously providing AS for a defending unit in a base cluster. The MP support reflects the maneuver commander's focus—destroying an attacking force, retaining or denying key terrain, moving away from an enemy force, or a combination of the aforementioned. However, despite the intent of the defensive operation being conducted, MP leaders must—

- Conduct MMS operations to help a force maneuver and mass. The MP must anticipate the shift from the defense to the offense and assist the forward movement of reserves.
- Conduct AS operations to deny information to enemy reconnaissance elements seeking out the exact location of the defending force. The MP are positioned where they can control key terrain or improve the defensive capability of bases and base clusters. The MP conduct aggressive R&S to locate and deny enemy access to critical logistical facilities.
- Protect sustainment resources while supporting the lateral, forward, and rearward movement of forces.
- Consider the type and size of the AO, the LOC security, the rear threat, and the number of EPWs/CIs and dislocated civilians to determine how they may affect the movement of forces.

> During Operation Desert Shield, EAC, corps, and division MP units conducted a myriad of missions to support both the XVIII Corps's (Airborne) and the VII Corps's movement to their TAA. The MP's main focus was on MMS and AS throughout the AO to ensure freedom of movement and to protect US and coalition forces against terrorists and other threats in the rear area.

10-6. The success of MP operations in the defense depends on the MP leader's situational awareness, the commander's intent, and the precise employment of MP resources when and where they are needed. In the defense, the MP perform their five functions to assist the defending commander by providing a lethal mobile force that allows him to concentrate his efforts and resources in fighting the next battle.

Chapter 11

Stability and Support Operations

ARFOR conduct stability operations in a complex, dynamic, and often asymmetric environment. Stability operations are usually nonlinear and noncontiguous. They are often time and manpower intensive. The purpose of support operations is to meet the immediate needs of designated groups for a limited time until civil authorities can accomplish these tasks without Army assistance.

FM 3-0

The Army's mission is to prepare for war and, when deterrence fails, to achieve a quick, decisive victory. The DOD projects that many of its future missions will be stability and support operations. However, these operations are not new to the Army. The Army has participated in stability and support operations in support of national interests throughout history. The Army has protected its citizens, assisted nations abroad, and served America in a variety of other missions. The pace, frequency, and variety of stability and support operations have quickened in the last three decades. As demonstrated in Operations Provide Comfort, Restore Democracy, and JTF Los Angeles, these types of operations present a challenge to all services.

OVERVIEW

11-1. Stability operations are conducted outside of the US and its territories to promote and protect US national interests by influencing political, civil, and military environments and by disrupting specific illegal activities. Stability operations may include both developmental and coercive actions. Developmental actions enhance a HN government's willingness and ability to care for its people. Coercive actions apply carefully prescribed limited force or the threat of forces to change the AO's environment.

11-2. Support operations provide essential supplies and services to relieve suffering and to help civil authorities prepare or respond to crises. In most cases, the ARFOR's efforts are focused on overcoming conditions created by man-made or natural disasters. The ultimate goal of support operations is to meet the immediate needs of designated groups (for a limited time) until civil authorities can accomplish these tasks without Army assistance.

11-3. Stability and support operations are distinctly different from offensive and defensive operations and must be analyzed differently. For instance, the application of the rules of engagement (ROE) will be unique to the operation. The restrictions on combat operations and the use of force must be understood and obeyed at all levels. MP leaders must clearly take the initiative and ensure that the applicable guidance is implemented and followed by all subordinates.

11-4. Stability and support operations have more diverse political considerations than are usually encountered in war. There are also complexities that must be addressed in joint, combined, and interagency stability and support operations. The Army's challenge is to execute these operations successfully in a joint or combined arena without degrading its capability to wage war successfully.

11-5. If the US is to fight and win future wars, it must develop, train, and employ its force structure in a manner that ensures success in every operation across the spectrum of military operations. Under these auspices, MP forces can provide commanders with especially valuable assets for stability and support operations. The MP have the unique capability of serving as a combat, CS, and CSS force. The domestic and international acceptability of the MP operational image frequently makes the MP the most appropriate force for stability and support operations. Additionally, the MP are flexible and capable of rapidly transitioning from one end of the spectrum to the other if the environment changes.

FORCE SUITABILITY

11-6. The MP's capabilities and their operational and supporting tasks are as integral to stability and support operations as they are to offensive and defensive operations. MP branch-/core-related skills are highly compatible with the capabilities required for stability and support operations. The MP provide a highly capable, politically acceptable force that is suitable for a variety of missions. They possess robust moving, shooting, and communicating capabilities; and they project an assist, protect, and defend image that is particularly important when tailoring a force that requires significant capabilities but a low political profile. The MP provide a flexible, but limited, economy-of-force organization. Mission requirements will always exceed available MP resources, especially in stability and support operations. For this reason, MP assets must be prioritized to deliver the greatest mission capability. The specific operations that MP units perform at a given time are determined by the echelon commander's needs, the mission requirements, and the availability of MP resources. As discussed in previous chapters, the echelon commander, through the command's PM, sets the priorities for MP operations. As in offensive and defensive operations, the PM knows that while the MP force can perform all MP functions, they may not be robust enough to perform all assigned MP missions all of the time. The PM must continuously make trade-offs between the numbers and kinds of MP missions that can be supported.

11-7. The MP's capabilities are further enhanced by their training, mind-set, and experience in dealing with people in highly stressful and confusing situations. Although capable of conducting combat operations when needed, the MP are highly practiced in de-escalation and in employing the minimum-essential force to contain potentially violent situations. This mind-set serves as the framework for MP law-enforcement training and is especially applicable in MP support for stability and support operations. Furthermore, this mind-set is exercised and reinforced daily in peacetime law-enforcement operations. This is a significant capability. MP soldiers learn and receive constant reinforcement training in controlling a situation on the spot before it escalates and cannot be controlled. The MP have a unique understanding of human nature and are adept at handling emotion-laden situations.

> During operations in Kosovo, as both Serbs and Albanians attempted to create unfavorable conditions and responses by US forces, the MP kept their cool under fire in the streets of Gnjilane and Kamenica. The MP's reaction led to a quick understanding by all sides that the US Army MP were fair and just and understood the tense situations.

11-8. The MP stress four essential training competencies—skill proficiency, human dimension and attitude, camaraderie and teamwork, and leadership. These competencies are the foundation for all MP operational performances. They equip the MP to exercise discretion in dealing with others, to protect and assist those in need, and to accomplish the varying demands of MP missions in stability and support operations and, when needed, in war.

STABILITY AND SUPPORT

11-9. Stability and support operations are performed in support of US authorities or in support of international bodies and foreign governments when so required by US authorities. The MP have long provided the essential support for stability and support operations; however, the frequency of these operations has increased significantly in recent years. Consequently, the MP have been deployed in greater numbers with greater frequency. MP units have a substantial history of successful, simultaneous, and continuous deployments in both a lead and a support role.

SUPPORT TO DOMESTIC CIVIL AUTHORITY

11-10. The purpose of support to domestic civil authorities is to—

- Render humanitarian assistance.
- Provide disaster relief.
- Restore order.
- Combat terrorism.
- Support counterdrug operations.

11-11. Military operations supporting domestic civil authorities are governed by the provisions of the Stafford Act, the Posse Comitatus Act, and other laws and regulations. Peacetime experience in civil-military affairs equips the MP to oversee, assist, and train other units and agencies for law enforcement, sensitive security, and operations to control dislocated civilians (within the legal limits of a particular stability or support operation). For this reason, the MP are often needed after other forces have redeployed, and they stay until government order is restored. Since active-component military missions are constrained by the Posse Comitatus Act, nonfederalized national guard (NG) units are often employed to conduct these operations (particularly counterdrug operations). The NG units can operate under gubernatorial authority within state jurisdiction without Posse Comitatus restrictions. The NG MP units are the only NG units trained, skilled, and experienced in providing law enforcement. They offer civilian authorities a unique, specialized capability not available from other NG units.

SUPPORT TO HUMANITARIAN ASSISTANCE AND DISASTER RELIEF

11-12. Humanitarian assistance and disaster-relief operations provide emergency assistance to victims of natural and man-made disasters and other endemic conditions. Natural disasters include earthquakes, floods, and hurricanes. Man-made disasters include riots and civil conflicts. Humanitarian assistance and disaster relief conducted by US forces alleviate urgent nonmilitary needs of a populace until the appropriate civilian agencies can provide the needed services without military support. These types of operations include, but are not limited to—

- Conducting searches and rescues.
- Evacuating, sheltering, sustaining, and protecting dislocated civilians.
- Making food, medical care and treatment, and other civilian-aid programs available.
- Reducing opportunities for criminal activity and restoring L&O.
- Maintaining other operations needed to ensure the well being of the affected population.
- Providing relief operations for man-made disasters. These operations are characterized by civil conflict, and they entail all of the requirements for other disasters but focus on the restoration of L&O.
- Providing force protection and security.

11-13. MP support can be an important asset in all disaster-relief operations. The MP are trained and equipped for decentralized operations. MP teams operate in highly mobile vehicles equipped with radios, which are invaluable in disaster-relief operations. Each MP platoon, company, and battalion headquarters is equipped with high-frequency radios. These capabilities allow the MP (even when widely dispersed) to maintain centralized communications; to provide L&O support; and to assist in search, rescue, and evacuation efforts. Because of their decentralized operations and communications equipment, the MP can also play a vital role in notification, ADC, and access and egress control. They can deploy with disaster-assessment teams to assist in advising and developing a disaster-assistance plan. The MP can also deploy to an affected area to—

- Assist in restoring order.
- Search for, rescue, and evacuate victims.
- Provide personnel and vehicular circulation controls.
- Provide ADC.
- Conduct detention operations.
- Provide physical and area security to protect life and property immediately following a disaster.
- Protect fire and emergency-services personnel.
- Gather, record, and report information.
- Provide force protection and security.

11-14. MP (I/R) battalions can support civil-affairs units by establishing I/R facilities to shelter, sustain, protect, and account for disaster-relief victims. These units can also train and assist government agencies (US and foreign) or police in the operation of facilities for dislocated civilians and detainees.

11-15. The MP possess capabilities critical to relief in man-made disasters characterized by civil conflict. The MP have the essential training and skills needed for relief operations that focus on restoring civil order. Restoration of civil order demands—

- **Objectivity and neutrality.** The objective of civil-disturbance control is to restore L&O, not to impose punishment or suppress peaceful dissent. Government forces quelling civil disorders must be emotionally objective and politically neutral. MP law-enforcement training stresses objectivity and neutrality, an important attribute for forces employed in stability or support operations.
- **Minimal use of force.** The application of more than minimum force may be detrimental to mission success. The MP are particularly suited for these situations, having been trained on the technique for the use and measured application of minimum forces to accomplish mission goals.
- **Public acceptance.** Government forces must avoid actions that will alienate the populace. Any action offensive to community sensitivities must be avoided. The MP receive training in crowd-control operations, interpersonal communications skills, and minimum use-of-force techniques—skills that are critical in gaining public acceptance.
- **Positive image.** In many disturbance operations, demonstrators and government forces compete for the approval and sympathy of the general population. The MP's assist-and-protect image provides the government with a distinct advantage in such situations.
- **Threat awareness.** Government forces must pay attention to and try to learn the organization, intentions, and activities of demonstrators. Equal attention must be paid to how the community will respond to the demonstrator's actions and the government's response. The MP interact daily with the populace, placing them in a position to develop criminal information and to determine the collective attitude of the populace.
- **Effective coordination.** Disturbance operations often involve many agencies from various levels of government. Thorough, timely, and coordinated preparation, planning, training, and execution are essential for mission success. MP civil-disturbance training (including planning and coordination measures) and their communications assets enhance coordination, particularly during mission execution.
- **Operational unity.** Multiple control forces representing different agencies with varying jurisdictional authority can pose C^2 problems. When unity of command cannot be achieved, operational unity must be achieved by colocating operational centers, integrating communications, and delineating organizational responsibilities. The MP are well trained in police and military operations, possessing a sound understanding of both civilian police and military measures, legal constraints, and effective coordination.

11-16. The MP operational tasks supporting man-made disasters vary according to the nature of the disaster and the operational environment. For example, in CONUS where the commitment of military forces to civil disturbances is considered a last resort and military involvement is limited by law, nonfederalized NG MP units have long been employed by civil authorities for their unique capabilities. MP forces can—

- Isolate and contain an area with barricades, roadblocks, and perimeter patrols. The MP can also assist civil-affairs units by enforcing pass and identification systems and public-utility controls.
- Provide mobile patrols to enhance the security of high-priority targets (such as buildings, utilities, and services that are critical to the community's economic and physical well being).
- Monitor, disperse, contain, or limit crowd movement through—
  - Observation points and patrols.
  - The communication of interest and intent to the crowd.
  - The channelization or diversion of the crowd.
  - Gaining the cooperation of the dissident leadership.
  - The proclamation of the illegal nature of the crowd's actions.
  - Show of force.
  - The use of crowd-control formations.
  - The apprehension of crowd members and leaders.
  - The application of minimum-force measures.
- Establish area control to prevent looting, to protect businesses and other likely targets, and to prevent arson. The MP accomplish this mission with saturation patrolling, including vehicle and foot patrols. With augmentation, the MP can perform air and water patrols. They also enforce populace control measures (such as ordinances to prevent gathering, permits to gather, restrictions on circulation, restrictions on the interference with government and public functions and personnel, restrictions on possessing weapons, and other measures instituted by civil authorities).
- Neutralize special threats (such as snipers and bombings) that are highly dangerous to both government forces and the community. Snipers pose a particularly grave danger in disaster-relief operations. The MP can take immediate protective actions, secure an area, isolate the threat and, when authorized, use SRTs to assist civilian or HN authorities in apprehending or neutralizing the sniper. The MP are trained to perform the necessary actions when an explosive device is discovered. In such an event, they can secure and evacuate an area, organize search teams, and isolate the site.

11-17. The MP can also support in rescue and recovery operations, which may include rescuing US or foreign nationals. Such operations are also employed to locate, identify, and recover sensitive materials deemed critical to national security. The operations may be conducted in benign or hostile environments. The MP provide OPSEC, and they process and account for civilians affected by rescue and recovery operations. The MWD teams can also help search for lost or hidden personnel. In stability and support operations, the MP can perform their police functions for rescue and recovery operations while keeping the perceived military signature low.

SUPPORT TO COUNTERDRUG OPERATIONS

11-18. Counterdrug operations are measures taken to disrupt, interdict, and destroy illicit drug activities. In a counterdrug role, the DOD may offer certain forms of support to HN counterdrug personnel, to US civilian law-enforcement agencies operating in a HN, to Department of State (DOS) counterdrug personnel, and to the Bureau of International Narcotics Matters. Without direction from the National Command Authority acting under constitutional or statutory authority, US forces engaged in counterdrug activities may not engage in direct law-enforcement activities (such as apprehension, search, seizure, and other similar activities). As discussed earlier, the Posse Comitatus Act authorizes specific DOD assistance in counterdrug activities, but US military forces usually may not participate directly in arrests, searches, seizures, and similar activities. Conversely, as discussed earlier, USC Title 32 authorizes state governors to use NG MP units with full authority to engage directly in law-enforcement activities. When authorized, US advisors may accompany counterdrug forces on operations.

11-19. MP operational contributions can include the use of force, drug identification, crime-scene protection, customs operations, surveillance of criminal activity, serious-incident reporting, and other related activities. Employing the MP for these operations ensures force suitability and avoids a mismatch in terms of unit capability and mission effectiveness.

11-20. Border-screening operations consist of three separate but related measures—mobile patrolling, the use of listening/observation posts, and airfield surveillance. These measures are part of the MP's AS and MMS functions used in tactical rear-area operations. The MP's operational effectiveness may be further increased through augmentation by a civilian law-enforcement officer who would be responsible for search, seizure, and arrest actions. Active-component units are permitted to assist in marijuana eradication in support of domestic law-enforcement agencies (DLEAs). Marijuana-eradication operations are an application of the MP AS, mobility, and L&O functions, thus providing realistic training while supporting an operation.

11-21. As discussed in previous chapters, the MP are well suited to augment customs operations. They routinely perform customs inspections for US forces in Korea and Germany. The NG MP regularly conduct vehicle, cargo, and container searches under the supervision of US Customs agents and at US ports of entry.

11-22. The USACIDC agents can provide support to counterdrug operations. The CID units identify the sources of illegal-drug distribution systems. When directed, the MP and the CID complement interagency counterdrug efforts to stop the production, flow, and distribution of drugs and to provide planning, training, equipment, and facilities to support domestic counterdrug operations. The CID also supports CRIMINTEL programs, internal security, and site surveys at borders where Army units will deploy. In addition to the aforementioned support, the DOD General Counsel has ruled that MWD teams can be loaned to DLEAs to assist in detecting smuggled contraband.

SUPPORT IN COMBATING TERRORISM

11-23. The Department of Justice (DOJ) is the lead federal agency for dealing with acts of terrorism within the US, its territories, and its possessions. Within the DOJ, the Federal Bureau of Investigation (FBI) has the lead. The FBI can train the police of friendly nations in antiterrorism and counterterrorism operations. Usually, US military forces act in a technical-advisory but not a law-enforcement capacity in combating acts of terrorism. It is often difficult to distinguish the acts of politically motivated terrorists from violent acts of criminals or individuals in society at large. In fact, all terrorist acts are criminal; however, they also may be part of an insurgency.

11-24. The MP are key players in the defensive planning, implementation, and control of combating-terrorism operations, whether in garrison or deployed to war or MOOTW. In whatever countries US forces operate, MP units conduct antiterrorism operations daily as part of their L&O and AS functions. The MP also respond to terrorist incidents. They help secure the area and search for evidence, and they help the lead agency. The MP provide advice and assistance in developing antiterrorism and counterterrorism programs for deployed US forces and DOS agencies and for police, paramilitary, and military forces of assisted nations (when so authorized). MP training, liaison, and joint patrols help strengthen HN-police organizations, improving their effectiveness and efficiency in combating terrorism. The MP also assist HN police in performing PRC operations.

11-25. The training of police (domestic and foreign) requires DA approval, and restrictions exist on some forms of training. Legal advice and coordination are a must before training can occur. The MP and the CID provide personnel and equipment for planning, advising, equipping, and training agencies to prevent, deter, and respond to terrorism. The training focuses on deterring terrorist acts such as bombings, kidnappings, assassinations, taking hostages, and hijackings. Related assistance may include training and equipment for explosive detection, management of hostage situations, physical security, protection of key personnel, and defensive/evasive driving methods. For counterterrorism operations on US military installations, the MP provide highly trained, specially equipped SRTs for situations requiring the selective use of force. Additionally, CID agents are highly trained in hostage negotiation, which increases the commander's options before resorting to force.

11-26. The MP and the CID assist in counterterrorism information-gathering activities through liaison and joint patrols with HN police. These activities are part of the PIO function. The MP have extensive contacts with civilian police, who in turn have contacts with the populace. These contacts can help surface criminal information concerning terrorist plans, recruitment, and support structures sooner than military sources. The CID can provide commanders with CRIMINTEL on a daily basis. At the same time, MP antiterrorism support includes measures to reduce the vulnerability of people and property. Such measures include—

- Personal awareness.
- Personal-protection operations and techniques.
- Crime prevention.
- Physical-security programs.

11-27. The MP and CID crime-prevention programs employ proactive measures to protect people and their property and US and HN property. Crime prevention is a continuous process of planning, implementing measures, evaluating effectiveness, modifying measures, and developing further information. The primary crime-prevention tool is the crime-prevention survey. It is used to examine the physical features of various critical sites. The MP physical-security programs help identify, reduce, eliminate, or neutralize conditions favorable to criminal, terrorist, and insurgent activities. The key measures used to accomplish this are physical-security inspections and surveys.

SUPPORT TO NONCOMBATANT EVACUATION OPERATIONS

11-28. Noncombatant evacuation operations (NEOs) relocate civilians who are endangered in a foreign nation. The NEOs are classified as benign (unopposed) or hostile (opposed). MP roles in support to NEOs differ accordingly. In a benign environment, NEOs are usually conducted with the help and full cooperation of the affected nation, with little or no opposition to evacuation. The use of force is limited to self-defense (as outlined by the ROE) and protection of the evacuees. In a hostile environment, NEOs are opposed by forces with specific intentions of preventing or destroying the operation. In this scenario, the MP will follow the ROE as established by higher headquarters.

11-29. MP operations in NEOs closely parallel EPW/CI evacuation operations. The objective of both is to safely collect, secure, protect, transport, process, and account for people, whether they are EPWs or noncombatants. However, due to the dissimilarities of these groups, the techniques employed differ.

11-30. The MP can secure selected facilities, reception and transfer points, or other locations required to collect, process, and evacuate people. They can provide route reconnaissance and mobile security teams to escort HRP, DOS personnel, evacuees, and sensitive material. If needed, MP units assist civil-affairs personnel with the movement and control of dislocated civilians. The MP may support civil affairs by setting up and operating evacuation sites, holding areas for foreign nationals denied evacuation, and reception and processing stations. This support can be provided by MP CS units for a short duration and by MP I/R units for a prolonged duration. Additionally, the MP will establish and maintain liaison with HN police to ensure a mutual understanding of jurisdiction, to coordinate efforts, and to provide criminal and tactical intelligence. In hostile environments requiring the employment of combat forces, the MP would be employed in their traditional CS role.

SUPPORT TO SECURITY-ASSISTANCE OPERATIONS

11-31. Security-assistance operations are one of the main tools of US foreign policy. A critical concern in these operations is the HN's ability to plan and manage its own defense resources. In security-assistance operations, the US provides defense materiel, military training, and defense-related services through grants, loans, credit, or cash sales. When authorized, the MP support security assistance with military and police training through the International Military Education and Training Program (IMETP). This program provides instruction and training to foreign military and civilian personnel on a grant/aid basis. Many foreign civil law-enforcement agencies request training in the following areas:

- Basic police procedures (such as patrolling and crime-scene protection).
- Physical security.
- Corrections.
- Civil-disturbance operations.
- Customs operations.
- Traffic control.
- Use of force.

11-32. The MP possess the expertise and experience to advise, train, and assist a HN's military and police forces. The MP's participation in the IMETP includes—

- Formal and informal instruction of foreign students.
- Technical education and tactical training with applicable training aids.
- Assistance to foreign police and military elements by MP mobile training teams (MTTs).
- Providing MP units or small, modular, functionally specific teams to support security-assistance needs. These teams can train HN police in both field and institutional facilities.
- Support to nation-assistance operations.

11-33. Nation-assistance operations support a nation's effort to promote development, ideally by using its national resources. This is achieved through education and the transfer of essential skills to the HN. The goals of such operations are to—

- Promote long-term regional stability.
- Develop sound and responsive democratic institutions.
- Develop supportive infrastructures.
- Promote an environment that allows for orderly political change and economic progress.

11-34. The MP support nation-assistance operations with training and joint MP and HN-police operations. The MP know and understand police operations, military operations, and C^2 procedures. Consequently, the MP are ideal for training the law-enforcement personnel in both field and institutional settings. The MP's nation-assistance goals are to—

- Enhance L&O in democratic societies.
- Improve efficiency and effectiveness.
- Promote the proper usage and maintenance of equipment.
- Establish a sound training base for police.

- Standardize procedures that enhance combined police operations.
- Promote friendship and goodwill toward the US.
- Preclude the need for US military interventions to counter acts of violence.

11-35. Combined police operations reinforce training and provide HN police with mobility, security, and communications to operate more effectively while conserving critical personnel resources. The MP provide support through training and the following:

- Law-enforcement, security, and criminal-information support to the HN police force (to include L&O operations and administration in a democratic society).
- Patrol and desk operations.
- Circulation-control operations.
- PRC and civil-disturbance operations.
- MWDs.
- Physical-security operations.
- Personnel-security operations.
- Area- and route-security operations.
- Counterdrug operations.
- Antiterrorism operations.
- Mass immigration operations.
- Customs and border operations.
- Confinement operations.
- Crime prevention.
- Community and police relations.
- C^2 of police operations.

11-36. When conducting nation-assistance operations, as well as other operations, the MP will often be tasked to support civil-affairs operations. Close coordination with the G5/S5 and liaison with the appropriate HN representatives have added emphasis to these types of operations. The scope of civil-affairs operations is influenced by the economic, social, psychological, and political background of the country and its people. Civil-affairs personnel will coordinate the MP's effort to restore stability, contribute to national development, and promote support for the host government. Civil-affairs personnel assist in preventing civilian interference with military operations and coordinate all other military and civil affairs (such as community relations, PRC, civil defense, and community- and area-security programs).

11-37. The MP can conduct ADC measures to assist the HN in the aftermath of a natural or man-made disaster. They help rescue and evacuate the injured for medical care, and they issue food, water, and essential supplies (See FM 41-10).

SUPPORT TO PEACEKEEPING OPERATIONS

11-38. Peacekeeping operations (PKOs) are military operations conducted with the consent of the belligerents in a conflict. The PKOs are intended to maintain negotiated truces and to facilitate diplomatic solutions. The US may participate in a PKO under the auspices of an international organization, in cooperation with other countries, or unilaterally. A PKO usually occurs in an ambiguous situation in which there is extreme tension and the possibility of violence. This violence may range from terrorism and sabotage to minor military conflicts involving known and unknown belligerents. MP forces may serve as an element of a UN peacekeeping force or as part of a multinational observer group. The overall operational control of a peacekeeping force is exercised by a multinationally staffed military peacekeeping command. However, the commander of each assigned national force retains command integrity of his unit.

11-39. The basic MP force structure and appropriate augmentation are situation dependent. The size and composition of the MP forces are determined by diplomatic negotiations. Personnel spaces are allocated based on the US contingents' respective missions. The principle of consent affects the composition and function of the peacekeeping force. Consent applies to the degree of interest the disputing countries have when contributing their forces to participate in the peacekeeping effort.

11-40. Each PKO is unique but is generally characterized by constraints in the use of force. In a PKO, force is normally limited to self-defense, the defense of others, and the protection of national security. Therefore, the soldiers in a PKO must possess the skills required for warfare, but they must also be trained so that they are equally able to act as intermediaries. In this aspect, the MP are a distinct asset. They are experienced in a case-by-case approach to the use of force. They are experienced in using discretion and in the de-escalation of potential violence. The MP training in EPW operations are especially useful in handling EPWs if the peacekeeping force supervises the exchange of prisoners of war.

11-41. The peacekeeping force employs a combination of the following techniques to control the potential for violence inherent to PKOs:

- Observation is the peacekeeping force's primary responsibility and basic requirement. It provides timely and accurate reports of any suspicious situation, incident, or occurrence.
- Information gathering is a continuous requirement. Peacekeepers must be constantly alert to what takes place around them and to any changes or inconsistencies in the behavior, attitude, and activities of military and civilian personnel.
- Surveillance and supervision operations help oversee the implementation of agreements. The operations require restraint, tact, and patience.
- Complaint investigations must be thorough and objective, and they must result in a fair assessment.
- Negotiation and mediation skills are often required of peacekeepers. If peacekeepers can resolve minor problems at the lowest level, they can prevent major issues from rising, which may prevent increased tensions or the resumption of fighting.
- Patrolling (mounted and dismounted) among the population publicizes presence. It also serves as a means of gathering information, informally enhancing supervision, and investigating complaints.

11-42. The above techniques are compatible with MP battlefield functions. Operationally, the MP conduct route and area R&S, operate mounted or dismounted patrols, operate listening/observation posts, provide humanitarian assistance, investigate possible terrorist and criminal acts and cease-fire and sanction violations, and gather information. The MP provide a flexible, wide-ranging means of information dissemination. Additionally, they provide crucial support for force protection. The MP and the CID employ active and passive measures to deter and defeat threats directed at service members, their families, DOD civilians, facilities, equipment, and very important persons (VIPs). The MP use emerging technologies in force protection to conduct security, detection, and identification operations and to warn of intruders.

11-43. As part of a PKO, the MP conduct crowd- and riot-control operations, including the extraction of mob leaders. The MP are an ideal force for controlling antagonistic masses engaged in rioting, looting, and demonstrating. The MP understand how to make the transition from the lower end of the use-of-force spectrum to the use of lethal force, if so required or directed. Finally, the MP deploying with peacekeeping forces can provide early-on force protection, including headquarters and initial aerial port/seaport of debarkation (APOD/SPOD) security. The MP's mobility, firepower, and communications provide critical reconnaissance, information-collection, and response-force capabilities and acceptability not found in combat forces.

11-44. When conducting crowd- and riot-control operations in a PKO or peace-enforcement operation (PEO), nonlethal weapons (NLWs) are an additional means of employing force for the purpose of limiting the probability of death or serious injury to noncombatants or belligerents. However, the use of lethal force must always remain an inherent right of individuals in instances when they, their fellow soldiers, or personnel in their charge are threatened with death or serious bodily harm. The NLWs add flexibility to the control of disturbances by providing an environment where military forces can permissively engage threatening targets with limited risk of noncombatant casualties and collateral damage (see FM 90-40).

11-45. The use of lethal force (employed under the standing ROE) will never be denied. At no time will forces be deployed without the ability to defend themselves against a lethal threat nor will they forego normal training, arming, and equipping for combat. Nonlethal options are a complement to, not a replacement for lethal force. They seek to expand a proactive response across the range of military operations.

11-46. The decision to use NLWs against an adversary during a confrontation should be delegated to the lowest possible level, preferably to the platoon or squad. However, this requires that all personnel (not just leaders) have a clear understanding of the ROE and the commander's intent.

11-47. Commanders and public-affairs officers must be prepared to address media questions and concerns regarding the use and role of NLWs. They must be prepared to address the role of NLWs, and they must make it clear that the presence of NLWs in no way indicates abandoning the option to employ deadly force in appropriate circumstances.

11-48. The PEOs are in the high end of the PKO's spectrum. These operations are just short of combat and require a different operational approach. MP operations to support PEOs correspond to MP battlefield functions supporting combat forces. The PEOs often pose special operational considerations for dealing with dislocated civilians. The MP provide PRC operations and I/R, when required.

ORGANIZATIONS AND CAPABILITIES

11-49. MP units are well suited for stability and support operations given their unique capabilities and low force signature. The MP brigade, battalion, company, and I/R units may be tailored for stability and support operations based on their unique METT-TC.

MP BRIGADE TF

11-50. In essence, the organization and capabilities of the MP organization in stability and support operations is the same as those for other operations. However, its augmentation is tailored based on METT-TC unique to stability and support operations and the commander's needs. The MP brigade has been the centerpiece of successful MP deployment during previous stability and support operations. The MP brigade provides comprehensive planning, C^2, and A/L support to assigned and attached subordinate units. The brigade also has the command and staff experience to coordinate and supervise HN security and law-enforcement activities.

11-51. The MP brigade headquarters provides senior MP leadership and the communications capabilities required to direct a variety of functional elements—CSS, PSYOP, public affairs, and combat-arms augmentation—in support of joint or unilateral TF operations. At authorized level of organization (ALO) 1, the MP brigade headquarters has an SJA section to coordinate the legal and jurisdictional issues inherent to stability and support operations. The brigade headquarters also has an S5 section and a liaison capability to coordinate support and assistance to HN agencies, friendly forces, other US forces, DOS personnel, and nongovernmental or private volunteer organizations.

11-52. The MP brigade TF (MPBTF), with its robust capabilities and low force signature, is well suited for the unique mission requirements and sensitivities of stability and support operations. The MPBTF can be employed as part of a combat TF in high-threat, opposed-entry operations or as the lead TF for a low-threat, unopposed-entry operation. The following are examples of previous MPBTFs and the operations in which they participated:

- **Operation Provide Comfort.** The 18th MP Brigade deployed to northern Iraq to provide humanitarian assistance and protection to the Kurds.
- **JTF Los Angeles.** The 49th MP Brigade (NG) provided civil-disturbance assistance to Los Angeles in the aftermath of the Rodney King trial.
- **Operation Hawkeye.** The 16th MP Brigade and units of the 89th MP Brigade provided security, protection, and humanitarian

assistance to the population of Saint Croix in the aftermath of Hurricane Hugo.

- **JTF Andrew.** The 16th MP Brigade provided humanitarian assistance in the aftermath of Hurricane Andrew.
- **Operation Sea Signal.** The 89th MP Brigade deployed to Guantanamo Bay, Cuba, to provide humanitarian assistance to Haitian migrants.
- **Operation Restore Democracy.** The 16th MP Brigade deployed to Haiti to provide humanitarian assistance.

11-53. Early deployments of the MPBTF headquarters or a mission-tailored portion of the headquarters positions a senior, effective MP C^2 headquarters to respond to all MP requirements. It can assist in the coordination with the media, governmental and nongovernmental agencies, and other forces, and it can coordinate civil-military aspects of the operation. The MP brigade headquarters can assimilate and employ additional forces without degrading operations. The augmentation of brigade headquarters is driven by METT-TC. With the brigade headquarters in place, subordinate or supporting units can be significantly expanded without the loss of C^2. In essence, little or no loss in operational momentum occurs when the C^2 element deploys first.

11-54. The MP LRP section of the MP brigade headquarters colocates and works with the higher headquarters' G3/Operations Directorate (J3) plans element. This allows the MP LRP to incorporate MP operations effectively into higher-level planning, and it enables an efficient transition to field operations. Also, because in peacetime this section functions as a PM cell, it provides the PM with efficient and effective technical support during stability and support operations. As the operations approach termination or enter sustainability and fewer forces are required, the brigade headquarters can redeploy, leaving a battalion TF in charge of MP forces.

MP BATTALION TF

11-55. The MP battalion (CS) provides command, control, and support for assigned MP companies (CS). MP battalions (CS) are best employed as part of a TF—either as part of a combat TF (high threat, opposed) or a support TF (low threat, unopposed). With sufficient augmentation, the battalion can provide the logistical, administrative, personnel, and operational support needed to lead a diverse TF. During stability and support operations, an MP battalion TF may be placed under the OPCON of a brigade-size TF (MP or non-MP) or a division TF. If placed under the OPCON of a division TF, both the division PM and the MP battalion commander will synchronize their efforts to maximize the employment of available divisional and nondivisional MP assets in support of the division commander's concept of operations.

11-56. As with the MPBTF, the MP battalion's contributions to stability and support operations are not a new concept. The following are examples of previous battalion TFs and the operations in which they were involved:

- **Cuban Refugee Movement.** The 519th, 716th, 720th, and 759th MP Battalions deployed for humanitarian-assistance operations.
- **Operations Restore Hope and Continued Hope.** The 720th MP Battalion provided humanitarian assistance in Somalia.

- **Operation Sea Signal.** The 720th MP Battalion deployed to Guantanamo Bay, Cuba, to conduct humanitarian-assistance operations for Haitian migrants. Later, the 716th and 759th MP Battalions also deployed to Guantanamo Bay to perform like functions for the Cubans.

MP COMPANY IN STABILITY AND SUPPORT OPERATIONS

11-57. The MP company is the MP Corps's primary mission performer. It executes a wide variety of missions. In these operations, MP companies—

- Provide mobility support for both vehicles and personnel.
- Provide AS, including security for critical personnel, sites, cargoes, and railways.
- Secure, safeguard, shelter, and control detainees and dislocated civilians.
- Restore order in civil disturbances and conduct PRC operations.
- Carry out limited L&O operations.
- Provide liaison, coordination, and training for all aspects of law enforcement to other agencies.
- Conduct police-intelligence collections and disseminate information to the populace, friendly forces, and other HN agencies.
- Support force protection and security.

11-58. The MP company possesses robust mobility, lethality, and communications. Each of the MP companies' three-man teams are equipped with either a high-mobility, multipurpose, wheeled vehicle (HMMWV), an up-armored HMMWV, or an armored security vehicle (ASV); an AN/VRC 91 radio; a Light-Vehicle Obscuration Smoke System (LVOSS); an MK-19 grenade machine gun; a squad automatic weapon (SAW); two M-16 or M-4 rifles; an M203 grenade launcher; and a 9-millimeter pistol as a side arm for each team member. In addition to conducting combat operations, the MP company commander has the capability to conduct limited L&O operations (desk and investigative operations). With a tremendous operational reach, the MP company will displace its platoons as far as possible to accomplish all of the MP functions.

MP I/R UNITS IN STABILITY AND SUPPORT OPERATIONS

11-59. The MP I/R units are specifically equipped and trained to support, safeguard, account for, and provide proper and humane treatment for EPWs/CIs, US military prisoners, or dislocated civilians. When conducting dislocated-civilian operations, the MP I/R units support and interact with civil-affairs and PSYOP units who are responsible for monitoring and handling dislocated civilians. Should more than one category of personnel have to be supported, the MP I/R units will establish separate compounds for each to preclude contact with other populations. When operation of the facility is going to be more than a short-term endeavor, the intent is to involve (as appropriate) NGOs, IHOs, HN authorities, and dislocated civilians in the administration of the facility. The tasks performed by MP I/R units for stability and support operations are basically the same as those for war, but the operations and procedures are modified according to the operational environment and the categories of people to be housed in the facility (see FM 19-40).

11-60. Criminal detainees (who are detained against their will) are subject to controls and procedures that closely parallel those used for EPWs/CIs. The primary emphasis is on guarding and preventing escape until their disposition is determined by recognized civilian judicial/law-enforcement authorities. When the operation is conducted outside of the US and its territories, criminal detainees are treated according to established legal procedures.

11-61. Dislocated civilians (who are being assisted rather than detained) are provided aid, shelter, and protection. The emphasis is on protecting them from harm by natural forces or hostile personnel. A special category of personnel arises when the MP I/R unit is required to house personnel that are dislocated civilians but who must be detained against their will. Such is the case of mass migrants who flee their countries and find themselves under US custody while formal proceeding guidance is being developed. In this case, MP I/R units must be sensitive to the situation and attempt to strike a balance between security, shelter, protection, and detention procedures.

11-62. Military detainees (who are detained against their will) are subject to controls and procedures that closely parallel those used for EPWs/CIs. In stability and support operations, US forces detain opposing-force military personnel. Without a formal declaration of war, opposing military personnel who are captured cannot be categorized as EPWs. However, they are afforded many of the EPW's rights and privileges under the Geneva convention. Such a situation occurred during Operation Just Cause when PDF personnel were detained by US forces.

11-63. In an operational environment in which hostile groups are engaged against one another, an I/R facility may be set up to protect one group from another. In this case, the facility's purpose is to shelter, sustain, account for, and protect its dislocated civilians from the hostile group. The MP I/R units and other supporting units concentrate on providing AS to protect the facility from direct fire. Other MP or combat forces provide protections beyond the direct-fire zone. Accountability for the dislocated civilians is coordinated with the SJA and civil affairs. They focus on maintaining a record of the people in the facility and their physical conditions. In operations in which no hostile groups are engaged (such as natural disasters), the I/R facility may be set up to provide shelter, sustain, and account for personnel. There may not be a need for external security personnel.

11-64. The MP I/R C^2 structure for stability and support operations is based on METT-TC. The nature and complexity of the mission, the number and types of detainees, and the operational duration should be considered. At one end of the operations spectrum, an MP brigade (I/R) may be required; while at the other end, an MP battalion (I/R) may be assigned to an MP brigade (CS).

Chapter 12

Force Protection

There is no more important responsibility than force protection. It is an inherent command responsibility and must be fully integrated into every unit's mission. A commander must continually review his unit's force-protection posture and crosswalk it with current and changing policy and threat levels. Force protection demands the personal involvement of commanders to ensure the best possible security consistent with the threat to personnel and mission-essential resources.

AR 525-13

Force protection consists of those actions that prevent or mitigate hostile actions against DOD personnel (to include family members), resources, facilities, and critical information. It coordinates and synchronizes offensive and defensive measures to enable the joint force to perform while degrading opportunities for the enemy. It includes air, space, and missile defense; NBC defense; antiterrorism; defensive-information operations; and security to operational forces and means.

OVERVIEW

12-1. The MP contribute significantly to the Army's overall force-protection effort. They are directly responsible for executing certain security-related measures (law enforcement and physical security) and for providing support to others (HRP security and antiterrorism), all of which contribute to force protection. Commanders use the Army's risk-management methodology and established security programs and capabilities within US forces to help prioritize and implement force-protection measures.

SUPPORT TO FORCE PROTECTION

12-2. MP support to force protection is accomplished under the leadership of the respective command's PM. It is accomplished mainly through an active role in physical security, HRP security, law enforcement, and antiterrorism.

PHYSICAL SECURITY

12-3. Physical security is an integral part of security of forces and means. Physical-security measures deter, detect, and defend against threats from terrorists, criminals, and unconventional forces. These measures include—

- Fencing and perimeter standoff space.
- Lighting and sensors.

- Vehicle barriers.
- Blast protection.
- Intrusion-detection systems and electronic surveillance.
- Procedural measures such as security checks, training and awareness, property accountability/inventory requirements, physical-security inspections of mission-essential or vulnerable areas (MEVAs), and physical-security surveys of installations.

12-4. These physical-security measures are not exclusively peacetime measures. They are designed to protect US personnel and equipment regardless of the mission or the geographical location.

12-5. The MP support the commander's security of forces and means programs by—

- Controlling or monitoring installation, base, or base-cluster access or entrance points.
- Monitoring intrusion-detection systems and providing a response force.
- Conducting physical-security inspections.
- Conducting perimeter security or site surveillance.
- Recommending the placement of walls, berms, gates, or barriers around designated MEVAs, high-value areas, or perimeters.
- Supporting the commander's risk-analysis effort.
- Conducting roving patrols, checkpoints, or roadblocks.
- Performing other physical-security measures as required by the commander.

HRP SECURITY

12-6. The HRP security program provides additional security to designated individuals (and their families) whom by virtue of their rank, assignment, symbolic value, vulnerabilities, or location or a specific threat are at a greater risk than the general population. This security program is applicable across the full spectrum of military operations.

12-7. The MP support the commander's HRP security measures by—

- Conducting protective-service details for HRP (normally conducted by USACIDC special agents).
- Responding to duress alarms and intrusion-detection systems.
- Conducting a PSVA (performed by the USACIDC).
- Employing MWDs.
- Coordinating with local, state, and federal agencies as well as joint, multinational, and HN law-enforcement agencies.
- Performing other HRP security measures as required by the commander.

LAW ENFORCEMENT

12-8. The Army's law-enforcement program supports the security of forces and means through the prevention, detection, response, and investigation of crimes. The MP, the USACIDC, and DOD police play a crucial role in the law-enforcement program. They provide a visible deterrent against a broad portion of the threat spectrum, investigate crimes, and provide an initial response force to threat incidents. As with the previous programs, the law-enforcement program is applicable anywhere across the spectrum of military operations.

12-9. The MP support law-enforcement operations by—

- Providing liaison teams with local, state, and federal agencies; HN police; and joint and multinational agencies.
- Employing SRTs and hostage-negotiation teams (normally provided by the USACIDC).
- Providing traffic enforcement, MSR regulation enforcement, and other route-control measures.
- Employing MWDs.
- Conducting MP investigations (criminal and noncriminal).
- Conducting patrolling, AS, and surveillance measures.
- Implementing applicable threat-condition (THREATCON) measures.
- Conducting and implementing other law-enforcement measures as required by the commander.

ANTITERRORISM

12-10. Terrorism is characterized as the unlawful use of violence or the threat of violence to coerce or intimidate a government or a society. Protection against a terrorist threat requires both an offensive counterterrorism capability and a defensive antiterrorism program. The primary focus of antiterrorism is on training and awareness and thorough planning. Antiterrorism is not a stand-alone program. It leverages other force-protection-related programs (such as physical security and law enforcement) to provide much of the physical defense against terrorism.

12-11. The MP support the commander's antiterrorist program by—

- Collecting CRIMINTEL (within the limits of AR 190-45).
- Responding to terrorist threats and incidents.
- Performing aggressive patrols and R&S operations.
- Detecting suspicious activities.
- Enforcing joint and cooperative agreements with the FBI, the USACIDC, MI, and other pertinent agencies (home or abroad).
- Conducting information collection and dissemination.
- Protecting critical assets, facilities, and personnel.
- Performing other antiterrorism measures as required by the commander.

Appendix A

Metric Conversion Chart

This appendix complies with current Army directives which state that the metric system will be incorporated into all new publications. Table A-1 is a conversion chart.

Table A-1. Metric Conversion Chart

Metric to English			English to Metric		
Multiply	**By**	**To Obtain**	**Multiply**	**By**	**To Obtain**
Length					
Centimeters	0.0394	Inches	Inches	2.54	Centimeters
Meters	3.28	Feet	Feet	0.0305	Meters
Meters	1.094	Yards	Yards	0.9144	Meters
Kilometers	0.621	Miles (stat)	Miles (stat)	1.5609	Kilometers
Kilometers	0.540	Miles (naut)	Miles (naut)	1.853	Kilometers
Millimeters	0.039	Inches	Inches	25.40	Millimeters
Area					
Square centimeters	0.1550	Square inches	Square inches	6.45	Square centimeters
Square meters	10.76	Square feet	Square feet	0.0929	Square meters
Square meters	1.196	Square yards	Square yards	0.836	Square meters
Volume					
Cubic centimeters	0.610	Cubic inches	Cubic inches	16.39	Cubic centimeters
Cubic meters	35.3	Cubic feet	Cubic feet	0.0283	Cubic meters
Cubic meters	1.308	Cubic yards	Cubic yards	0.765	Cubic meters
Milliliters	0.0338	US liq ounces	US liq ounces	29.6	Milliliters
Liters	1.057	US liq quarts	US liq quarts	0.946	Liters
Liters	0.264	US liq gallons	US liq gallons	3.79	Liters
Weight					
Grams	0.0353	Ounces	Ounces	28.4	Grams
Kilograms	2.20	Pounds	Pounds	0.454	Kilograms
Metric tons	1.102	Short tons	Short tons	0.907	Metric tons
Metric tons	0.984	Long tons	Long tons	1.016	Metric tons

Appendix B

Doctrine and Training Literature

Users of this manual must understand the elements of doctrinal literature and their relationship to each other. The commonly used terms doctrine and tactics, techniques, and procedures (TTP) are interrelated and mutually supportive; yet, each term has its own usage, level of detail, and place in the hierarchy of doctrinal publications. Because there is a close interrelationship, publications overlap in what they provide to the user. Figure B-1 illustrates the relationship of these doctrinal terms and publications.

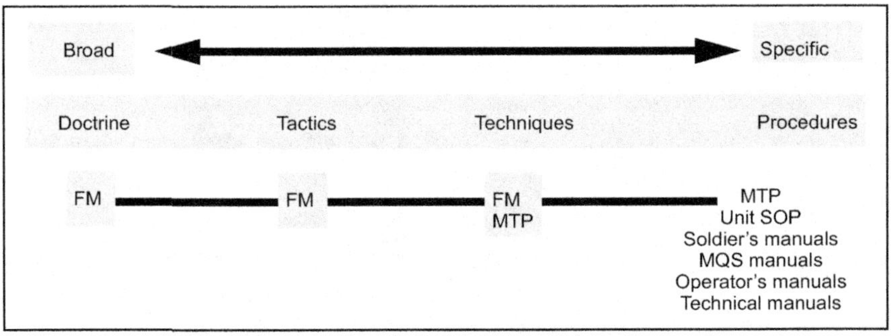

Figure B-1. Doctrinal and Training Literature

DOCTRINE

B-1. Doctrine is the body of fundamental principles by which the military guide their actions in support of national objectives. It is authoritative, but requires judgment in application. Doctrine provides the Army's fighting philosophy, establishes a common language, and structures the manner within which units plan and conduct combat operations.

B-2. Doctrine is not absolute; it is applied by the leader to meet the circumstances of the situation being faced. Doctrine provides for this flexibility with broad fundamentals and principles for conducting operations. Leaders well grounded in doctrine understand how to synchronize the elements of combat power and how to apply TTP using the assets available to accomplish the mission.

B-3. Doctrine is described in FMs. FM 3-0 defines the systems of war fighting practiced by the US Army. FMs 100-15 and 71-100 expand on this foundation. This manual describes the CS doctrine of the MP; it is the capstone manual for MP operations. It defines the role, operational requirements, functions, and C^2 relationships of MP units. This manual is not oriented on a specific organizational structure. Leaders at all levels apply this doctrine to their unique organizational and operational situations and use various tactics and techniques to accomplish the doctrine.

TACTICS

B-4. Tactics describe how the leader carries out doctrine (relating to the arrangement of forces for battle). Tactics include the—

- Employment of units in combat (to accomplish a mission assigned by the commander).
- Ordered arrangement and maneuver of units in relation to each other and to the enemy in order to use their full potential (a description of how the commander should arrange his forces and maneuver to accomplish a mission).

B-5. This is the part of doctrinal literature used for training in preparation for combat. Like doctrine, tactics presented in manuals are applied with judgment by the leader. The FMs prescribe "how to fight" and the mission training plans (MTPs) describe "what to train."

TECHNIQUES

B-6. Techniques are the methods of performing any act, especially the detailed methods used by troops or commanders in performing assigned tasks. Techniques describe the basic methods of using equipment and personnel. Techniques give details on how commanders actually carry out assignments. They improve the force's efficiency by ensuring the uniformity of action or by ensuring that the actions of various individuals and elements complement those of other individuals or elements.

B-7. The FMs and MTPs provide a description of techniques using the personnel and equipment available in actual organizations. These techniques show (at a more detailed level) how to use available assets to carry out tactics.

PROCEDURES

B-8. Procedures are the lowest level of detail. They address "how to" at the task level. Procedures are a particular course or mode of action that describe how to perform a certain task. Procedures include the—

- Standing methods used by units to accomplish tasks.
- Operating procedures for weapons and equipment.
- Crew drills.
- Staff actions and coordination requirements.
- Methods of target engagement by direct- and indirect-fire weapons.

B-9. Procedures are building blocks of individual- and collective-task accomplishment that serve as the foundation of tactics and techniques. Procedures are explained in unit standing operating procedures (SOPs), MTPs, soldier's manuals, operator's manuals, technical manuals (TMs), and similar publications.

Appendix C

Battlefield Workload Analysis

With minor adjustments in the capability block, the BWA can be used to determine MP requirements for the division. A BWA work sheet is shown in Figure C-1. A separate matrix is prepared for each EAC and corps MP brigade. Figure C-2, page C-3, is a sample of a BWA matrix for a 3-division corps MP CS company.

Mission/Function[1]	Capability[1]	Workload[2]	Rqmt	HN[18]	US Rqmt	Risk[19]	Mission Rqmt
Divisions supported[3]	1						
MMS	360 km MSR and ASR						
Refugees on control routes[4]	150,000 per day						
AS—corps rural terrain[5]	2,000 sq km						
AS—EAC rural terrain[5]	3,000 sq km						
AS—urban terrain[5]	800 sq km (urban)						
AS—USAF main opns base[5]	1,000 sq km/USAF MOB						
AS—convoy escort[6]	# tms per convoy	# convoys					
AS—ammo security[7]	1.3 modular ammo co						
	2 modular ammo co						
AS—pipeline security[8]	400 km						
AS—EAC pipeline co[8]	3						
AS—trains[9]	7						
AS—small critical sites[10]	12						
AS—major HQ security[11]	1						
AS—EAC port security[12]	0.66 of port						
AS—combat operations	When required, assets are diverted from other missions.						
L&O—incident response[13]	75,000 nondivision troops						
I/R—US prisoner detention[14]	700						
I/R—EPW escort/corps[15]	1,900 walking						
	2,500 by vehicle	Min: 1 plt					
	3,800 by train						
I/R—EPW holding/corps[16]	2,000	Min: 1 plt					
Other assigned missions[17]	Mission	Force size					
		Totals[20]					
Battalion HHD requirement = 0.199 x number of companies[21]							

Figure C-1. Sample BWA Work Sheet

NOTE: Workload ÷ capability = doctrinal requirement - HN offset = US requirement - acceptable risk = mission requirement

[1]Mission/capabilities are as stated in section I of TOEs 19477L000 and 19677L000. The unit capability for each mission is what the unit can accomplish when performing no other mission (for example, if it is only doing MSRs, the unit can cover 360 kilometers in a 24-hour period). Unless otherwise specified, the capabilities are the same at corps and EAC.

[2]Workload is the amount (MSR or area) or number of supported units/facilities.

[3]Corps support to each division is covered by allocating one MP company per supported division. Any additional support requirements should be based on a detailed division PM/MP mission analysis and included under other assigned missions.

[4]Unless specifically tasked, the Army normally relies on HN support to manage refugees. Redirecting refugees from MSRs and keeping them off of MSRs is part of (as is straggler control and TCP) the mission and MSR workload. When assigned the mission, the capability is for control of refugees on established refugee control routes that are in addition to the MSRs.

[5]Compute each square kilometer mission requirement separately. Do not compute the same terrain in more than one computation (for example, the area around the Air Force's main operating base (MOB) is not included with either rural or urban square kilometers).

[6]Convoy-escort capability is not stated in the TOE and is an AS function rather than an MMS function. Determine the average number of teams per mission and the average number of simultaneous missions. Multiply the number of teams required per mission by 0.028, then multiple the result by the number of simultaneous missions to determine the total requirement. State the number of teams per convoy and the number of convoys, then convert the force size to a 3-decimal-place company equivalent. Each additional team, squad, and platoon equivalent equals 0.028, 0.083, and 0.25 respectively.

[7]Security for modular ammo companies (ASPs) supporting divisions is three platoons and is normally conducted directly behind the division (corps forward). However, METT-TC may require the ammo units to move into the DRA. In this case, corps units should be attached to the division for the duration. Security for corps storage areas (CSAs) and theater storage areas (TSAs) is one MP company per two ammo companies. Security is provided on an area-support basis in the form of screening, with limited fixed posts.

[8]The EAC petroleum terminal/pipeline operating company manages the EAC pipeline that is normally laid adjacent to MSRs. Pipeline security is inherently provided by the MP operating on the MSRs. Unless the pipeline is not near the MSR, MP support is only allocated against the EAC petroleum terminal/pipeline operating company's bulk storage facilities. Security is provided on an area-support basis in the form of screening, with limited fixed posts.

[9]Train security is normally an EAC mission. This capability equates to simultaneously providing five MP teams to each of seven trains. Actual workloads and requirements are computed in the same manner as a convoy escort.

[10]Small critical facilities are normally defined as key MSR bridges and tunnels and communication nodes that may require additional security.

[11]The major headquarters are corps, COSCOM, TSC, unified command and, if directed, the JTF. The ASCC depends on the MP to provide security for two CPs. The ASCC requirement is an additional three platoons to support a second CP.

[12]The requirement for one seaport is 1.5 companies.

[13]Teams provide incident response, crime-scene security, and traffic control. They also assist L&O detachment teams and CID investigators.

[14]For deliberate planning in support of an OPLAN, personnel to be confined will normally be held by either the unit or the EAC I/R MP battalion performing the confinement mission. This does not preclude a requirement for the echelon PM to operate a field detention facility until prisoners are released back to their units or evacuated to the EAC confinement facility.

[15]The primary mode is vehicle (back-haul doctrine). Regardless of the mode, a minimum of one platoon (0.25 requirement) per supported division is allocated for EPW escort from the division central collection point (DCCP) to the corps holding area (CHA). Because the actual number of EPWs is unknown, coupled with no time/distance study for each supported division, one platoon per division provides a minimum capability for continuous evacuation within the stated doctrinal timeline (evacuate to the CHA within 12 hours of arrival at the DCCP).

[16]At a minimum, one platoon (0.25 requirement) is allocated to operate the CHA (provides the capability to hold 500 EPWs at any given time).

[17]Add any additional continuous missions assigned by the echelon commander not covered above. Examples include VIP escort/security for other than the CG and the deputy commanding general (DCG) of major headquarters, screening exposed corps flanks, and increasing teams per mission/function which decreases doctrinal capability (for example, permanent TCPs or checkpoints). Each additional team, squad, or platoon equivalent equals 0.028, 0.083, and 0.25 respectively.

Figure C-1. Sample BWA Work Sheet (continued)

[18]For each mission/function, any workload either fully or partially accomplished by the HN is subtracted from the doctrinal mission requirement. Enter the MP company equivalent offset and subtract from the doctrinal requirements to determine the US requirement.

[19]For each mission/function, if the operational/MP commander will accept mission risk, subtract the MP company equivalent to determine the mission requirement.

[20]Add columns to determine the total doctrinal company requirement, HN offsets, US requirement, acceptable risk, and mission requirement to support the operation/AO. Round the requirement to the nearest whole number (0.4 or fewer rounds down, 0.5 or greater round up).

[21]Determine battalion HHD requirements by multiplying the company requirements by 0.199 and rounding.

Figure C-1. Sample BWA Work Sheet (continued)

Mission/Function	Capability	Workload	Rqmt	HN	US Rqmt	Risk	Mission Rqmt
Divisions supported	1	3	3.000		3.000		3.000
MMS	360 km MSR and ASR	1,050	2.917		2.917		2.917
Refugees on control routes	150,000 per day	25,000	0.167	0.167	0.000		0.000
AS—corps rural terrain	2,000 sq km	6,500	3.250		3.250	0.500	2.750
AS—EAC rural terrain	3,000 sq km						
AS—urban terrain	800 sq km (urban)	1,500	1.875		1.875	0.500	1.625
AS—USAF main opns base	1,000 sq km/USAF MOB	1,000	1.000		1.000		1.000
AS—convoy escort	5 tms per convoy	4	0.560		0.560		0.560
AS—ammo security	1.3 modular ammo co	3	2.308		2.308		2.308
	2 modular ammo co	1	0.500		0.500		0.500
AS—pipeline security	400 km						
AS—EAC pipeline co	3						
AS—trains	7						
AS—small critical sites	12	9	0.750		0.750	0.250	0.500
AS—major HQ security	1	2	2.000		2.000		2.000
AS—EAC port security	0.66 of port						
AS—combat operations	When required, assets are diverted from other missions.						
L&O—incident response	75,000 nondivision troops	45,000	0.600		0.600	0.300	0.300
I/R—US prisoner detention	700						
I/R—EPW escort/corps	1,900 walking						
	2,500 by vehicle	Min: 1 plt/div	0.750		0.750		0.750
	3,800 by train						
I/R—EPW holding/corps	2,000	Min: 1 plt/	0.250		0.250		0.250
Other assigned missions	PSD for Corps CofS	1 sqd	0.083		0.083		0.083
	Flank screen	1 plt	0.250		0.250		0.250
		Totals	20	0.177	20	1.3	19
Battalion HHD requirement = 0.199 x number of companies			4		4		4
NOTE: Workload ÷ capability = doctrinal requirement - HN offset = US requirement - acceptable risk = mission requirement							

Figure C-2. Sample BWA Matrix

Appendix D

Automated Information Systems in Support of Battle Command

As the MP Corps continues to advance into the twenty-first century, it must continue to pursue and stay in tune with the Army's initiative to acquire advanced technologies and operational concepts. This will give MP soldiers on the battlefield an information advantage that is critical for successful performance of their functions. The following information opens a window into the near future and explains some of the automation initiatives that will enhance the Army's and the MP Corps's soldiers.

ARMY INFORMATION SYSTEMS

D-1. The Global Command and Control System (GCCS) is the midterm solution and the bridge to the concepts outlined in the concept of command, control, communications, computers, and intelligence for the warrior (C4IFTW). The GCCS's core consists of the basic functions required by the war fighter to plan, execute, and manage military operations. The Army's selected C^2 system is the Army Battle Command System (ABCS).

D-2. The ABCS integrates the C^2 systems found at each echelon—from the ground-force component commander at the theater or JTF level to the individual soldier or the weapons platform. Whether ARFOR are deployed for land combat or are conducting peace operations, providing humanitarian assistance, or giving aid to civil authorities, the ABCS supports the mission by integrating the battle-space automation systems and communications that functionally link strategic and tactical headquarters. The ABCS is interoperable with joint and multinational C^2 systems at upper echelons across the full range of C^2 functionality, and it is vertically and horizontally integrated at the tactical and operational levels. The ABCS has three major components—

- The Global Command and Control System—Army (GCCS-A).
- The Army Tactical Command and Control System (ATCCS).
- The FBCB2.

GLOBAL COMMAND AND CONTROL SYSTEM—ARMY

D-3. The GCCS-A is the Army's component of the joint GCCS. It was built from application programs developed by the Army Worldwide Military Command and Control System [WMCCS] Information System (AWIS), namely the—

- Strategic Theater Command and Control System (STCCS).

- Theater Automated Command and Control Information Management System (TACCIMS).
- EAC portion of the Combat Service Support Control System (CSSCS) (CSSCS/EAC).

D-4. The GCCS-A is required in the MP LRP, S1, S3/NBC, and S4 sections within the MP brigade at EAC; in the Assistant Chief of Staff, G1 (Personnel) (G1), the G3, and the Assistant Chief of Staff, G4 (Logistics) (G4) sections within the MP command (I/R); and in the S1, S3, and S4 sections within the MP brigade (I/R).

ARMY TACTICAL COMMAND AND CONTROL SYSTEM

D-5. The ATCCS embodies the commander and his staff and the doctrine, procedures, and tools used to command and control forces on the tactical battlefield. The ATCCS's functional subsystems provide tools to support the C^2 process across the spectrum of low- to high-intensity conflicts worldwide; for MOOTW; and for combat, CS, and CSS functional areas at corps-and-below units. The ATCCS provides a common set of C^2 tools and functions for commanders that extends the ATCCS to the battalion level. The component's ATCCS's are the—

- Maneuver Control System (MCS).
- All-Source Analysis System (ASAS).
- Advanced Field-Artillery Tactical Data System (AFATDS).
- Air- and Missile-Defense Planning and Control System (AMDPCS).
- CSSCS.

D-6. The FBCB2 provides situational awareness and C^2 to the lowest tactical echelons. It facilitates a seamless flow of battle-command information and situation-awareness information across the battle space at brigade and below on Army platforms.

FORCE XXI BATTLE COMMAND—BRIGADE AND BELOW

D-7. The FBCB2 system, a battle-command information system, provides the technology to complete the ABCS information flow process from brigade to platform and across all platforms within the brigade TF. The FBCB2 enables commanders to remotely operate and maintain ABCS database connectivity (regardless of the command vehicle) and to digitally control and monitor their subordinate units' statuses and positions. Interoperable FBCB2 systems will enhance the battle commander's ability to have a detailed picture of the battlefield (situational awareness), to maximize force survivability, and to facilitate battlefield synchronization to reduce or destroy the enemy's capability to fight. The FBCB2 relies on integrating, processing, and moving information in a digital form. To support projections of future conflicts with an accompanying increase in the tempo of battle, the FBCB2 must be able to obtain information contained in databases from higher, lower, and adjacent elements without formally requesting that information through command channels. The war fighter requires the ability to pull data from all battlefield-system databases to keep current with the rapid changes caused by the dynamics of battle. The FBCB2 is required for divisional MP company commanders, PMs, platoon leaders, teams, and squads.

MANEUVER CONTROL SYSTEM

D-8. The MCS is the ABCS's horizontal and vertical integrator. The MCS will enable commanders and their staffs to collect, coordinate, and act swiftly on near-real-time battlefield information. It will quickly, accurately, and automatically move tactical information on the battlefield. This allows a commander to readily access information and display current situation reports (SITREPs) that assess enemy strength and movement as well as the status of friendly forces. The MCS will then aid the battle staff in rapidly disseminating the commander's orders. It is intended to provide two major functional capabilities—maneuver functional-area control and force-level information. The MCS is the ATCCS's functional subsystem that—

- Supports the combined-arms force commander directly by providing automated support for planning, coordinating, controlling, and using maneuver functional-area assets and tasks.
- Coordinates and synchronizes supporting arms in conducting operational planning, field operations, and training.

D-9. The MCS is required for division PM operations sections, company CPs at echelons corps and below (ECB), MP battalion commanders, S3 sections at corps, MP brigade commanders, and corps MP LRP and S3 sections.

ALL-SOURCE ANALYSIS SYSTEM—REMOTE WORK STATION

D-10. The ASAS—remote workstation (RWS) subsystem will operate within the existing intelligence and electronic warfare (IEW) battlefield operating system (BOS) and ABCS architectures to provide automation support to G2/S2 intelligence staff elements and other designated intelligence organizations from the maneuver battalion through EAC. It will operate at the security level appropriate to its echelon, mission, and functions. It assists the G2/S2 in providing IEW support to the commander and other staff elements in planning, coordinating, and conducting current and future tactical operations. It interfaces with the sensitive compartmented information (SCI) analysis and control element (ACE) subsystem to provide operational guidance and to exchange IEW information and products. It also operates as the technical control portion of the IEW C^2 node of the ABCS to provide current IEW and enemy-situation information to the force-level information database for access and use by ABCS users. Two RWSs are also provided to the brigade analysis control team (ACT) of the DS MI company to process data input from DS collection systems, to filter intelligence data from broadcast systems, and to pass tactically relevant enemy-situation information or situation-development data to the brigade S2's RWS. The ASAS-RWS is required in the S2 section of the CID group, the S2/S3 section of the CID battalion, the G2 section of the MP command (I/R), the S2 section of the MP battalion (I/R), and the S2 section of the MP battalion and brigade at all echelons.

COMBAT SERVICE SUPPORT CONTROL SYSTEM

D-11. The CSSCS will function as an integral part of the ATTCS. It is designated as an information system with the capability to rapidly collect, analyze, and disseminate logistical, medical, and personnel information to accelerate and improve decision making by CSS and force commanders. It also enables the rapid exchange of C^2 information with other battlefield functional areas (BFAs) through automated interfaces. The system will process unclassified through secret information. The CSSCS will be employed in all divisions, corps, EAC, separate brigades, and armored cavalry regiments. The MP require CSSCS functionality in all S1 and S4 sections in battalion or higher organizations except where the GCCS-A is required as stated above.

TRANSPORTATION COORDINATOR AUTOMATED MOVEMENT SERVER

D-12. The Transportation Coordinator Automated Movement Server (TC-AIMS II) is used by transportation agents and deploying units of each service to automate the processes of planning, organizing, coordinating, and controlling deployment and redeployment operations worldwide during peace as well as contingencies. The TC AIMS II will link all DOD-component unit movements with the installation transportation officer/traffic-management officer (ITO/TMO). This will provide the capability to plan and execute deployments rapidly to meet CINC-required arrival and departure dates. The TC AIMS II will also provide the ITO's/TMO's day-to-day traffic management functions and the theater transportation managers with the capabilities to use theater common lift assets effectively. The TC AIMS II will also allow CSS organizations and supporting commands to sustain the force and perform effective movement control, coordination, distribution, and sustainment once in theater. This is also defined as transportation movement and support of DOD personnel and cargo during all phases of military operations in all environments (including RSOI and battlefield operations). The TC AIMS II's functionality is required at all division PM operations and S3 sections within MP battalions and brigades to facilitate MSR regulation enforcement. Additionally, MP companies, battalions, brigades, and commands require the TC AIMS II to plan, organize, coordinate, and control deployments and redeployments.

GLOBAL COMBAT SUPPORT SYSTEM—ARMY

D-13. The GCSS-A is an integrated logistical support system with accompanying hardware that will connect the service's supply, finance, transportation, maintenance, and personnel databases. Under this system, the Army will integrate 13 Legacy tactical logistics systems and will eventually interface with the service's wholesale logistics systems as well as the consolidated logistics systems of the other services and the DOD. The system is critical for modernizing automation, for reducing logistical footprints in theaters, and for expanding war fighters' capabilities. The system will also replace the Standard Army Management Information System (STAMIS). The GCSS-A is required at all MP division company CPs and MP battalion and brigade S4 sections. Additionally, it is required at all supply functions within MP detachments that support I/R missions.

MILITARY POLICE AUTOMATED SYSTEMS

D-14. In an effort to maintain an information advantage and to acquire advance technologies and operational concepts, the MP Corps has acquired the following systems:

- The Military Police Management Information System (MPMIS) and its subsystems
- The CID's automated systems and their subsystems.

MILITARY POLICE MANAGEMENT INFORMATION SYSTEM

D-15. The MPMIS is a DA-mandated reporting system that does not interface with the ABCS. The MPMIS provides automated tools for MP records management and reporting. It reduces the administrative burden on the PM's staff. It increases the awareness of problem areas by managing crime information. The MPMIS consists of the following elements:

- **Offense Reporting Systems Windows (ORS-2W).** The ORS-2W provides automation support to track and report complaints, criminal incidents, and traffic incidents. It incorporates the CRC, the federally mandated National Incident-Based Reporting System (NIBRS), the DOD-mandated Defense Incident-Based Reporting System (DIBRS), the military community (MILCOM), and the installation's data requirements. The ORS-2W's functionality is required at PM activities at all echelons of command within the Army.
- **Correctional Reporting System (CRS-3).** The CRS-3 provides automation support to track and report inmates and corrections staff at all Army-controlled correctional facilities.
- **Security Management System (SMS).** The SMS provides automated support to assist physical-security inspectors and force-protection officers in reporting inspections and surveys. It tracks intrusion-detection devices and MWD training and health requirements.
- **Vehicle Registration System (VRS).** The VRS provides installations with an automated means to track vehicle, weapon, and bicycle registrations.
- **Prisoner-of-War Information System (PWIS-3).** The PWIS-3 provides the automated means of tracking an EPW's status from capture through release. This system is required at all EPW management activities.

CID AUTOMATED SYSTEMS

D-16. The Headquarters (HQ) USACIDC mandates CID automated systems. Similar to the MPMIS, it does not interface with the ABCS. The system facilitates data-collection efforts within the command. The CID's automated systems consist of the—

- **Automated Criminal Investigative Reporting System (ACIRS).** The ACIRS supports management and data-collection functions at the CRC and HQ USACIDC. It supports the following:
  - Maintaining investigative data.
  - Conducting agent investigative activities.
  - Reporting investigative data.
  - Managing the investigative processes.
  - Conducting criminal-activity analyses.

- **CRIMINTEL.** Data sources for CRIMINTEL include MP reports, CRIMINTEL reports and bulletins, reports of investigation, crime-prevention surveys, case files, collection folders, raw-data folders, significant data files, and target-analysis files. The CRIMINTEL interfaces with the ACIRS, but it does not share CRIMINTEL products.

Appendix E

Police Information Assessment Process

The PIAP differs somewhat from the tactical IPB. While the IPB allows the maneuver commander to see the effects that a variety of factors may have on his forces, the PIAP looks beyond this approach to determine how other relevant information may impact his forces. This information may be obtained from police or criminal actions or incidents encountered during the performance of MP functions.

IMPLEMENTATION

E-1. The PIAP is not a substitute for the IPB. MP leaders must continue to rely on the IPB and to use its estimates as a starting point for the PIAP. However, the PIAP is a dynamic and continuous cycle that complements the IPB (Figure E-1). The following paragraphs represent the doctrinal approach to implementing the PIAP. These steps provide the basic foundation to the process. They are not exclusive; the successful development of the PIAP will depend on the MP leader's ability to apply the process to his specific environment, METT-TC, and the commander's priorities.

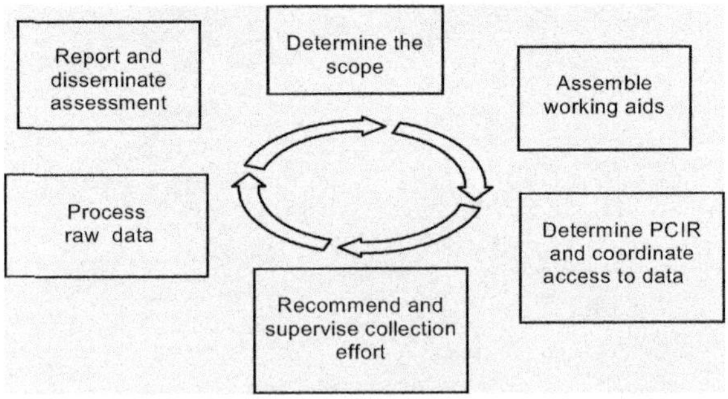

Figure E-1. The PIAP

DETERMINE THE SCOPE OF THE PIAP

E-2. While determining the scope of the PIAP might sound simple, it is one of the most crucial elements of the entire process. It entails identifying what you want to be the end result of your PIAP. To help determine the scope of the PIAP, MP leaders must perform the following subtasks:

- **Conduct a mission analysis.** A detailed mission analysis will ensure that all available resources are focused toward one goal. The more specific the mission is, the more focused and productive the effort. This involves more than just looking at the mission statement, the commander's intent, and the concept of operations. It involves translating the mission and situation into easily understood, manageable goals. The following are examples of these missions:
  - Identifying criminal threats from local national to US Army logistics operations within the port of Mogadishu, Somalia.
  - Determining the impact of gang violence on US forces operating in the vicinity of Camp Bondsteel, Kosovo.
  - Determining how rioters' violence will affect force protection before deploying forces to Los Angeles, California.
  - Determining the status of public service agencies (such as the police, the fire department, and the emergency medical services [EMS]) in Homestead, Florida, after the city was struck by Hurricane Andrew.

NOTE: As you can see from the examples, these missions are quite specific. In all likelihood, it would be necessary to establish several different PIAP missions to adequately protect Army elements operating in the MP's AOR.

- **Prioritize missions.** MP leaders must prioritize each PIAP mission since MP resources will probably be limited. The commander's intent is the key factor for determining priorities. When the commander's most important mission has been determined, MP leaders must decide how they can contribute to the overall success of that mission.
- **Determine the AO.** Generally, the AO is a geographical area (including the airspace above) usually defined by lateral, forward, and rear boundaries assigned to a commander. MP leaders must know where major units are located and how the characteristics of the location may impact ongoing operations. For example, units in direct contact with the enemy or isolated from the local population are less likely to be the target of a PIAP mission than those units adjacent to great population centers or in rest and relaxation centers.
- **Determine key terrain.** Key terrain is any locality or area that affords a marked advantage to the retaining side.
- **Determine the AOI.** An AOI is the area of concern to the commander. This includes the area of influence and the areas adjacent thereto and extending into enemy territory and to the objectives of current or planned operations. A thorough understanding of the AOI may help predict potential threat actions. For example, how far are local agitators and ringleaders willing to travel to interfere with US operations in Pristina, Kosovo? Would they go to other US compounds?
- **Determine required information products/sources.** The desired end-state product must be decided. Should it be a report? An updated SITREP? An upgrade in force-protection measures? Or a combination of products? It should also be determined who can provide the best

information (the CID; MI; the local police chief; or joint, interagency, or multinational agencies).

ASSEMBLE WORKING AIDS

E-3. During this step, MP leaders assemble the necessary graphic and nongraphic aids and information that will help them accomplish their mission. The CID and the MP may form an analysis team to track and analyze police information that might assist tactical operations or population-control operations. Working aids include, but are not limited to—

- **Maps.** Besides the standard military maps that depict the military AOs, the MP must attempt to procure indigenous maps of AOs. Detailed city and street maps will prove to be invaluable even if they are in a foreign language.
- **Language aids.** Phrase books and translation dictionaries cannot take the place of intensive language training, but they are better than nothing. These are essential even if the MP have assigned interpreters.
- **Open-source information.** This can cover an extensive range of invaluable information, but it should be available from a variety of sources. Examples include—
  - Demographic information on the HN population.
  - Societal information on the HN.
  - The locale and disposition of HN police forces.
  - Historical crime data of the area.
  - Environmental information (terrain and weather factors).
  - Local newspaper articles.
  - Internet sources on all of the above.

NOTE: Sources for this type of information can come from intelligence estimates provided by the higher headquarters' S2/G2, other military units, or the HN.

- **OPSEC information.** This information includes the disposition of US forces (to include MP, MI, CID, and other services' investigating units) deploying or being used within the AO.
- **Other working aids.** These may include items such as computers, printers, boards, and acetate that will be necessary depending on the missions the MP are undertaking.

DETERMINE INFORMATION REQUIREMENTS AND COORDINATE ACCESS TO DATA

E-4. Once the mission analysis is completed, the next step is to determine if there are information gaps. An information gap is a missing piece of information that is critical to the analysis. MP leaders must identify and prioritize the gaps using the time, the available resources, and the commander's intent. Once an information gap is identified, it becomes PCIR. Examples of PCIR are as follows:

- Where is the nearest police station, and how did they handle a certain problem?

- What is the composition of the two gangs operating in our AO, and how can they influence US operations?
- Why is river-crossing point Bravo a danger to US crossing forces? Why are local rioters disrupting only US operations on the east bank?

E-5. The next step is to determine if PCIR can be answered with data already available. To do this, the available military and nonmilitary sources must be identified. Some sources may include—

- **USACIDC.** The USACIDC is responsible for programs such as combating terrorism, CRIMINTEL, personal-security assessment, crime analysis, and LOGSEC threat assessment. Some of the information gathered by the USACIDC (although releasable under existing controls and restraints) may help with the PIAP mission. In fact, USACIDC special agents may already be working in the AO and may be available to help the MP with the collection effort.
- **MI element.** Like the USACIDC, the MI element operating in the AO may have come across a piece of information (police or criminal) that may help accomplish the PIAP mission. The MP must make daily contact with the S2/G2 and coordinate access to invaluable information. As with the USACIDC, available intelligence information may only be releasable under strict controls and restrictions imposed by the commander.
- **Other MP units.** Coordination with an adjacent MP unit may prove to be beneficial, especially if the other MP unit has conducted a similar PIAP mission or has conducted operations in the AOI.
- **HN law-enforcement agencies.** Local law-enforcement agencies will provide the essential populace knowledge otherwise not available to US forces. Chiefs of police, sheriffs, and other key community leaders may help MP leaders fill the information need of the PIAP mission.
- **Joint, interagency, and multinational forces.** Close and effective liaison with these elements will result in quick access to information that can produce the same benefits as the agencies listed above.

NOTE: If, after consulting with all available sources, the PCIR is not resolved, the MP leader must decide the best collection effort to solve the PCIR.

RECOMMEND AND SUPERVISE THE POLICE/CRIMINAL COLLECTION EFFORT

E-6. The collection effort is the means by which specific PCIR will be met. This implies developing a collection strategy, tasking specific collectors, and supervising the collection effort.

- **Develop a collection strategy.** After a through study of the availability, capability, and disposition of the potential collecting resources, MP leaders select which asset is better suitable to perform the mission. Are organic MP the best collectors for this mission? Or do we need to request support from USACIDC units? What unique or organic capabilities do each bring to the effort? Which MP function is the best? Do we conduct MMS, AS, or L&O operations? Part of the collection strategy includes coordinating with the S2/G2, the SJA, the

CID, and other agencies before launching the collection effort. This coordination will eliminate duplication of effort, interference with an ongoing effort, or stepping out of legal limits.

- **Task or request specific collectors.** The missions that will be tasked to respective collectors must be determined. Information collectors can be tasked with more than one mission at a time. However, it is imperative that their tasks be prioritized based on mission requirements and time available. The appropriate tasking or request chain must be used to request an MP team, a platoon, or USACIDC special agents.
- **Supervise the collection effort.** The collectors must be provided with reporting guidelines. How often should they report? Should they report "no information?" Specific reporting instructions should be provided, including how, when, and where they report. A collection tasking chart should also be used (Figure E-2). As more and more collectors are being used, it becomes important to track their missions, capabilities, and success. A brief note on the reporting instructions of each collector should be included.

PIAP Missions	Unit Tasked	Echelon Employed	POC	Collection Strategy Employed	Date Collection Started	Location	Date Collection Ended
PCIR 1							
PCIR 2							
PCIR 3							
PCIR 4							

Figure E-2. Collection Tasking Chart

PROCESS POLICE/CRIMINAL RAW DATA

E-7. This processing step converts raw data into police intelligence through analysis. The key to processing the data is to understand it. For example, as collectors report raw data it may often appear to be meaningless information. Rarely will one collector receive all of the necessary information to answer the PCIR at one time. Instead, a variety of information must be fitted together to form a coherent, even if incomplete, picture.

E-8. Every piece of data is important. This is the cardinal rule of the entire process. Until the data is analyzed for reliability and compared with other data, all reports must be treated equally. This is important to remember, for the trend will be to use data that either fits your expectations or gives the most complete picture.

E-9. A report of "no activity" can be as important as reports with information. If the collector is capable of collecting the information and is actively trying to collect, reports of "no activity" can be critical. First, if there is no indication of a change, they can be used to reprioritize PIAP missions. Second, they can be used to retask the collectors to other areas. It is important to check the mission's original intent periodically to ensure that it is still valid and merits further efforts.

E-10. To determine the reliability of the data, the MP leader must evaluate the source. Does the information come from the police chief, local youth, or other military forces? One of the best ways to determine the reliability of the information is to have more than one collector reporting on the same PCIR. If more than one collector reports the same information, it can normally be presumed to be accurate.

E-11. Information can be analyzed in a variety of ways—chronologically, geographically, by the impact on the force, by the type of activity or the modus operandi, or by the association of participants. The MP leader must evaluate the information after it is analyzed to determine whether the information collected answers the PCIR or whether it helps accomplish the PIAP mission. If the information does not answer these questions, he must decide whether it is necessary to task other collectors, retask the same collectors, or wait until the situation develops.

REPORT AND/OR DISSEMINATE POLICE AND CRIMINAL ASSESSMENT

E-12. Reporting and disseminating is the most important aspect of the PIAP. Collection and analysis is valueless unless the data is communicated to commanders who can best use the information.

E-13. Reporting requirements must be established based on the unit's SOP and other established processes. MP leaders must decide if the information collected must be sent using an established priority format or if it can wait to be sent using SITREPs. Does the information collected affect the force-protection measures or does it immediately impact the tactical scenario? Do we need to alert units operating in the vicinity of AA Fox to avoid sector 2? Additionally, MP leaders must ensure that any information released follows existing constraints and regulations.

E-14. It is also advantageous to provide feedback to the collectors. Collectors often receive no feedback on their performance. Informing them as to which part of their information was vital and beneficial can often increase their productivity in future operations.

SAMPLE SCENARIO

E-15. The following example helps illustrate the implementation of all six steps of the PIAP:

E-16. While deployed in support of Operation Control Chaos, US forces assigned to TF Blue Thunder are tasked with stability and support operations in the city of Corbina. This city, although currently implementing a peace treaty, was the site of extreme civil unrest. Even though the overall situation is not as chaotic as it was before US intervention, there are some sectors that are still in turmoil. Common crimes (such as robberies, assaults, looting, and vandalism) still occur daily and there is fear that this condition may spread out of control throughout the city. The TF commander knows that in order to bring peace to the entire city, he must attempt to stabilize the situation in some of these bad sectors. He decides that the best way to deal with this situation is to assign a sector (sector 51) to an MP battalion in an attempt to control the situation. If the MP battalion succeeds, he may use the same concept throughout the city. He is aware of the five MP functions and how they can help bring this sector under some control. Under this framework, the TF commander issues the following mission: *On order, the MP battalion will conduct stability and support operations in the vicinity of sector 51, the city of Corbina, in support of TF Blue Thunder.* Upon receipt of this mission, the battalion commander gathered his staff and initiated the planning process. Before the MP staff could initiate the PIAP, they made contact with the S2/G2 and received an updated IPB estimate. Now that the IPB estimates are on-hand, the staff initiates the PIAP.

DETERMINE THE SCOPE OF THE PIAP

E-17. The staff performs the following steps to determine the scope of the PIAP:

- **Conduct a mission analysis.** The staff looks at the given mission, the higher echelon commander's intent, and the concept of operations. After considering all of these factors, they produce a restated mission—*On order, the battalion will conduct MP operations in sector 51 to determine the type of threat and to bring this sector under control in support of TF Blue Thunder.*
- **Prioritize PIAP missions.** Since there is only one mission, this subtask is not applicable.
- **Determine the AO and the AOI.** The staff determines that the AO is the entire sector 51 and that the AOIs are the sectors adjacent to 51.
- **Determine key terrain.** The staff determines that the key terrains are the market place, the plaza, and the city park. They select these areas because most of the incidents occurred in these populated areas. Stabilization in these key areas may be advantageous to mission success.
- **Determine required information products/sources.** The end result is the stabilization of sector 51. Minimize the crime incidents and restore L&O. Information should come from the local police chief, local community leaders, civil affairs, PSYOP, and the CID.

ASSEMBLE WORKING AIDS

E-18. The staff assembles the following working aids in support of the PIAP:

- **Maps.** The staff acquires local maps that detail the market place, the plaza, and the city park. Although the maps are in a foreign language, the staff is able to identify these three prominent features and surrounding areas.
- **Language aids.** In addition to dictionaries, the staff is able to get two translators from civil affairs.

- **Open-source information.** Open-source information includes a variety of encyclopedias and country books. Newspaper reports and historical crime data is available via the internet or through MI sources. The emphasis is in obtaining information on the history and composition of the local police force.
- **Classified intelligence.** The staff receives a briefing from the S2/G2 describing the operations of SOF units in the area.
- **Other working aids.** Computers, charts, templates, and acetate are available.

DETERMINE INFORMATION REQUIREMENTS AND COORDINATE ACCESS TO DATA

E-19. After a careful analysis of all available information, the following information gaps were turned into PCIR:

- Why is the police force not actively engaged in L&O?
- What is the major threat to civilians in each area?
- Are local gangs involved?
- Is their local judicial system in place?
- Where are prisoners housed?

E-20. The staff consults with the USACIDC, MI, other MP units, and other sources but determines that the information provided is not clear or sufficient. Since the PCIR are not resolved, the staff decides to recommend that the commander initiate a collection effort.

RECOMMEND AND SUPERVISE THE POLICE/CRIMINAL COLLECTION EFFORT

E-21. The staff performs the following steps to recommend and supervise the police/criminal collection effort:

- **Develop a collection strategy.** Upon coordinating with the S2/G2, the SJA, and the CID, the MP commander decides that the best collectors will come from within organic assets. He decides that he will employ the MP to go out and seek the needed information. The battalion commander will employ a company commander to interview the local police chief and an MP/interpreter will patrol around the designated AOI. The MP commander will need additional interpreters to help him with his collection effort.
- **Task or request specific collectors.** The MP staff requests an interpreter through the normal chain of command. The TF G3 assigns six interpreters for a period of 90 days.
- **Supervise the collection effort.** All collectors are instructed to report not later than (NLT) 1800 every day. The report will be in the form of an "end of day" outbrief to the S2 and the battalion commander. A collection tasking chart (Figure E-3) will also be completed.

PROCESS POLICE/CRIMINAL RAW DATA

E-22. After receiving all of the above information, the S2 concludes that the inability of the police force to do their job has had a domino effect on all areas of peace and order. This information is confirmed through other reports from information collected by the CID (through their own efforts) and the CA. The information also reveals that the gangs have intentions of attacking US forces if they interfere with their "business." The S2 determines that although the situation may seem to be chaotic, once the locals are able to trust the police forces, everything should return to normal.

REPORT AND/OR DISSEMINATE POLICE AND CRIMINAL ASSESSMENT

E-23. Since the collection effort identifies a potential threat to US forces, the commander decides to notify higher headquarters and increase his own force-protection measures. He requires MP elements patrolling in designated areas to be no less than a squad. If patrolling dismounted, a HMMWV or an ASV will trail the squad. The MP commander uses the information provided during the PIAP to develop his implementation plan. This plan includes joint patrols, training, and construction of a joint MP/local national police station.

	PIAP Mission: Bring Sector 51 Under Control							
PCIR	**Unit Tasked**	**Echelon Employ-ed**	**POC**	**Collection Strategy Employed**	**Date Collection Started**	**Location**	**Date Collection Ended**	**PCIR Result**
Why is the police force not actively engaged in L&O operations?	A Co	2 teams	CPT Smith	Direct interview	31 Jan 00	Sector 51, Police HQ	2 Feb 00	The police force does not know how to interact with the population.
What is the major threat to civilians in each area?	A Co	6 teams	1LT Thomas	Patrol and interview	31 Jan 00	Market place, plaza, and park	7 Feb 00	Muggings, robberies, and stabbings are the major threats to civilians.
Are local gangs involved?	A Co	2 teams	CPT Smith	Direct interview	31 Jan 00	Sector 51, Police HQ	2 Feb 00	Two rival gangs exist.
Is a local judicial system in place?	A Co	2 teams	1LT Jones	Direct interview	31 Jan 00	Courthouse	3 Feb 00	A judicial system exists, but the judge is afraid to sentence anyone.
Where are the prisoners housed?	A Co	4 teams	1LT Lucas	AS and MMS	31 Jan 00	Local prison	2 Feb 00	The local prison is abandoned.

SAMPLE

Figure E-3. Sample Collection Tasking Chart

Glossary

1LT	first lieutenant
A/L	administrative and logistics
AA	assembly area
AATF	air-assault task force
ABCS	Army Battle Command Systems
ACE	analysis and control element
ACIRS	Automated Criminal Investigative Reporting System
ACT	analysis control team
ADA	air-defense artillery
ADC	area damage control
AFATDS	Advanced Field-Artillery Tactical Data System
ALO	authorized level of organization
AMDPCS	Air- and Missile-Defense Planning and Control System
AO	area of operation
AOE	Army of Excellence
AOI	area of interest
AOR	area of responsibility
APOD	aerial port of debarkation
APOE	aerial port of embarkation

AR	armored
AR	Army regulation
ARFOR	Army forces
AS	area security
ASAS	All-Source Analysis System
ASCC	Army service component command
ASG	area support group
ASP	ammunition supply point
ASR	alternate supply route
ASV	armored security vehicle
ATCCS	Army Tactical Command and Control System
attn	attention
AWIS	Army WMCCS Information System
BCT	brigade combat team
bde	brigade
BFA	battlefield functional area
BLD	brigade liaison detachment
bn	battalion
BOS	battlefield operating system
BRT	brigade reconnaissance troop
BWA	battlefield workload analysis

C^2	command and control
C^3	command, control, and communications
C4I	command, control, communications, computers, and intelligence
C4IFTW	command, control, communications, computers, and intelligence for the warrior
CAS	close air support
CASE	corps-area support element
CCIU	computer-crime investigative unit
cdr	commander
CG	commanding general
CHA	corps holding area
CI	civilian internee
CID	Criminal Investigation Division
CIDR	Criminal Investigation Division Regulation
CINC	commander in chief
CIP	Criminal Intelligence Program
cmd	command
CMOC	civil-military operations center
co	company
COA	course of action
CofS	Chief of Staff
commo	communications

COMMZ communications zone

conf confinement

CONUS continental US

COSCOM corps support command

CP command post

CPT captain

CRC Crime Records Center

CRIMINTEL criminal intelligence

CRS-3 Correctional Reporting System

CS combat support

CSA corps storage area

CSE corps support element

CSG corps support group

CSM command sergeant major

CSS combat service support

CSSCS Combat Service Support Control System

cust customs

CZ combat zone

DA Department of the Army

DC District of Columbia

DCCP division central collection point

DCG	deputy commanding general
DIBRS	Defense Incident-Based Reporting System
DISCOM	division support command
div	division
DLEA	domestic law-enforcement agency
DMMC	division materiel management center
DNA	deoxyribonucleic acid
DOD	Department of Defense
DOJ	Department of Justice
DOS	Department of State
DPM	deputy provost marshal
DRA	division rear area
DS	direct support
DSA	division support area
DSB	division support battalion
DSE	division support element
DSS	decisive, shaping, and sustaining
DZ	drop zone
EAC	echelons above corps
EAD	echelons above division
ECB	echelons corps and below

EG escort guard

EMS emergency medical services

EPW enemy prisoner of war

ERP engineer regulation point

ES escort

FA field artillery

FBCB2 Force XXI Battle Command—Brigade and Below

FBI Federal Bureau of Investigation

Feb February

FEBA forward edge of the battle area

FIU field investigative unit

FM field manual

FOB forward operating base

FOIA Freedom of Information Act

FORSCOM US Army Forces Command

FSB forward-support battalion

FSC forward-support company

G1 Assistant Chief of Staff, G1 (Personnel)

G2 Assistant Chief of Staff, G2 (Intelligence)

G3 Assistant Chief of Staff, G3 (Operations and Plans)

G4 Assistant Chief of Staff, G4 (Logistics)

G5 Assistant Chief of Staff, G5 (Civil Affairs)

G6 Assistant Chief of Staff, G6 (Communications)

GAC ground-assault convoy

GCCS Global Command and Control System

GCCS-A Global Command and Control System—Army

GD guard

GO government organization

GS general support

HHB headquarters and headquarters battery

HHC headquarters and headquarters company

HHD headquarters and headquarters detachment

HMMWV high-mobility, multipurpose, wheeled vehicle

HN host nation

HQ headquarters

HRP high-risk personnel

HVA high-value asset

I/R internment and resettlement

IBCT initial/interim brigade combat team

IC information center

IEW intelligence and electronic warfare

IG inspector general

IHO	international humanitarian organization
IMETP	International Military Education and Training Program
IO	international organization
IPB	intelligence preparation of the battlefield
IRIC	internment/resettlement information center
ISB	intermediate staging base
ITO	installation transportation officer
J3	Operations Directorate
Jan	January
JFC	joint-force command
JOA	joint-operations area
JP	joint publication
JRA	joint rear area
JSOA	joint special operations area
JTF	joint task force
km	kilometer(s)
L&O	law and order
LID	light infantry division
liq	liquid
LOC	lines of communication
LOGSEC	logistics security

LRP long-range plan

LSTA logistics-security threat assessment

LVOSS Light-Vehicle Obscuration Smoke System

LZ landing zone

M/O marksman/observer

M/S mobility/survivability

MACOM major Army command

MAJ major

MCRP Marine Corps reference publication

MCS Maneuver Control System

MDMP military decision-making process

mech mechanized

MEDEVAC medical evacuation

METL mission-essential task list

METT-TC mission, enemy, terrain, troops, time available, and civilian considerations

MEVA mission-essential or vulnerable area

MFPU major fraud procurement unit

MI military intelligence

MILCOM military community

min minimum

MKT mobile kitchen trailer

MLRS Multiple-Launch Rocket System

MMS maneuver and mobility support

MOB main operating base

MOOTW military operations other than war

MP military police

MPBTF military police brigade task force

MPMIS Military Police Management Information System

MQS military qualification standards

MSR main supply route

MTOE modified table of organization and equipment

MTP mission training plan

MTT mobile training team

MWD military working dog

NATO North Atlantic Treaty Organization

naut nautical

NBC nuclear, biological, and chemical

NCO noncommissioned officer

NEO noncombatant evacuation operation

NG national guard

NGO nongovernmental organization

NIBRS National Incident-Based Reporting System

NLT	not later than
NLT	nonlethal weapon
NWP	Naval warfare publication
OCONUS	outside the continental US
OPCON	operational control
OPLAN	operation plan
opns	operations
OPSEC	operations security
ORS-2W	Offense Reporting Systems Windows
PASE	port-area support element
PCIR	police/criminal information requirements
PDF	Panamanian Defense Force
PEO	peace-enforcement operation
PI	police intelligence
PIAP	police information assessment process
PIO	police intelligence operations
PIR	priority intelligence requirements
PKO	peacekeeping operation
plt	platoon
PM	provost marshal
POC	point of contact

POD port of debarkation

POE port of embarkation

POL petroleum, oil, and lubricants

PRC populace and resource control

PSD protective security detail

PSU protective-service unit

PSVA personal-security vulnerability assessment

PSYOP psychological operations

pub publication

PWIS-3 Prisoner-of-War Information System

PZ pickup zone

R&S reconnaissance and surveillance

RAOC rear-area operations center

RC reserve component

ROC rear operations center

ROE rules of engagement

rqmt requirement

RSOI reception, staging, onward movement, and integration

RSTA reconnaissance, surveillance, and target acquisition

RWS remote workstation

S1 Adjutant (US Army)

S2 Intelligence Officer (US Army)

S3 Operations and Training Officer (US Army)

S4 Supply Officer (US Army)

S5 Civil-Affairs Officer (US Army)

S6 Communications Officer (US Army)

SALUTE size, activity, location, unit, time, and equipment

SAW squad automatic weapon

SCI sensitive compartmented information

sec section

SEP separate

SFC sergeant first class

SITREP situation report

SJA staff judge advocate

SMS Security Management System

SOC security-operations cell

SOF special-operations forces

SOFA status of forces agreement

SOI signal operating instructions

SOP standing operating procedures

SPOD seaport of debarkation

SPOE seaport of embarkation

SPOTREP	spot report
spt	support
sq	square
sqd	squad
SRT	special-reaction team
STAMIS	Standard Army Management Information System
stat	statute
STCCS	Strategic Theater Command and Control System
TA	theater Army
TAA	tactical assembly area
TAACOM	Theater Army Area Command
TAC	tactical
TACCIMS	Theater Automated Command and Control Information Management System
TAI	traffic accident investigation
TASE	TSC-area support element
TC-AIMS	Transportation Coordinator Automated Movement Server
TCF	tactical combat force
TCP	traffic-control post
TF	task force
THREATCON	threat condition
TM	technical manual

TMO	traffic-management officer
tms	teams
TO	theater of operations
TOE	tables of organization and equipment
TRADOC	US Army Training and Doctrine Command
TRP	traffic-regulation plan
TSA	theater storage area
TSC	theater support command
TSE	tactical support element
TTP	tactics, techniques, and procedures
UMT	unit ministry team
UN	United Nations
US	United States
USACIDC	US Army Criminal Investigation Command
USACIL	US Army Criminal Investigations Laboratory
USAF	US Air Force
USAMPS	US Army Military Police School
USC	US Code
USCG	United States Coast Guard
USCS	US Customs Service
USDA	US Department of Agriculture

VIP very important person

VRS Vehicle Registration System

WMCCS Worldwide Military Command and Control System

Bibliography

AR 15-6. *Procedures for Investigating Officers and Boards of Officers.* 11 May 1988.

AR 190-12. *Military Police Working Dogs.* 30 September 1993.

AR 190-45. *Law Enforcement Reporting.* 30 September 1988.

AR 190-47. *The Army Corrections System.* 15 August 1996.

AR 195-2. *Criminal Investigation Activities.* 30 October 1985.

AR 525-13. *Antiterrorism / Force Protection (AT / FP): Security of Personnel, Information, and Critical Resources.* 10 September 1998.

CIDR 195-1. *Special Agent Operations Manual.* October 1994.

DA Form 2028. *Recommended Changes to Publications and Blank Forms.* 1 February 1974.

FM 3-0. *Operations.* To be published within the next 6 months.

FM 19-1. *Military Police Support for the AirLand Battle.* 23 May 1988.

FM 19-4. *Military Police Battlefield Circulation Control, Area Security, and Enemy Prisoner of War Operations.* 7 May 1993.

FM 19-10. *The Military Police Law and Order Operations.* 30 September 1987.

FM 19-40. *Enemy Prisoners of War, Civilian Internees and Detained Persons.* 27 February 1976.

FM 34-2. *Collection Management and Synchronization Planning.* 8 March 1994.

FM 34-130. *Intelligence Preparation of the Battlefield.* 8 July 1994.

FM 41-10. *Civil Affairs Operations.* 14 February 2000.

FM 63-4. *Combat Service Support Operations—Theater Army Area Command.* 24 September 1984.

FM 71-3. *The Armored and Mechanized Infantry Brigade.* 8 January 1996.

FM 71-100. *Division Operations.* 28 August 1996.

FM 90-13-1. *Combined Arms Breaching Operations.* 28 February 1991.

FM 90-40. *Multiservice Procedures for the Tactical Employment of Nonlethal Weapons (NLW) {MCRP 3-15.8; NWP 3-07-31; USCG Pub 3-07.31}.* 6 October 1998.

FM 100-1. *The Army.* 14 June 1994.

FM 100-5. *Operations.* 14 June 1993.

FM 100-8. *The Army in Multinational Operations.* 24 November 1997.

FM 100-10. *Combat Service Support.* 3 October 1995.

FM 100-15. *Corps Operations.* 29 October 1996.

FM 100-17-3. *Reception, Staging, Onward Movement, and Integration.* 17 March 1999.

FM 101-5. *Staff Organization and Operations.* 31 May 1997.

FM 101-5-1. *Operational Terms and Graphics (MCRP 5-2A).* 30 September 1997.

JP 3-10. *Joint Doctrine for Rear Area Operations.* 28 May 1996.

USC Title 18. *Crimes and Criminal Procedure.* 23 January 2000.

USC Title 32. *National Guard.* 5 January 1999.

Index

Change Number 1

Headquarters
Department of the Army
Washington, DC, 17 May 2004

Intelligence

Contents

Distribution Restriction: Approved for public release; distribution is unlimited.

*This publication supersedes FM 34-1, 27 September 1994.

Preface

FM 2-0 is the Army's keystone manual for military intelligence (MI) doctrine. It describes—

- The fundamentals of intelligence operations.
- The operational environment (OE).
- Intelligence in unified action.
- The Intelligence Battlefield Operating System (BOS).
- Intelligence considerations in strategic readiness.
- The intelligence process.
- MI roles and functions within the context of Army operations.

This manual conforms to the overarching doctrinal precepts presented in FM 3-0.

This manual provides doctrinal guidance for the Intelligence BOS actions in support of commanders and staffs. It also serves as a reference for personnel who are developing doctrine; tactics, techniques, and procedures (TTP); materiel and force structure; and institutional and unit training for intelligence operations.

This manual provides MI guidance for all commanders, staffs, trainers, and MI personnel at all echelons. It forms the foundation for MI and the Intelligence BOS doctrine development, and applies equally to the Active Component (AC), United States Army Reserve (USAR), and Army National Guard (ARNG). It is also intended for commanders and staffs of joint and combined commands, US Naval and Marine Forces, units of the US Air Force, and the military forces of multinational partners.

Headquarters, US Army Training and Doctrine Command is the proponent for this publication. The preparing agency is the US Army Intelligence Center and School. Send written comments and recommendations on DA Form 2028 (Recommended Changes to Publications and Blank Forms) directly to: Commander, ATZS-FDT-D (FM 2-0), 550 Cibeque Street, Fort Huachuca, AZ 85613-7017. Send comments and recommendations by e-mail to *ATZS-FDC-D@hua.army.mil.* Follow the DA Form 2028 format or submit an electronic DA Form 2028.

Unless otherwise stated, masculine nouns and pronouns do not refer exclusively to men.

This manual contains Army tactical task (ART) description taken verbatim from the *Army Universal Task List* (AUTL). ART task descriptions are followed by a reference to FM 7-15 in parentheses and the ART number; for example (See FM 7-15, ART 5.1.1.).

PART ONE

Intelligence in the Operational Environment

Part One discusses MI's role in peace, conflict, and war. Supporting the warfighter with effective intelligence is the primary focus of Military Intelligence. Intelligence provides commanders and decisionmakers with the requisite information facilitating their situational understanding so that they may successfully accomplish their missions in full spectrum operations.

Chapter 1 describes the operational environment and the roles of MI within the operational environment. It introduces the Intelligence warfighting function, the intelligence tasks, and the intelligence process, which are the mechanisms through which MI supports the warfighter. This chapter also introduces the intelligence disciplines, which are explained in detail in Part Three of this manual.

Chapter 2 describes the interaction of MI within the nation's intelligence community structure, providing an overview of the intelligence community at the national level and the unified action level—joint, multinational, and interagency aspects of full spectrum operations. This chapter also discusses the concepts and components of intelligence reach.

Chapter 1

Intelligence and the Operational Environment

ROLE OF INTELLIGENCE

1-1. The commander requires intelligence about the enemy and the area of operations (AO) prior to engaging in operations in order to effectively execute battles, engagements, and other missions across the full spectrum of operations. Intelligence assists the commander in visualizing the AO, organizing the forces, and controlling operations to achieve the desired tactical objectives or end-state. Intelligence supports force protection (FP) by alerting the commander to emerging threats and assisting in security operations.

1-2. The unit may need to deal with multiple threats. The commander must understand how current and potential enemies organize, equip, train, employ, and control their forces. Intelligence provides an understanding of the enemy, which assists in planning, preparing, and executing military operations. The commander must also understand the operational environment and its effects on both friendly and enemy operations. The commander receives mission-oriented intelligence on enemy forces and the

AO from the G2/S2. The G2/S2 depends upon the intelligence, surveillance, and reconnaissance (ISR) effort to collect and provide information on the enemy and AO.

1-3. One of the most significant contributions that intelligence personnel can accomplish is to accurately predict future enemy events. Although this is an extremely difficult task, predictive intelligence enables the commander and staff to anticipate key enemy events or reactions and develop corresponding plans or counteractions. Intelligence analysis conclusions, assessments, or products that attempt to define, describe, present, or portray the future situation or conditions of the enemy, terrain, weather, and civil considerations.

1-4. The most important purpose of intelligence is to influence decisionmaking. Commanders must receive the intelligence, understand it (because it is tailored to the commander's requirements), believe it, and act on it. Through this doctrinal concept, intelligence drives operations.

THE INTELLIGENCE WARFIGHTING FUNCTION

1-5. *The intelligence warfighting function is the related tasks and systems that facilitate understanding of the operational environment.* (FM 3-0) It includes tasks associated with intelligence, surveillance, and reconnaissance (ISR) operations, and is driven by the commander. Intelligence is more than just collection. It is a continuous process that involves analyzing information from all sources and conducting operations to develop the situation. The intelligence warfighting function includes the following tasks:

- Support to force generation.
- Support to situational understanding.
- Perform intelligence, surveillance, and reconnaissance.
- Support to targeting and information operations.

1-6. The intelligence warfighting function is one of six warfighting functions—movement and maneuver, intelligence, fires, sustainment, command and control, and protection. *A warfighting function is a group of tasks and systems (people, organizations, information, and processes) united by a common purpose that commanders use to accomplish missions and training objectives* (FM 3-0). (See FM 3-0, chapter 4, for a detailed discussion of the warfighting functions.) The intelligence warfighting function is a flexible force of personnel, organizations, and equipment that, individually or collectively, provide commanders with the timely, relevant, and accurate intelligence required to visualize the battlefield, assess the situation, and direct military actions. Additionally, the intelligence warfighting function is—

- A complex system that operates worldwide, from "mud-to-space," in support of an operation, to include the ability to leverage theater and national capabilities.
- Requires cooperation and division of labor internally, higher, lower, adjacent, and across components and the coalition.

1-7. The intelligence warfighting function not only includes assets within the MI branch but also includes the assets of all branches or warfighting functions that conduct intelligence warfighting function tasks. Every Soldier, as a part of a small unit, is a potential information collector and an essential component to help reach situational understanding. Each Soldier develops a special level of awareness simply due to exposure to events occurring in the AO and has the opportunity to collect and report information by observation and interaction with the population (see paragraph 3-3).

1-8. Planning and executing military operations will require intelligence regarding the threat (traditional, irregular, catastrophic, and disruptive) and the AO. The intelligence warfighting function generates intelligence and intelligence products that portray the enemy and aspects of the environment. These intelligence products enable the commander to identify potential courses of action (COAs), plan operations, employ forces effectively, employ effective tactics and techniques, and implement protection.

1-9. The intelligence warfighting function is always engaged in supporting the commander in offensive, defensive, stability, and, when directed, civil support operations. Intelligence provides products that are timely, relevant, accurate, and predictive. Hard training, thorough planning, meticulous preparation, and aggressive execution posture the Army for success. In the current environment we must maintain intelligence readiness to support operations on "no notice." This support is comprehensive and reaches across full spectrum operations and levels of war to produce the intelligence required to successfully accomplish the mission. A combination of space, aerial, seaborne, and ground-based systems provide the most comprehensive intelligence possible. During force projection operations, the intelligence warfighting function supports the commander with accurate and responsive intelligence from predeployment through redeployment.

1-10. The intelligence warfighting function architecture provides specific intelligence and communications structures at each echelon from the national level through the tactical level. These structures include intelligence organizations, systems, and procedures for collecting, processing, analyzing, and delivering intelligence and other critical information in a useable form to those who need it, when they need it. Effective communications connectivity and automation are essential components of this architecture.

INTELLIGENCE TASKS (METL)

1-11. The personnel and organizations within the Intelligence warfighting function conduct four primary intelligence tasks that facilitate the commander's visualization and understanding of the threat and the AO. These tasks are interactive and often take place simultaneously. (Refer to FM 7-15 for the complete subordinate task listing.) Figure 1-1 shows these tasks tailored to the commander's needs.

INTELLIGENCE TASKS	COMMANDER'S FOCUS	COMMANDER'S DECISIONS
Support to Situational Understanding - Perform IPB - Perform Situation Development - Provide Intelligence Support to Force Protection - Conduct Police Intelligence Operations	**Plan a mission.** **Execute the operation.** **Secure the force.**	**Which COA will I implement?** **Which enemy actions are expected?**
Support to Strategic Responsiveness - Perform I&W - Ensure Intelligence Readiness - Conduct Area Studies of Foreign Countries - Support Sensitive Site Exploitation	**Orient on contingencies.**	**Should I increase the unit's level of readiness?** **Should I implement the OPLAN?**
Conduct Intelligence, Surveillance, and Reconnaissance - Perform Intelligence Synchronization - Perform ISR Integration - Conduct Tactical Reconnaissance - Conduct Surveillance	**Plan the mission.** **Prepare.** **Execute.** **Assess.**	**Which DPs, HPTs, etc., are linked to the enemy actions?** **Are the assets available and in position to collect on the DPs, HPTs, etc.?** **Have the assets been repositioned for contingency mission?**
Provide Intelligence Support to Effects - Provide Intelligence Support to Targeting - Provide Intelligence Support to IO - Provide Intelligence Support to Combat Assessment	**Destroy/suppress/ neutralize targets.** **Reposition intelligence or attack assets.**	**Is my fire (lethal or non-lethal) and maneuver effective?** **Should I refire the same targets?**

Figure 1-1. Intelligence Tailored to the Commander's Needs.

SUPPORT TO SITUATIONAL UNDERSTANDING

1-12. This task centers on providing information and intelligence to the commander, which facilitates the commander's understanding of the enemy and the environment. It supports the command's ability to make sound decisions. Support to situational understanding comprises four subtasks: perform intelligence preparation of the battlefield (IPB), perform situation development, provide intelligence support to FP, and conduct police intelligence operations.

Perform Intelligence Preparation of the Battlefield

1-13. The G2/S2 is the staff proponent for IPB. IPB is the staff planning activity undertaken by the entire staff to define and understand the battlefield and the options it presents to friendly and threat forces. IPB includes input from the whole staff. There is only one IPB in each headquarters with inputs from all affected staff cells; they are not separate warfighting function or staff section IPBs throughout the headquarters. It is a systematic process of analyzing and visualizing the operational environment in a specific geographic area for a specific mission or in anticipation of a specific mission. By applying IPB, the commander and staff gain the information necessary to selectively apply and maximize combat power at critical points in time and space. IPB is most effective when it integrates each staff element's expertise into the final products. To conduct effective IPB, the G2/S2 must—

- Produce IPB products that support the staff's preparation of estimates and the military decision-making process (MDMP).

- Identify characteristics of the AO, including the information environment, that will influence friendly and threat operations.
- Establish the area of interest (AOI) in accordance with the commander's guidance.
- Identify gaps in current intelligence holdings.
- Determine multiple enemy COAs (ECOAs) by employing predictive analysis techniques to anticipate future enemy actions, capabilities, or situations.
- Establish a database that encompasses all relevant data sets within and related to the operational environment.
- Determine the enemy order of battle (OB), doctrine, and TTP. Identify any patterns in enemy behavior or activities.
- Accurately identify and report hazards within the AO, including the medical threat and toxic industrial material (TIM).
- Accurately identify threat capabilities, high-value targets (HVTs), and threat models.
- Integrate IPB information into COA analysis and the MDMP.
- Update IPB products as information becomes available.

Perform Situation Development

1-14. Situation development is a process for analyzing information and producing current intelligence about the enemy and environment during operations. The process helps the intelligence officer recognize and interpret the indicators of enemy intentions, objectives, combat effectiveness, and potential ECOAs. Situation development—

- Confirms or denies threat COAs.
- Provides threat locations.
- Explains what the threat is doing in relation to the friendly force operations.
- Provides an estimate of threat combat effectiveness.

1-15. Through situation development, the intelligence officer is able to quickly identify information gaps, and explain enemy activities in relation to the unit's own operations, thereby assisting the commander in gaining situational understanding. Situation development helps the commander make decisions and execute branches and sequels. This reduces risk and uncertainty in the execution of the plan. The intelligence officer maintains, presents, and disseminates the results of situation development through intelligence input to the common operational picture (COP) and other intelligence products.

Provide Intelligence Support to Force Protection

1-16. Provide intelligence in support of protecting the tactical force's fighting potential so that it can be applied at the appropriate time and place. This task includes the measures the force takes to remain viable and functional by protecting itself from the effects of or recovery from enemy activities. FP consists of those actions taken to prevent or mitigate hostile actions against Department of Defense (DOD) personnel (to include Department of Army

[DA] civilians, contractors, uniformed personnel, and family members), resources, facilities, and critical information. These actions—

- Conserve the force's fighting potential for application at the decisive time and place.
- Incorporate coordinated and synchronized offensive and defensive measures.
- Facilitate the effective employment of the joint force while degrading the capabilities of and opportunities for the threat.

1-17. Intelligence to FP consists of monitoring and reporting the activities, intentions, and capabilities of adversarial groups and determining their possible COAs. Detecting the adversary's methods in today's OE requires a higher level of situational understanding, informed by current and precise intelligence. This type of threat drives the need for predictive intelligence based on analysis of focused information from intelligence, law enforcement, and security activities.

1-18. Intelligence analysis in support of FP employs analytical methodologies and tools to provide situational understanding and to predict the adversary's actions. Modified or standard time-event charts, association matrixes, activity matrixes, link diagrams, and overlays are beneficial in monitoring the actions of the adversary. Overlays may include (but are not limited to) threat training camps, organizations, finances, personalities, industrial sites, information systems, decisionmaking infrastructures, specific activities, and locations of previous attacks.

Conduct Police Intelligence Operations

1-19. Police intelligence operations (PIO) are a military police (MP) function that supports, enhances, and contributes to the commander's force protection program, COP, and situational understanding. The PIO function ensures that information collected during the conduct of other MP functions—maneuver and mobility support, area security, law and order, and internment and resettlement—is provided as input to the intelligence collection effort and turned into action or reports. PIO has three components: (See FM 7-15, ART 1.1.4)

- Collect police information.
- Conduct Police Information Assessment Program (PIAP).
- Develop police intelligence products.

Notes:

US Code, Executive Orders, DOD Directives, and Army Regulations contain specific guidance regarding prohibition on the collection of intelligence information on US citizens, US corporations, and resident aliens. These laws and regulations include criminal penalties for their violation. Any PIO directed against US citizens should undergo competent legal review prior to their initiation.

The inclusion of the PIO task branch in the Intelligence warfighting function does not change the intelligence process described in this manual.

SUPPORT TO STRATEGIC RESPONSIVENESS

1-20. Intelligence support to strategic responsiveness supports staff planning and preparation by defining the full spectrum of threats, forecasting future threats, and forewarning the commander of enemy actions and intentions. Support to strategic responsiveness consists of four subtasks: Perform I&W, ensure intelligence readiness, conduct area studies of foreign countries, and support sensitive site exploitation. (See FM 7-15, ART 1.2)

Perform Indications and Warnings

1-21. This activity provides the commander with forewarning of enemy actions or intentions; the imminence of threat actions. The intelligence officer develops I&W in order to rapidly alert the commander of events or activities that would change the basic nature of the operations. It enables the commander to quickly reorient the force to unexpected contingencies and shape the battlefield. (See FM 7-15, ART 1.2.1)

1-22. The G2/S2 at the operational and strategic levels develops I&W in order to rapidly alert the commander of events or activities that would change the basic nature of the operations so the commander can initiate the appropriate action in a timely manner. I&W reduce the risk of enemy actions that are counter to planning assumptions. I&W enable the commander to quickly reorient the force to unexpected events and to shape the battlefield by manipulating enemy activities.

Ensure Intelligence Readiness

1-23. Intelligence readiness operations support contingency planning and preparation by developing baseline knowledge of multiple potential threats and operational environments. These operations and related intelligence training activities engage the Intelligence warfighting function to respond effectively to the commander's contingency planning intelligence requirements. (See FM 7-15, ART 1.2.2) While still in garrison, intelligence defines the full spectrum of threats and forecasts future threats and dangers. Intelligence readiness operations accomplish the following:

- Provide intelligence to support contingency-based training and staff planning.
- Identify, consider, and evaluate all potential threats to the entire unit.
- Provide a broad understanding of the operational environment of the contingency area, which is developed through continuous exchange of information and intelligence with higher echelon and joint intelligence organizations.

Conduct Area Studies of Foreign Countries

1-24. Study and understand the cultural, social, political, religious, and moral beliefs and attitudes of allied, host nation (HN), or indigenous forces to assist in accomplishing goals and objectives. (See FM 7-15, ART 1.2.3)

Note: The inclusion of this task does not change the support to strategic responsiveness provided by MI organizations described in this manual.

Support Sensitive Site Exploitation

1-25. Sensitive site exploitation consists of a related series of activities inside a sensitive site captured from an adversary. A sensitive site is a designated, geographically limited area with special military, diplomatic, economic, or information sensitivity for the United States. This includes factories with technical data on enemy weapon systems, war crimes sites, critical hostile government facilities, areas suspected of containing persons of high rank in a hostile government or organization, terrorist money laundering, and document storage areas for secret police forces. These activities exploit personnel, documents, electronic data, and material captured at the site, while neutralizing any threat posed by the site or its contents. While the physical process of exploiting the sensitive site begins at the site itself, full exploitation may involve teams of experts located around the world. (See FM 7-15, ART 1.2.4)

Note: The inclusion of this task does not change the support to strategic responsiveness provided by MI organizations described in this manual.

CONDUCT INTELLIGENCE, SURVEILLANCE, AND RECONNAISSANCE

1-26. With staff participation, the intelligence officer synchronizes intelligence support to the ISR effort by focusing the collection, processing, analysis, and intelligence products on the critical needs of the commander. The operations officer, in coordination with the intelligence officer, tasks and directs the available ISR assets to answer the commander's critical information requirements (CCIRs). Through various detection methods and systematic observation, reconnaissance and surveillance obtains the required information. A continuous process, this task has four subtasks: perform intelligence synchronization, perform ISR integration, conduct tactical reconnaissance, and conduct surveillance. (See FM 7-15, ART 1.3, and FM 3-90).

Perform Intelligence Synchronization

1-27. The intelligence officer, with staff participation, synchronizes the entire collection effort to include all assets the commander controls, assets of lateral units and higher echelon units and organizations, and intelligence reach to answer the commander's priority intelligence requirements (PIRs) and information requirements (IRs). (See FM 7-15, ART 1.3.1)

1-28. The intelligence officer, with staff participation, supports the G3/S3 in orchestrating the entire ISR effort to include all assets the commander controls, assets of lateral units and higher echelon units and organizations, and intelligence reach to answer the CCIRs (PIRs and friendly force information requirements [FFIRs]) and other intelligence requirements. Intelligence synchronization activities include the following:

- Conducting requirements management (RM): anticipate, develop, analyze, validate, and prioritize intelligence requirements. Recommend PIRs to the commander. Manage the commander's intelligence requirements, requests for information (RFIs) from subordinate and lateral organizations, and tasks from higher headquarters. Eliminate satisfied requirements and add new requirements as necessary.

- Developing indicators.
- Developing specific IRs (specific information requirements [SIRs]).
- Converting the SIRs into intelligence tasks or ISR tasks. (See Figure 1-2 for the ISR task development process.) The S2/G2 assigns intelligence production and reach tasks to subordinate intelligence elements or personnel, submits RFIs to higher and lateral echelons, and coordinates with (or assists) the G3/S3 to develop and assign ISR tasks.
- Comparing the ISR tasks to the capabilities and limitations of the available ISR assets (in coordination with the operations officer).
- Forwarding SIRs that cannot be answered by available assets to higher or lateral organizations as RFIs.
- Assessing collection asset reporting and intelligence production to evaluate the effectiveness of the ISR effort.
- Maintaining situational understanding to identify gaps in coverage and to identify the need to cue or redirect ISR assets.
- Updating the intelligence synchronization plan. The G2/S2 manages and updates the intelligence synchronization plan as PIRs are answered and new requirements arise.

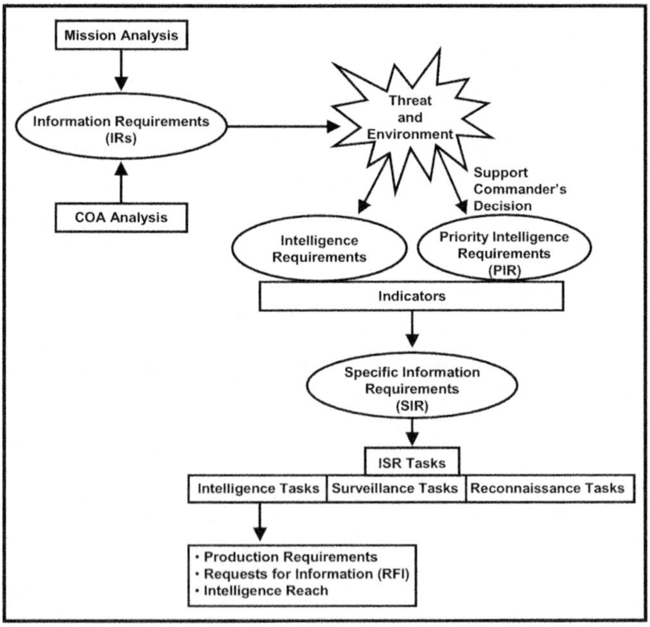

Figure 1-2. ISR Task Development Process.

1-29. **Intelligence Synchronization Considerations.** The G2/S2 generally follows six considerations in planning intelligence synchronization and ISR activities: anticipate, integrate, prioritize, balance, control, and reach. Refer to FM 2-01 for more information regarding intelligence synchronization.

- **Anticipate.** The intelligence staff must recognize when and where to shift collection or identify new intelligence requirements. The overall intent of this principle is to identify a new or adjust an existing requirement and present it to the commander for approval before waiting for the commander or staff to identify it.
- **Integrate.** The intelligence staff must be fully and continuously integrated into the unit's orders production and planning activities to ensure early identification of intelligence requirements. Early and continuous consideration of collection factors enhances the unit's ability to direct collection assets in a timely manner, ensures thorough planning, and increases flexibility in selecting assets.
- **Prioritize.** Prioritize each intelligence requirement based on its importance in supporting the commander's intent and decisions. Prioritization, based on the commander's guidance and the current situation, ensures that limited ISR assets and resources are directed against the most critical requirements.
- **Balance.** ISR capabilities complement each other. The intelligence staff should resist favoring or becoming too reliant on a particular unit, discipline, or system. Balance is simply planning redundancy, when required, eliminating redundancy when not desired, and ensuring an appropriate mix of ISR assets or types. The intelligence synchronization matrix (ISM) is useful in determining or evaluating balance.
- **Control.** To ensure timely and effective responses to intelligence requirements, a unit should first use ISR assets it controls. These assets usually are more responsive to their respective commander and also serve to lessen the burden on the ISR assets of other units, agencies, and organizations.
- **Reach.** Although usually not as responsive as a unit's own assets, intelligence reach may be the only way to satisfy an intelligence requirement. If at all possible, one should not depend solely on intelligence reach to answer a PIR.

1-30. **Develop IRs**. The intelligence staff develops a prioritized list of what information needs to be collected and produced into intelligence. Additionally, the intelligence staff dynamically updates and adjusts those requirements in response to mission adjustments and changes. This list is placed against a latest time intelligence is of value to ensure intelligence and information are reported to meet operational requirements. (See FM 7-15, ART 1.3.1) Figure 1-3 shows a comparison of IR, PIR, and intelligence requirements.

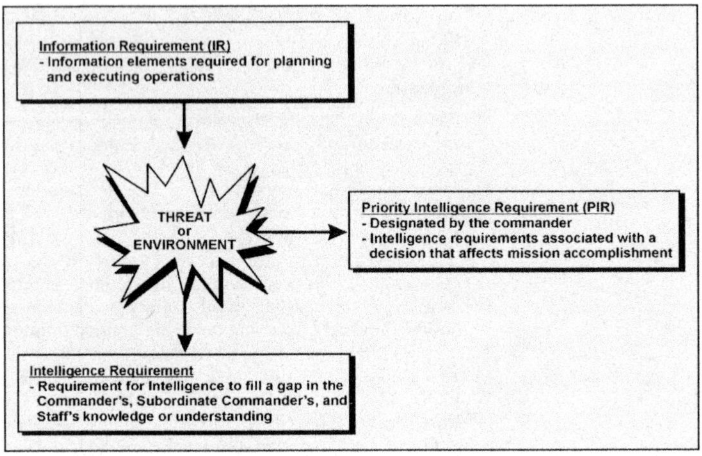

Figure 1-3. IR, PIR, and Intelligence Requirements Comparison.

1-31. An effective discussion of ISR has to include an understanding of the CCIRs. The CCIRs are elements of information required by commanders that directly affect decisionmaking and dictate the successful execution of military operations. The commander decides what information is critical based on the commander's experience, the mission, the higher commander's intent, and the staff's input (initial IPB, information, intelligence, and recommendations). Refer to FM 3-0 for more information regarding CCIRs.

1-32. Based on the CCIRs, two types of supporting IRs are generated: PIRs and FFIRs. However, commanders may determine that they need to know whether one or more essential elements of friendly information (EEFI) have been compromised or that the enemy is collecting against a designated EEFI. In those cases, commanders may designate that question as one of their CCIRs. Figure 1-4 shows the CCIR composition.

1-33. IRs are all of the information elements required by the commander and staff for the successful planning and execution of operations; that is, all elements necessary to address the factors of mission, enemy, terrain and weather, troops and support available, time available, civil considerations (METT-TC). Vetting by the commander or a designated representative turns an IR into either a PIR or an intelligence requirement. IRs are developed during COA analysis based on the factors of METT-TC.

1-34. PIRs are those intelligence requirements for which a commander has an anticipated and stated priority during the task of planning and decisionmaking. PIRs are associated with a decision based on an enemy action or inaction or the AO that will affect the overall success of the commander's mission. **The commander designates intelligence requirements tied directly to decisions as CCIR (PIR and FFIR).**

Answers to the PIRs help produce intelligence essential to the commander's situational understanding and decisionmaking. For information on PIR development, see FM 2-01.

1-35. The G2/S2 recommends to the commander those IRs produced during the MDMP that meet the criteria for PIR. They do not become CCIR (PIR and FFIR) until approved by the commander. Additionally, The commander may unilaterally designate PIRs. The IRs that are not designated by the commander as PIRs become intelligence requirements. The intelligence requirement is a gap in the command's knowledge or understanding of the AO or threat that the Intelligence warfighting function must fill.

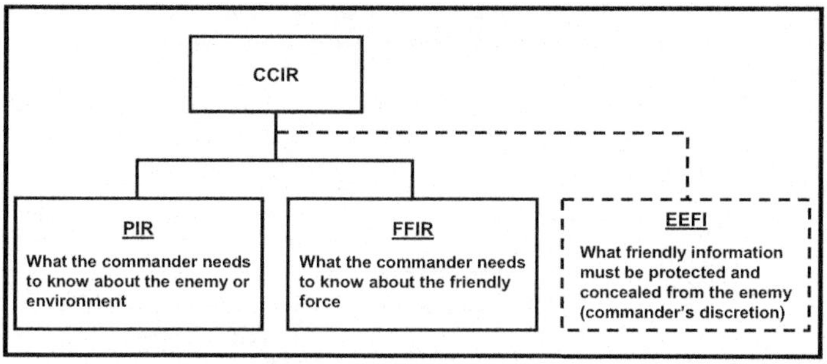

Figure 1-4. Commander's Critical Information Requirements Composition.

1-36. The G3/S3 then tasks the unit's assets to answer both the PIR and intelligence requirements through the ISR plan. PIR should—

- Ask only one question.
- Support a decision.
- Identify a specific fact, event, activity (or absence thereof) which can be collected.
- If linked to an ECOA, indicate an ECOA prior to, or as early as possible in, its implementation.
- Indicate the latest time the information is of value (LTIOV). The LTIOV is the absolute latest time the information can be used by the commander in making the decision the PIR supports. The LTIOV can be linked to time, an event, or a point in the battle or operation.

1-37. **Friendly Force IRs.** The staff also develops FFIRs which, when answered, provide friendly force information that the commander and staff need to achieve situational understanding and to make decisions.

1-38. **Essential Elements of Friendly Information**. EEFI establish information to protect, not information to obtain. However, commanders may determine that they need to know whether one or more EEFI have been compromised or that the enemy is collecting against a designated EEFI. In those cases, commanders may designate that question as one of their CCIRs, which generates PIRs and/or FFIRs. For example, a commander may determine that if the enemy discovers the location and movement of the friendly reserve, the operation is at risk. In this case, the location and movement of the friendly reserve are EEFI. The commander designates determining whether the enemy has discovered the location and movement of the friendly reserve as one of the CCIR. That CCIR, in turn, generates PIR and FFIR to support staff actions in determining whether the EEFI has been compromised.

1-39. **Develop the Intelligence Synchronization Plan.** The entire unit staff develops their IRs and determines how best to satisfy them. The staff uses reconnaissance and surveillance assets to collect information. The intelligence synchronization plan includes all assets that the operations officer can task or request and coordination mechanisms to ensure adequate coverage of the AOIs. (See FM 7-15, ART 1.3.1.2)

1-40. The intelligence synchronization plan, often presented in a matrix format as an ISM, aids in synchronizing the entire ISR effort with the overall operation and the commander's decisions and/or decision points (DPs). The intelligence synchronization plan is often produced in conjunction with the ISR plan. However, before performing intelligence synchronization and finalizing the intelligence synchronization plan, the G2/S2 must have—

- The CCIR (PIR and FFIR).
- A prioritized list of the remaining intelligence requirements.
- Evaluated ISR assets and resources.
- All of the assigned ISR tasks.

Perform ISR Integration

1-41. The operations officer, in coordination with the intelligence officer and other staff members, orchestrates the tasking and directing of available ISR assets to answer the CCIR. The operations officer, with input from the intelligence officer, develops tasks from the SIRs which coincide with the capabilities and limitations of the available ISR assets and the latest time information is of value (LTIOV). Intelligence requirements are identified, prioritized, and validated and an ISR plan is developed and synchronized with the scheme of maneuver. (See FM 7-15, ART 1.3.2, and FM 3-90)

1-42. The G3/S3, in coordination with the G2/S2 and other staff members, orchestrates the tasking and directing of available ISR assets to answer the CCIRs (PIRs and FFIRs) and IRs. The result of this process is the forming of the ISR plan. The ISR plan provides a list of all the ISR tasks to be accomplished. The G2/S2 and the G3/S3 develop tasks from the SIRs. These tasks are then assigned based on the capabilities and limitations of the available ISR assets and the LTIOV.

1-43. **Develop the ISR Plan.** The operations officer is responsible for developing the ISR plan. The entire unit staff analyzes each requirement to determine how best to satisfy it. The staff will receive orders and RFIs from both subordinate and adjacent units and higher headquarters. The ISR plan includes all assets that the operations officer can task or request and coordination mechanisms to ensure adequate coverage of the area of interest. (See FM 7-15, ART 1.3.2.1, and FM 3-90)

1-44. **The ISR Plan**. The ISR plan is usually produced as the ISR Annex to an OPORD (Annex L, Intelligence Surveillance, and Reconnaissance). Refer to FM 5-0 for specific and authoritative information on the ISR Annex. ISR is a continuous combined arms effort led by the operations and intelligence staffs in coordination with the entire staff that sets reconnaissance and surveillance in motion. The PIRs and other intelligence requirements drive the ISR effort. The commander takes every opportunity to improve situational understanding and the fidelity of the COP about the enemy and terrain through the deployment of ISR assets. Commanders integrate reconnaissance and surveillance to form an integrated ISR plan that capitalizes on their different capabilities. The ISR plan is often the most important part of providing information and intelligence that contributes to answering the CCIRs. For the G2/S2, an effective ISR plan is critical in answering the PIR. Also see FM 3-55 for more information on the ISR plan.

1-45. The ISR plan is not an MI-specific product—the G3/S3 is the staff proponent of the ISR plan— it is an integrated staff product executed by the unit at the direction of the commander. The G2/S2, however, must maintain situational understanding in order to recommend to the commander and G3/S3 changes or further development of the ISR plan. Based on the initial IPB and CCIRs, the staff—primarily the G2/S2—identifies gaps in the intelligence effort and develops an initial ISR plan based on available ISR assets. The G3/S3 turns this into an initial ISR Annex that tasks ISR assets as soon as possible to begin the collection effort.

1-46. The G3/S3, assisted by the G2/S2, uses the ISR plan to task and direct the available ISR assets to answer the CCIRs (PIRs and FFIRs) and intelligence requirements. Conversely, the staff revises the plan as other intelligence gaps are identified if the information is required to fulfill the CCIRs or in anticipation of future intelligence requirements. With staff participation, the G2/S2 intelligence officer synchronizes the ISR effort through a complementary product to the ISR plan—the intelligence synchronization plan.

1-47. **Execute and Update the ISR Plan**. The operations officer updates the ISR plan based on information received from the intelligence officer. The operations officer is the integrator and manager of the ISR effort through an integrated staff process and procedures. As PIRs are answered and new information requirements arise, the intelligence officer updates intelligence synchronization requirements and provides the new input to the operations officer who updates the ISR plan. He works closely with all staff elements to ensure the unit's organic collectors receive appropriate taskings. This ISR reflects an integrated collection strategy and employment, production and dissemination scheme that will effectively answer the commander's PIR. (See FM 7-15, ART 1.3.2.2, and FM 3-90)

Conduct Tactical Reconnaissance

1-48. To obtain, by visual observation or other detection methods, such as signals, imagery, measurement of signature or other technical characteristics, human interaction and other detection methods about the activities and resources of an enemy or potential enemy, or to secure data concerning the meteorological, hydrographic, or geographic characteristics and the indigenous population of a particular area. This task includes the conduct of NBC reconnaissance and the tactical aspects of SOF special reconnaissance. The five subtasks are—

- Conduct a Zone Reconnaissance.
- Conduct an Area Reconnaissance.
- Conduct a Reconnaissance in Force.
- Conduct a Route Reconnaissance.
- Conduct a Reconnaissance Patrol.

Note: This task branch includes techniques by which ART 1.1.4.1 (Collect Police Information) may be performed. (FM 7-15)

1-49. Reconnaissance is a mission undertaken to obtain by visual observation or other detection methods, information about the activities and resources of an enemy or potential enemy, and about the meteorological, hydrographic, or geographic characteristics of an AO. MI personnel and organizations can conduct reconnaissance through obtaining information derived from signals, imagery, measurement of signatures, technical characteristics, human interaction, and other detection methods. When performing reconnaissance, it is important to—

- Orient the reconnaissance asset on the named area of interest (NAI) and/or reconnaissance objective in a timely manner.
- Report all information rapidly and accurately.
- Complete the reconnaissance mission not later than (NLT) the time specified in the order.
- Answer the requirement that prompted the reconnaissance task.

Conduct Surveillance

1-50. To systematically observe the airspace, surface, or subsurface areas, places, persons, or things in the AO by visual, aural (audio), electronic, photographic, or other means. Other means may include but are not limited to space-based systems, and using special NBC, artillery, engineer, SOF, and air defense equipment. (See FM 7-15, ART 1.3.4, and FM 3-90)

Note: This task is a technique by which ART 1.1.4.1 (Collect Police Information) may be performed.

1-51. Conducting surveillance is systematically observing the airspace, surface, or subsurface areas, places, persons, or things in the AO by visual, aural, electronic, photographic, or other means. Surveillance activities include—

- Orienting the surveillance asset on the NAI and/or the surveillance objective in a timely manner.
- Reporting all information rapidly and accurately.
- Completing the surveillance mission NLT the time specified in the order.
- Answering the requirement that prompted the surveillance task.

PROVIDE INTELLIGENCE SUPPORT TO EFFECTS

1-52. The task of providing the commander information and intelligence support for targeting of the threat's forces, threat organizations, units and systems through lethal and non-lethal fires to include electronic attack and information operations. This task includes three subtasks: provide intelligence support to targeting, provide intelligence support to information operations, and provide intelligence support to combat assessment. (See FM 7-15, ART 1.4)

Provide Intelligence Support to Targeting

1-53. The intelligence officer, supported by the entire staff, provides the commander information and intelligence support for targeting the threat's forces and systems through employment of direct, indirect lethal and non-lethal fires. It includes identification of threat capabilities and limitations. The Intelligence warfighting function plays a crucial role by providing intelligence throughout the steps of the targeting process: decide, detect, deliver, assess. Support to targeting includes identifying threat capabilities and limitations. The G2/S2 supports the unit's lethal and non-lethal targeting effort through target development, target acquisition (TA), electronic warfare (EW), and combat assessment. This task has two subtasks:

- *Provide Intelligence Support to Target Development* is the systematic analysis of the enemy forces and operations to determine HVTs, systems, and system components for potential attack through maneuver, fires, or information.
- *Provide Intelligence Support to Target Detection* establishes procedures for dissemination of targeting information. The targeting team develops the sensor/attack system matrix to determine the sensor required to detect and locate targets. The intelligence officer places the following requirements into the integrated ISR plan.
  - Requires reconnaissance and surveillance operations to identify, locate, and track high-payoff targets (HPTs) for delivery of lethal or non-lethal effects.
  - Includes employing fires, offensive IO, and other attack capabilities against enemy C2 systems as part of the unit's FS plan and IO objectives.

Provide Intelligence Support to Information Operations

1-54. IO are actions taken to affect adversary information, influence other's decisionmaking processes and information systems while protecting one's own information and information systems. Overall operational continuity and mission success requires close, mutual coordination and synchronization of intelligence plans and operations with IO elements and related activities. Refer to Appendix A, this manual, for more information. This task has three subordinate tasks: (See FM 7-15, ART 1.4.2)

- Provide Intelligence Support to Offensive IO.
- Provide Intelligence Support to Defensive IO.
- Provide Intelligence Support to Activities Related to IO.

Provide Intelligence Support to Combat Assessment

1-55. Combat assessment is the determination of the overall effectiveness of force employment during military operations. The objective of combat assessment is to identify recommendations for the course of military operations. It answers the question, "Were the objectives met by force employment?" Although the assessment is primarily an intelligence responsibility, it requires input from and coordination with operations and fire support staffs. Combat assessment consists of conducting physical damage, functional damage, and target system assessments. (See FM 7-15, ART 1.4.3)

1-56. The staff determines how combat assessment relates to a specific target by conducting physical damage, functional damage, and target system assessments.

- **Conduct Physical Damage Assessment (PDA).** A PDA is an estimate of the extent of physical damage to a target based upon observed or interpreted damage. This post-attack target analysis is a coordinated effort among all units.
- **Conduct Functional Damage Assessment (FDA).** The FDA estimates the remaining functional or operational capability of a targeted facility or object. The staff bases FDA on the assessed physical damage and estimates of the threat's ability to recuperate (to include the time required to resume normal operations). Multiple echelons typically conduct this all-source analysis. The targeting or combat assessment cell integrates the initial target analyses with other sources, including intelligence, and then compares the original objective with the current status of the target to determine if the objective has been met.
- **Conduct Target System Assessment (TSA).** The TSA is an estimate of the overall impact of force employment against an adversary's target system. The unit, supported by higher echelon assets, normally conducts this assessment. The analyst combines all combat assessment reporting on functional damage to targets within a target system and assesses the overall impact on that system's capabilities. This process lays the groundwork for future recommendations for military operations in support of operational objectives.

1-57. **Munitions Effects Assessment (MEA)**. MEA takes place concurrently and interactively with combat assessment since the same signatures used to determine the level of physical damage also give clues to munitions effectiveness. MEA is primarily the responsibility of operations and FS personnel, with input from the G2/S2. After the same weapon is used to attack several targets of a specific type, MEA should be accomplished to evaluate weapon performance. MEA analysts seek to identify through systematic trend analysis any deficiencies in weapon system and munitions performance or combat tactics by answering the question, "Did the weapons employed perform as expected?" Using a variety of input (targeting analysts, imagery analysts, structural engineers, and mission planners) analysts prepare a report assessing munitions performance. If combat troops capture attacked targets, it is then possible to collect detailed information on the target. Reports should detail weapon performance against specified target types. This information could have a crucial impact on future operations.

1-58. **Re-attack Recommendation**. Re-attack recommendations follow directly from both battle damage assessment (BDA) and MEA efforts. Basically re-attack recommendations answer the question, "Have we achieved the desired effects against our targeted objectives?" Evolving objectives, target selection, timing, tactics, weapons, vulnerabilities, and munitions are all factors in the new recommendations, combining both operations and intelligence functions.

1-59. **Targeting Meeting**. The role of the G2/S2 in targeting meetings is critical to ensuring intelligence supports the targeting process. The G2/S2 must come to the targeting meeting prepared. Additionally, the G2/S2 must understand the steps of the targeting process and tailor the intelligence products and G2/S2 participation according to the targeting process steps of: decide, detect, deliver, and assess.

1-60. There are two major areas in which the G2/S2 must prepare in order to support the targeting meeting.

- **Analysis:** Determine what the enemy is doing (current situation) and anticipated threat models; ECOA situation templates, ECOA sketches and statements, HVTs, and other associated products.
- **IPB Products:** Updated from the initial IPB effort, including analytical results (see above bullet), and tailored to the targeting requirements.

1-61. **Decide**. At the targeting meeting, the G2/S2 should be prepared to do the following:

- Brief the current and projected future enemy situation with an event template.
- Brief any combat assessments.
- Brief HVTs for potential selection of HPTs.
- Brief the current PIR portion of CCIRs, the PIRs of the CCIRs for the next phase or in accordance with the same timeframe of the targeting meeting.
- Brief the current ISM or intelligence synchronization plan.
- Participate in wargaming.
- Refine the initial ISR plan (in conjunction with the FS and G3/S3 representatives).

1-62. **Detect**. Based on information from the decide portion of the targeting meeting, the G2/S2 should be prepared to do the following:

- Identify specific ISR assets.
- Develop IRs and propose pertinent IRs to the commander for designation as PIRs.
- Complete the initial ISR plan; ensure it addresses the HPTs.

1-63. **Electronic Warfare**. EW provides targeting information to the commander based on transmitter location data. There are three major subdivisions within EW: electronic support (ES), electronic protection (EP), and electronic attack (EA).

- **ES** – Also referred to as electronic warfare support (EWS) in JP 1-02, involves actions to search for, intercept, identify, and locate or localize sources of intentional and unintentional radiated electromagnetic energy. In addition to supporting the overall ISR effort, ES is needed to produce intelligence required to support EA missions.
- **EP** – Involves passive and active means taken to protect personnel, facilities, and equipment from any effects of friendly or enemy employment of EW that degrades, neutralizes, or destroys friendly combat capabilities.
- **EA** – Involves the use of electromagnetic energy, directed energy, or anti-radiation weapons to attack personnel, facilities, or equipment with the intent of degrading, neutralizing, or destroying enemy combat capability and is considered a form of fires. EA includes actions taken to prevent or reduce an enemy's effective use of the electromagnetic spectrum (EMS), such as jamming and electromagnetic deception.

1-64. **Deliver**. The G2/S2 should be prepared to do the following:

- Brief how the Intelligence warfighting function will work to detect and track the target through the entire targeting process.
- Brief the implementation of the intelligence synchronization and ISR plans, including the LTIOV and communications structure.

1-65. **Assess**. Depending on the factors of METT-TC and the PIRs, the same ISR assets that detect and track the targets may be required to support combat assessments in order to determine if re-attack is required. Conversely, if re-attack is a decision that the commander will make based upon the effects on the enemy, combat assessment should be reflected in the PIR, and supported in the corresponding ISR plan.

THE OPERATIONAL ENVIRONMENT

1-66. The operational environment can be defined as a composite of the conditions, circumstances, and influences that affect the employment of capabilities and bear on the decision of the commander. The operational environment encompasses physical areas and factors of the air, land, maritime, and space domains. It also includes the information environment and enemy, adversary, friendly, and neutral systems.

OPERATIONAL VARIABLES (PMESII-PT)

1-67. Analysis of the operational environment—in terms of political, military, economic, social, infrastructure, and information with the addition of physical environment and time (PMESII-PT) variables—provides relevant information that senior commanders use to understand, visualize, and describe the operational environment. As a set, these operational variables are often abbreviated as PMESII-PT. A description of each of the PMESII-PT variables follows:

- **Political**. Describes the distribution of responsibility and power at all levels of governance or cooperation.
- **Military**. Explores the military capabilities of all relevant actors in a given operational environment.
- **Economic**. Encompasses individual behaviors and aggregate phenomena related to the production, distribution, and consumption of resources.
- **Social**. Describes the cultural, religious, and ethnic makeup within an operational environment.
- **Information**. Describes the nature, scope, characteristics, and effects of individuals, organizations, and systems that collect, process, disseminate, or act on information.
- **Infrastructure**. Is composed of the basic facilities, services, and installations needed for the functioning of a community or society.
- **Physical Environment**. Defines the physical circumstances and conditions that influence the execution of operations throughout the domains of air, land, sea, and space.
- **Time**. Influences military operations within an operational environment in terms of the decision cycles, operational tempo, and planning horizons. (FM 3-0)

1-68. For the intelligence portion, when identifying threats based on the systems approach to PMESII-PT, there are four primary components of the operational environment. These are the threat, terrain, weather, and civil considerations.

1-69. While the operational variables are directly relevant to campaign planning, they are too broad to be applied directly to tactical planning. That does not mean they are not of value at all levels; they are fundamental to developing the understanding of the operational environment necessary to conduct planning at any level, in any situation. The degree to which each operational variable provides useful information depends on the situation and echelon. For example, social and economic variables often receive close analysis as part of enemy and civil considerations at brigade and higher levels. They may affect the training and preparation of small units. However, they are not relevant to a small-unit leader's mission analysis. That leader may only be concerned with such questions as "Who is the tribal leader for this village?" "Is the electrical generator working?" "Does the enemy have antitank missiles?"

MISSION VARIABLES

1-70. Upon receipt of a warning order (WARNO) or mission, Army leaders narrow their focus to six mission variables. Mission variables are those aspects of the operational environment that directly affect a mission. They outline the situation as it applies to a specific Army unit. The mission variables are mission, enemy, terrain and weather, troops and support available, time available, and civil considerations (METT-TC). These are the categories of relevant information used for mission analysis. Army leaders use the mission variables to synthesize operational variables and tactical-level information with local knowledge about conditions relevant to their mission. The METT-TC categories are listed below:

- **Mission**. The mission is the task, together with the purpose, that clearly indicates the action to be taken and the reason.
- **Enemy**. Relevant information regarding the enemy may include the following:
  - Threat characteristics (previously order of battle [OB] factors).
    - Doctrine (or known execution patterns).
    - Personal habits and idiosyncrasies.
    - Equipment, capabilities, and vulnerabilities.
  - Threat COAs.
- **Terrain and Weather**. Terrain and weather are natural conditions that profoundly influence operations. Terrain and weather are neutral; they favor neither side unless one is more familiar with—or better prepared to operate in—the environment.
- **Troops and Support Available**. The number, type, capabilities, and condition of available friendly troops and support available. These include the resources available from joint, interagency, intergovernmental, multinational, host nation (HN), commercial (via contracting), and private organizations.
- **Time Available**. Time is critical to all operations. Controlling and exploiting time is central to initiative, tempo, and momentum. By exploiting time, commanders can exert constant pressure, control the relative speed of decisions and actions, and force exhaustion on enemy forces.
- **Civil Considerations**. Civil considerations comprise six characteristics expressed in the memory aid ASCOPE:
  - Areas.
  - Structures.
  - Capabilities.
  - Organizations.
  - People.
  - Events.

Note. For additional information on ASCOPE and the IPB process, see FM 2-01.3. Understanding the operational environment requires understanding the civil aspects of the AO. Civil considerations reflect how the manmade infrastructure, civilian institutions, and the attitudes and activities of the civilian leaders, populations, and organizations within an AO influence the conduct of military operations.

1-71. METT-TC enables leaders to synthesize operational level information with local knowledge relevant to their missions and tasks in a specified AO. Tactical and operational leaders can then anticipate the consequences of their operations before and during execution. See FM 3-0 for a detailed discussion of PMESII-PT and METT-TC.

THE INTELLIGENCE PROCESS

1-72. Intelligence operations generally include the five functions that constitute the intelligence process: **plan, prepare, collect, process,** and

produce. Additionally, there are three common tasks that occur across the five functions of the intelligence process: **analyze, disseminate,** and **assess**. The three common tasks are discussed after the last function, produce. The intelligence process functions are not necessarily sequential; this is what differentiates the Army's intelligence process from the Joint intelligence cycle. The intelligence process provides a common model with which to guide one's thinking, discussing, planning, and assessing about the threat or AOI environment. The intelligence process generates information about the threat, the AOI, and the situation, which allows the commander and staff to develop a plan, seize and retain the initiative, build and maintain momentum, and exploit success.

PLAN

1-73. The plan step consists of the activities that identify pertinent IRs and develop the means for satisfying those requirements. The CCIRs (PIRs and FFIRs) drive the planning of the ISR effort. The intelligence officer supports the G3/S3 in arranging the ISR effort, based on staff planning, to achieve the desired collection effects. Planning activities include, but are not limited to—

- Conducting IPB.
- Managing requirements.
- Submitting RFIs and using intelligence reach to fill information gaps.
- Evaluating reported information.
- Establishing the intelligence communications and dissemination architecture.
- Developing, managing, and revising the intelligence synchronization plan and the ISR plan as mission requirements change.

PREPARE

1-74. The failure of MI units or personnel to accomplish their tasks or missions can often be attributed to their failure to prepare. The prepare step includes those staff and leader activities which take place upon receiving the operation order (OPORD), operation plan (OPLAN), warning order (WARNO), or commander's intent in order to improve the unit's ability to execute tasks or missions.

1-75. The most habitual and egregious preparation failures committed by leaders (as evidenced by performance at the combat training centers) is not the conduct of specific tasks, but the failure to adequately coordinate for more generic combat requirements such as—

- Friendly forward unit liaison.
- Departing or reentering friendly lines or AOs.
- Fire support: enroute, at mission location, and return.
- Casualty evacuation procedures.
- Passwords (running, forward of friendly lines), recognition signals, call signs, and frequencies.
- Resupply.
- Movement through friendly AOs.

COLLECT

1-76. The collect step involves collecting and reporting information in response to ISR tasking. ISR assets collect information and data about threat forces, activities, facilities, and resources as well as information concerning the environmental and geographical characteristics of a particular AO. A successful ISR effort results in the timely collection and reporting of relevant and accurate information. This collected information forms the foundation of intelligence databases, intelligence production, and the G2's/S2's situational understanding. The requirements manager evaluates the reported information for its responsiveness to the CCIRs (PIRs and FFIRs).

PROCESS

1-77. Processing involves converting collected data and information into a form that is suitable for analyzing and producing intelligence. Examples of processing include enhancing imagery, translating a document from a foreign language, converting electronic data into a standardized report that can be analyzed by a system operator, and correlating information.

1-78. Processing data and information is performed unilaterally and cooperatively by both humans and automated systems.

PRODUCE

1-79. The produce step involves evaluating, analyzing, interpreting, synthesizing, and combining information and intelligence from single or multiple sources into intelligence or intelligence products in support of known or anticipated requirements. Production also involves combining new information and intelligence with existing intelligence in order to produce intelligence in a form that the commander and staff can apply to the MDMP and that supports and helps facilitate situational understanding. During the produce step, the intelligence staff manipulates information by—

- Analyzing the information to isolate significant elements.
- Evaluating the information to determine accuracy, timeliness, usability, completeness, precision, and reliability.
- Combining the information with other relevant information and previously developed intelligence.
- Applying the information to estimate possible outcomes.
- Presenting the information in a format that will be most useful to its eventual user.

1-80. The intelligence staff deals with numerous and varied production requirements based on PIRs and intelligence requirements; diverse missions, environments, and situations; and presentation requirements. Through analysis, collaboration, and intelligence reach, the G2/S2 and the staff use the collective intelligence production capability of higher, lateral, and subordinate echelons to meet the production requirements. Proficiency in these techniques and procedures facilitates the intelligence staff's ability to answer command and staff requirements regardless of the factors of METT-TC.

1-81. The three common tasks, discussed below, can occur throughout the intelligence process.

Analyze

1-82. The intelligence staff analyzes intelligence and information about the enemy's capabilities, friendly vulnerabilities, and the battlefield environment as well as issues and problems that arise within the intelligence process itself to determine their nature, origin, and interrelationships. This analysis enables commanders, staffs, and leaders to determine the appropriate action or reaction and to focus or redirect assets and resources to fill information gaps or alleviate pitfalls. It is also within the analyze function that intelligence analysts sort through large amounts of collected information and intelligence to obtain only that information which pertains to the CCIRs (PIRs and FFIRs), maintenance of the COP, and facilitates the commander's situational understanding.

Disseminate

1-83. Disseminating is communicating relevant information of any kind from one person or place to another in a usable form by any means to improve understanding or to initiate or govern action. Disseminating intelligence entails using information management techniques and procedures to deliver timely, relevant, accurate, predictive, and usable intelligence to the commander. Determining the product format and selecting the means to deliver it are key aspects of dissemination. Information presentation may be in a verbal, written, interactive, or graphic format. The type of information, the time allocated, and the individual preference of the commander all influence the information format.

Assess

1-84. Assessment plays an integral role in all aspects of the intelligence process. Assessing includes evaluating the effectiveness of intelligence in supporting the operation. Assessing the situation and available information begins upon receipt of the mission and continues throughout the intelligence process. The continual assessment of intelligence operations and ISR efforts, available information and intelligence, and various aspects of the factors of METT-TC are critical to ensure—

- The G2/S2 answers all CCIRs (PIRs and FFIRs).
- The G2/S2 and G3/S3 redirect ISR assets to support changing requirements.
- Using information and intelligence properly.

INTELLIGENCE DISCIPLINES

1-85. Intelligence disciplines are categories of intelligence functions. The Army's intelligence disciplines are All-Source Intelligence, Human Intelligence (HUMINT), Geospatial Intelligence (GEOINT), Imagery Intelligence (IMINT), Signals Intelligence (SIGINT), Measurement and Signatures Intelligence (MASINT), Open-Source Intelligence (OSINT), Technical Intelligence (TECHINT), and Counterintelligence (CI). For more information regarding the intelligence disciplines, see Part Three of this manual, as well as the respective manuals, which covers each individual intelligence discipline.

ALL-SOURCE INTELLIGENCE

1-86. All-source intelligence is defined as the intelligence products, organizations, and activities that incorporate all sources of information and intelligence, including open-source information, in the production of intelligence. All-source intelligence is a separate intelligence discipline, as well as the name of the task used to produce intelligence from multiple intelligence or information sources.

HUMAN INTELLIGENCE

1-87. HUMINT is the collection of foreign information—by a trained HUMINT Collector—from people and multimedia to identify elements, intentions, composition, strength, dispositions, tactics, equipment, personnel, and capabilities. It uses human sources as a tool, and a variety of collection methods, both passively and actively, to collect information.

GEOSPATIAL INTELLIGENCE

1-88. Title 10 U.S. Code § 467 establishes GEOINT. The definition of geospatial intelligence is—

> *The exploitation and analysis of imagery and geospatial information to describe, assess, and visually depict physical features and geographically referenced activities on the Earth. GEOINT consists of imagery, imagery intelligence, and geospatial information.*

1-89. The Army implementation of GEOINT is a result of the Army's organization, manning, and training. There are multiple types of data and information that various Army units and organizations collect, provide, and analyze in order to support the GEOINT enterprise. The two primary GEOINT service providers in the Army are MI units and organizations and Engineer (topographic) units and organizations. MI units and organizations provide imagery and IMINT to the enterprise. Engineer (topographic) units and organizations provide geospatial data and information to the enterprise. Therefore, while some of the collection, analysis, and exploitation of imagery and geospatial information occur within the intelligence warfighting function; some of the collection, analysis, and exploitation of imagery and geospatial information occur outside intelligence.

IMAGERY INTELLIGENCE

1-90. IMINT is intelligence derived from the exploitation of imagery collected by visual photography, infrared, lasers, multi-spectral sensors, and radar. These sensors produce images of objects optically, electronically, or digitally on film, electronic display devices, or other media.

SIGNALS INTELLIGENCE

1-91. SIGINT is a category of intelligence comprising either individually or in combination all communications intelligence (COMINT), electronic intelligence (ELINT), and foreign instrumentation signals intelligence (FISINT), however transmitted. SIGINT is derived from communications, electronics, and foreign instrumentation signals.

MEASUREMENT AND SIGNATURES INTELLIGENCE

1-92. MASINT is technically derived intelligence that detects, locates, tracks, identifies, and/or describes the specific characteristics of fixed and dynamic target objects and sources. It also includes the additional advanced processing and exploitation of data derived from IMINT and SIGINT collection. MASINT collection systems include but are not limited to radar, spectroradiometric, electro-optical (E-O), acoustic, radio frequency (RF), nuclear detection, and seismic sensors, as well as techniques for gathering chemical, biological, radiological, and nuclear (CBRN), and other material samples.

OPEN-SOURCE INTELLIGENCE

1-93. The *National Defense Authorization Act for Fiscal Year 2006* states:

> *Open source intelligence is produced from publicly available information that is collected, exploited, and disseminated in a timely manner to an appropriate audience for the purpose of addressing a specific intelligence requirement.*

1-94. Expressed in terms of the Army intelligence process, OSINT is relevant information derived from the systematic collection, processing, and analysis of publicly available information in response to intelligence requirements.

1-95. The Army does not have a specific military occupational specialty (MOS), additional skill identifier (ASI), or special qualification identifier (SQI) for OSINT. With the exception of the Asian Studies Detachment, the Army does not have base tables of organization and equipment (TOE) for OSINT units or staff elements. OSINT missions and tasks are imbedded within existing missions and force structure or accomplished through task organization.

TECHNICAL INTELLIGENCE

1-96. TECHINT is intelligence derived from the collection and analysis of threat and foreign military equipment and associated materiel for the purposes of preventing technological surprise, assessing foreign scientific and technical (S&T) capabilities, and developing countermeasures designed to neutralize an adversary's technological advantages.

COUNTERINTELLIGENCE

1-97. CI counters or neutralizes intelligence collection efforts through collection, counterintelligence investigations, operations, analysis, and production, and functional and technical services. CI includes all actions taken to detect, identify, track, exploit, and neutralize the multidiscipline intelligence activities of friends, competitors, opponents, adversaries, and enemies; and is the key intelligence community contributor to protect US interests and equities. CI assists in identifying EEFI, identifying vulnerabilities to threat collection, and actions taken to counter collection and operations against US forces.

UNIFIED ACTION

1-98. Unified action is the synchronization, coordination, and/or integration of the activities of governmental and nongovernmental entities with military operations to achieve unity of effort. (JP 1) Joint operations focus and maximize the complementary and reinforcing effects and capabilities of each service. Joint force commanders (JFCs) synchronize the complementary capabilities of the service components that comprise the joint force.

1-99. The employment of MI in campaigns and major operations must be viewed from a joint perspective, and the "mud-to-space" intelligence concept must establish a fully interoperable and integrated joint intelligence capability. ARFOR intelligence assets work with multinational and interagency partners to accomplish their missions. Ideally, multinational and interagency intelligence partners provide cultures, perspectives, and capabilities that reinforce and complement Army MI strengths and capabilities. Close intelligence coordination is the foundation of successful unified action. (See chapter 2.)

1-100. A critical part of current operations is the execution of the joint doctrinal concept of persistent surveillance. Joint doctrine defines persistent surveillance as a collection strategy that emphasizes the ability of some collection systems to linger on demand in an area to detect, locate, characterize, identify, track, target, and possibly provide battle damage assessment and re-targeting in real or near-real time. Persistent surveillance facilitates the formulation and execution of preemptive activities to deter or forestall anticipated adversary courses of action (JP 2-0).

THE NATURE OF LAND OPERATIONS

1-101. Landpower is the ability—by threat, force, or occupation—to gain, sustain, and exploit control over land, resources, and people (FM 3-0). Army operations reflect expeditionary and campaign capabilities that constantly adapt to each campaign's unique circumstances. Expeditionary capabilities require forces organized to be modular, versatile, and rapidly deployable. Rapidly deployed expeditionary force packages provide immediate options for seizing or retaining the operational initiative and also allow the conduct of sustained operations for as long as necessary. Army forces are organized, trained, and equipped for endurance. The Army's preeminent challenge is to balance expeditionary agility and responsiveness with the endurance and adaptability needed to complete a campaign successfully.

1-102. The capability to prevail in close combat is indispensable and unique to land operations. The outcome of battles and engagements depends on Army forces' ability to prevail in close combat. Many factors inherent in land combat combine to complicate the situation. These factors include chaos, complexity, insufficient intelligence, errors in understanding or planning, difficult terrain, the civilian population, and an adaptive and lethal enemy. The axiom "intelligence drives operations" continues to be true especially for land operations; operations and intelligence are inextricably linked.

1-103. Four considerations are preeminent in generating expeditionary capabilities and Army force packages.

- **Scope.** Considers and strives to understand an enemy throughout the depth of an operational area. Commanders rely on intelligence in order to use maneuver, fires, and other elements of combat power to defeat or destroy enemy forces.
- **Duration**. Forces routinely conduct missions prior to, during, and after the commitment of land combat forces. Intelligence is always engaged.
- **Terrain**. Missions occur among a complex variety of natural and manmade features. Employing forces in the complexity of the ground environment requires thorough planning.
- **Permanence.** Forces are integrated with, or assigned to, land combat forces as they seize or secure ground.

1-104. Several attributes of the land environment affect the application of landpower. These attributes include—

- The requirement to deploy and employ Army forces rapidly.
- The requirement for Army forces to operate for protracted periods.
- The nature of close combat.
- Uncertainty, chance, friction, and complexity.

1-105. Reconnaissance, Surveillance, and Target Acquisition (RSTA)/ISR is the means the Army uses to implement the joint doctrinal concept of persistent surveillance in support of tactical operations. Dependable technology and responsive intelligence lessen the effects of uncertainty, chance, friction, and complexity. Complex and dynamic Army tactical operations require extensive ISR capabilities to satisfy the commander's information requirements to detect, locate, characterize, identify, track, and target HPTs, and to provide combat assessment in real time within a very fluid operational environment.

1-106. *Reconnaissance, Surveillance, and Target Acquisition/ Intelligence, Surveillance, and Reconnaissance* is a full spectrum combined arms mission that integrates ground and air capabilities to provide effective, dynamic, timely, accurate, and assured combat information and multidiscipline actionable intelligence for lethal and non-lethal effects and decisions in direct support of the ground tactical commander.

FORCE PROJECTION OPERATIONS

1-107. Force projection is the military component of power projection. It is a central element of the national military strategy. Army organizations and installations, linked with joint forces and industry, form a strategic platform to maintain, project, and sustain ARFOR, wherever they deploy. Force projection operations are inherently joint and require detailed planning and synchronization. Force projection encompasses a range of processes—mobilization, deployment, employment, sustainment, and redeployment—discussed below.

1-108. The Army must change its mindset from depending on an "intelligence buildup" to performing intelligence readiness on a daily basis in order to meet the requirements for strategic responsiveness. MI personnel, even in garrison at the lowest tactical echelons, must use their analytic and other systems and prepare for possible operations on a daily basis.

1-109. Built on a foundation of intelligence readiness, the Intelligence warfighting function provides the commander with the intelligence needed to plan, prepare, and execute force projection operations. Successful intelligence during force projection operations relies on continuous collection and intelligence production before and during the operation. In a force projection operation, higher echelons will provide intelligence to lower echelons until the tactical ground force completes entry and secures the lodgment area. The joint force J2 must exercise judgment when providing information to subordinate G2s/S2s so not to overwhelm them. Figure 1-5 is an example of force projection intelligence.

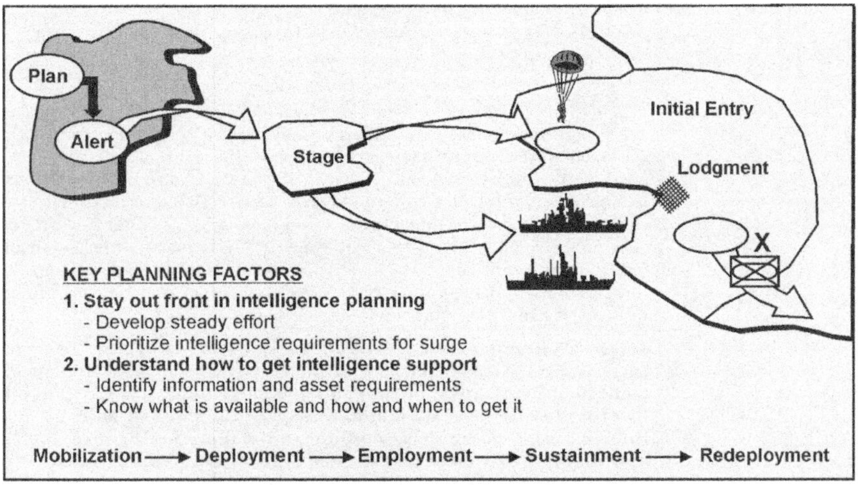

Figure 1-5. Force Projection Intelligence.

1-110. The G2/S2 must anticipate, identify, consider, and evaluate all threats to the entire unit throughout force projection operations. This is critical during the deployment and entry operations stages of force projection. During these stages, the unit is particularly vulnerable to enemy actions because of its limited combat power and knowledge of the AO. Intelligence personnel must, therefore, emphasize the delivery of combat information and intelligence products that indicate changes to the threat or AO developed during predeployment IPB.

1-111. The G2/S2 should—

- Review available databases on assigned contingency AOIs, conduct IPB on these AOIs, and develop appropriate IPB products.
- Comply with higher headquarters standing operating procedures (SOPs) and manuals for specific intelligence operations guidance.
- Coordinate for and rehearse electronic message transfers (for example, Internet Protocol addresses, routing indicators) using the same communications protocols with theater, higher headquarters, subordinate, and lateral units that the unit would use when deployed.
- Plan, train, and practice surging intelligence functions on likely or developing contingency crises.
- Prepare and practice coordination from predeployment through redeployment with other elements and organizations (for example, HUMINT, IMINT, SIGINT, MASINT, IO, staff weather officer [SWO], civil affairs [CA], psychological operations [PSYOP], and special operations forces [SOF] units, to include databases and connectivity).
- Include the following as a part of daily (sustainment) operations:
  - US Army Reserve and other augmentation.
  - A linguist plan with proficiency requirements. (Alert linguists through early entry phases of deployment.)
  - Training (individual and collective).
- Establish formal or informal intelligence links, relationships, and networks to meet developing contingencies.
- Forward all RFIs to higher headquarters in accordance with SOPs.
- Establish statements of intelligence interests (SIIs), other production, and I&W requirements.

1-112. To draw intelligence from higher echelons and focus intelligence downward, based on the commander's needs, the G2/S2 must—

- Understand the J2's multiple echelon and broadcast dissemination capability to ensure near-real time (NRT) reporting to all deployed, in transit, or preparing to deploy forces.
- Maintain or build intelligence databases on the environment and threats for each probable contingency.
- State and record the CCIR (as a minimum, list the PIRs and ISR tasks or requests).

1-113. Until the unit's collection assets become operational in the AO, the G2/S2 will depend upon intelligence from the ARFOR or JTF to answer the unit's intelligence needs. The following paragraphs describe the intelligence and ISR considerations during force projection.

MOBILIZATION

1-114. Mobilization is the process by which the armed forces or part of them are brought to a state of readiness for war or other national emergency. It assembles and organizes resources to support national objectives. Mobilization includes activating all or part of the US Army Reserve, and assembling and organizing personnel, supplies, and materiel. A unit may be brought to a state of readiness for a specific mission or other national emergency. This process, called mobilization, is where specific US Army

Active or Reserve units, capabilities, and personnel are identified and integrated into the unit. During mobilization, the G2/S2 must—

- Monitor intelligence reporting on threat activity and I&W data.
- Manage IRs and RFIs from their unit and subordinate units to include updating ISR planning.
- Establish habitual training relationships with their US Army Active and Reserve augmentation units and personnel as well as higher echelon intelligence organizations as identified in the existing OPLAN. Support the US Army Reserve units and augmentation personnel by preparing and conducting intelligence training and threat update briefings and by disseminating intelligence.
- Identify ISR force requirements for the different types of operations and contingency plans (CONPLANs).
- Identify individual military, civilian, and contractor augmentation requirements for intelligence operations. The Army, and the Intelligence warfighting function in particular, cannot perform its missions without the support of its Department of the Army Civilians (DACs) and contractors. The force increasingly relies on the experience, expertise, and performance of non-uniformed personnel and has fully integrated these non-uniformed personnel into the warfighting team.

1-115. During mobilization the G2/S2, in conjunction with the rest of the staff, must ensure the adequacy of training and equipping of US Army Active and Reserve MI organizations and individual augmentees to conduct intelligence operations.

1-116. The G2/S2 supports peacetime contingency planning with IPB products and databases on likely contingency areas. The G2/S2 establishes an intelligence synchronization plan that will activate upon alert notification. For smooth transition from predeployment to entry, the G2/S2 must coordinate intelligence synchronization and communications plans before the crisis occurs. The intelligence synchronization plan identifies the intelligence requirements supporting those plans, to include—

- ISR assets providing support throughout the AOI.
- Command and support relationships of ISR assets at each echelon.
- Report and request procedures not covered in unit SOPs.
- Sequence of deployment of ISR personnel and equipment. Early deployment of key ISR personnel and equipment is essential for force protection and combat readiness. Composition of initial and follow-on deploying assets is influenced by the factors of METT-TC, availability of communications, and availability of lift.
- Communications architecture supporting both intelligence staffs and ISR assets.
- Friendly vulnerabilities to hostile intelligence threats and plans for conducting FP measures. The staff must begin this type of planning as early as possible to ensure adequate support to FP of deploying and initial entry forces.
- Monitor time-phased force and deployment data (TPFDD) and recommend changes in priority of movement, unit, or capability to enable ISR operations.

1-117. The G2/S2 must continually monitor and update the OPLANs to reflect the evolving situation, especially during crisis situations. National intelligence activities monitor regional threats throughout the world and can answer some intelligence requirements supporting the development of OPLANs.

1-118. Upon alert notification, the G2/S2 updates estimates, databases, IPB products, and other intelligence products needed to support command decisions on force composition, deployment priorities and sequence, and the AOI. Units reassess their collection requirements immediately after alert notification. The G2/S2 begins verifying planning assumptions within the OPLANs. CI and ISR personnel provide FP support and antiterrorism measures.

1-119. Throughout mobilization, unit intelligence activities will provide the deploying forces with the most recent intelligence on the AO. The intelligence staff will also update databases and situation graphics. The G2/S2 must—

- Fully understand the unit, ARFOR, and joint force intelligence organizations.
- Revise intelligence and intelligence-related communications architecture and delete or integrate any new systems and software with the current architecture.
- Support 24-hour operations and provide continuous intelligence.
- Plan all required intelligence reach procedures.
- Determine transportation availability for deployment and availability when deployed.
- Determine all sustainability requirements.
- Determine intelligence release requirements and restrictions; releasability to multinational and HN sources.
- Review status of forces agreements (SOFAs), rules of engagement (ROE), international laws, and other agreements, emphasizing the effect that they have on intelligence collection operations. (Coordinate with the staff judge advocate [SJA] on these issues.)
- Ensure ISR force deployment priorities are reflected in the TPFDD to support ISR operations based upon the factors of METT-TC.
- Ensure intelligence links provide the early entry commander vital access to multi-source army and joint intelligence collection assets, processing systems, and databases.
- Review the supporting unit commanders' specified tasks, implied tasks, task organization, scheme of support, and coordination requirements with forward maneuver units. Address issues or shortfalls and direct or coordinate changes.
- Establish access to national HUMINT, IMINT, SIGINT, MASINT, and CI databases, as well as automated links to joint service, multinational, and HN sources.

DEPLOYMENT

1-120. Deployment is the movement of forces and materiel from their point of origin to the AO. This process has four supporting components:

predeployment activities, fort to port, port to port, and port to destination; these components are known collectively as reception, staging, onward movement, and integration (RSO&I) activities. Success in force projection operations hinges on timely deployment. The size and composition of forces requiring lift are based on the factors of METT-TC, availability of pre-positioned assets, the capabilities of HN support, and the forward presence of US forces. Force or tactical tailoring is the process used to determine the correct mix and sequence of deploying units.

1-121. During deployment, intelligence organizations at home station or in the rear area take advantage of modern SATCOM, broadcast technology, and automated data processing (ADP) systems to provide graphic and textual intelligence updates to the forces enroute. Enroute updates help eliminate information voids and, if appropriate, allow the commander to adjust the plan prior to arrival in theater in response to changes in the OE or enemy actions.

1-122. Intelligence units extend established networks to connect intelligence staffs and collection assets at various stages of the deployment flow. Where necessary, units establish new communications paths to meet unique demands of the mission. The theater and corps analysis and control elements (ACEs) play a critical role in making communications paths, networks, and intelligence databases available to deploying forces.

1-123. Space-based systems are key to supporting intelligence during the deployment and the subsequent stages of force projection operations by—

- Monitoring terrestrial AOIs through ISR assets to help reveal enemy location and disposition, attempting to identify the enemy's intent.
- Providing communications links between forces enroute and in the continental United States (CONUS).
- Permitting MI collection assets to accurately determine their position through the Global Positioning System (GPS).
- Providing timely and accurate data on meteorological, oceanographic, and space environmental factors that might affect operations.
- Providing warning of theater ballistic missile launches.
- Providing timely and accurate weather information to all commanders through the Integrated Meteorological System (IMETS).

1-124. Situation development dominates intelligence operations activities during initial entry operations. The G2/S2 attempts to identify all threats to arriving forces and assists the commander in developing FP measures. During entry operations, echelons above corps (EAC) organizations provide intelligence. This support includes providing access to departmental and joint intelligence and deploying scalable EAC intelligence assets. The entire effort focuses downwardly to provide tailored support to deploying and deployed echelons in response to their CCIRs (PIRs and FFIR).

1-125. Collection and processing capabilities are enhanced, as collection assets build up in the deployment area, with emphasis on the build-up of the in-theater capability required to conduct sustained ISR operations. As the build-up continues, the G2/S2 strives to reduce total dependence on extended split-based intelligence from outside the AO. As assigned collection assets arrive into the theater, the G2/S2 begins to rely on them for tactical intelligence although higher organizations remain a source of intelligence.

1-126. As the ARFOR enter the theater of operations, the joint force J2 implements and, where necessary, modifies the theater intelligence architecture. Deploying intelligence assets establishes liaison with staffs and units already present in the AO. Liaison personnel and basic communications should be in place prior to the scheduled arrival of parent commands. ISR units establish intelligence communications networks.

1-127. CONUS and other relatively secure intelligence bases outside the AO continue to support deployed units. Systems capable of rapid receipt and processing of intelligence from national systems and high capacity, long-haul communications systems are critical to the success of split-based support of a force projection operation. These systems provide a continuous flow of intelligence to satisfy many operational needs.

1-128. The G2/S2, in coordination with the G3/S3, participates in planning to create conditions for decisive operations. The G2/S2 also adjusts collection activities as combat strength builds. During entry operations the G2/S2—

- Monitors FP indicators.
- Monitors the ISR capability required to conduct sustained intelligence operations.
- Monitors intelligence reporting on threat activity and I&W data.
- Develops measurable criteria to evaluate the results of the intelligence synchronization plan.
- Assesses—
  - Push versus pull requirements of intelligence reach.
  - Effectiveness of the intelligence communications architecture.
  - Reporting procedures and timelines.
  - Intelligence to OPLANs and OPORDs, branches, and sequels (to include planning follow-on forces).

EMPLOYMENT

1-129. Employment is the conduct of operations to support a JFC commander. Employment encompasses an array of operations, including but not limited to—

- Entry operations (opposed or unopposed).
- Shaping operations (lethal and non-lethal).
- Decisive operations (combat or support).
- Postconflict operations (prepare for follow-on missions or redeployment).

Entry Operations

1-130. Enemies often possess the motives and means to interrupt the deployment flow of ARFOR. Threats to deploying forces may include advanced conventional weaponry (air defense, mines, etc.) and WMD. Sea and air ports of debarkation (PODs) should be regarded as enemy HPTs because they are the entry points for forces and equipment. PODs are vulnerable because they are fixed targets with significant machinery and equipment that is vulnerable to attack; in addition to military forces and materiel, HN support personnel, contractors, and civilians may all be working there. An enemy attack, or even the threat of an enemy attack, on a POD can have a major impact on force projection momentum. Commanders at all levels require predictive intelligence so that they may focus attention on security actions that reduce vulnerabilities. To avoid, neutralize, or counter threats to entry operations, the commanders rely on the ability of the G2/S2 to support future operations by accurately identifying enemy reactions to US actions, anticipating their response to our counteractions and predicting additional ECOAs.

1-131. Predictive intelligence also supports the decisions the commander and staff must make about the size, composition, structure, and deployment sequence of the force in order to create the conditions for success. Commanders rely on predictive intelligence to identify potential friendly decisions before the actual event. While thorough planning develops friendly COAs to meet possible situations, the nature of an operation can change significantly before execution. The G2/S2 must provide timely, accurate, and predictive intelligence to ensure the commander can retain the initiative to implement the plan or make decisions before losing the opportunity to do so.

Shaping Operations

1-132. Shaping operations create and preserve conditions for the success of the decisive operation. Shaping operations include lethal and non-lethal activities conducted throughout the AO. They support the decisive operation by affecting enemy capabilities and forces, or by influencing enemy decisions. G2/S2 intelligence analysis and ISR activities support the development and execution of shaping operations by identifying threat centers of gravity and decisive points on the battlefield. The G2/S2 also ensures the intelligence process focuses on the CCIRs (PIRs and FFIRs). It is critical that the G2/S2 provide timely, accurate, and predictive intelligence to the commander to facilitate the execution of shaping operations. Predictive intelligence should provide sufficient time for the commander to understand how the enemy will react to US COAs so that appropriate shaping operations can be implemented.

Decisive Operations

1-133. Decisive operations are those that directly accomplish the task assigned by the higher headquarters. Decisive operations conclusively determine the outcome of major operations, battles, and engagements. Continuously synchronized ISR activities, coupled with predictive intelligence results and products during all stages of force projection, combine to ensure the commander is prepared to employ the right forces with the right support to conduct decisive operations at the most appropriate place and time. In addition to coordinating with the G3/S3 to synchronize all ISR activities, the G2/S2 must answer the CCIRs (PIRs and FFIRs) and provide continuous assessments of the enemy's current situation and predict the enemy's subsequent COA, branches, and/or sequels.

Postconflict Operations

1-134. Upon cessation of hostilities or truce, deployed forces enter a new stage of force projection operations. Postconflict operations focus on restoring order, reestablishing HN infrastructure, preparing for redeployment of forces, and planning residual presence of US forces. While postconflict operations strive to transition from conflict to peace, there remains a possibility of resurgent hostilities by individuals and forces. The ISR effort, particularly predictive intelligence analysis, remains just as critical in postconflict operations as in the other employment operations. ISR operations support the postconflict emphasis on restoration operations exemplified by commanders redirecting their CCIRs (PIRs and FFIRs) and IRs to support units conducting these missions. These operations might include—

- Engineer units conducting mine clearing or infrastructure reconstruction operations.
- Medical and logistics units providing humanitarian relief.
- MP units providing law and order assistance.
- CA units reestablishing local control and preparing for the orderly transition of local governments and to support civil-military operations (CMO).

SUSTAINMENT

1-135. Sustainment involves providing and maintaining levels of personnel and materiel required to sustain the operation throughout its duration. It is essential to generating combat power. CSS may be split-based between locations within and outside CONUS. These operations include ensuring units have the MI assets required to accomplish the mission, such as personnel (including linguists), communications systems, ISR systems, and appropriate maintenance support.

REDEPLOYMENT

1-136. Redeployment is the process by which units and materiel reposture themselves in the same theater; transfer forces and materiel to support another JFC's operational requirements; or return personnel and materiel to the home or demobilization station upon completion of the mission. Redeployment operations encompass four phases:

- Recovery, reconstitution, and pre-deployment activities.
- Movement to and activities at the port of embarkation (POE).
- Movement to the POD.
- Movement to home station.

1-137. As combat power and resources decrease in the AO, FP and I&W become the focus of the commander's intelligence requirements. This in turn drives the selection of those assets that must remain deployed until the end of the operation and those that may redeploy earlier. The S2—

- Monitors intelligence reporting on threat activity and I&W data.
- Continues to conduct intelligence to FP.
- Requests ISR support (theater and national systems) and intelligence in support of redeployment.

1-138. After redeployment, MI personnel and units recover and return to predeployment activities. ISR units resume contingency-oriented peacetime intelligence operations. US Army Reserve ISR units demobilize and return to peacetime activities. G2/S2s must—

- Monitor intelligence reporting on threat activity and I&W data.
- Update or consolidate databases.
- Maintain intelligence readiness.
- Provide their input into the Force Design Update (FDU) process to refine modified table of organizations and equipment (MTOE) and evaluate the need for Individual Mobilization Augmentee (IMA) personnel.

Chapter 2

Intelligence and Unified Action

UNIFIED ACTION

2-1. As discussed in JP 1-02 and Chapter 1, unified action describes the wide scope of actions (including the synchronization of activities with governmental and NGO agencies) taking place within unified commands, subordinate unified commands, or JTFs under the overall direction of the commanders of those commands. Under unified action, commanders integrate joint, single-service, special, and supporting intelligence operations with interagency, nongovernmental, and multinational operations. The ARFOR often brings unique ISR capabilities to unified action operations.

2-2. In a unified action, the intelligence staff takes on additional responsibilities, relationships, and infrastructure to execute multi-service and multinational intelligence operations. This chapter discusses the synchronization of Army intelligence efforts with joint and other national and international partners to achieve unity of effort and to accomplish the commander's intent.

THE LEVELS OF WAR

2-3. The levels of war are doctrinal perspectives that clarify the links between strategic objectives and tactical actions. Although there are no finite limits or boundaries between them, the three levels of war are strategic, operational, and tactical (see Figure 2-1).

- The strategic level is that level at which a nation, often as one of a group of nations, determines national and multinational security objectives and guidance and develops and uses national resources to accomplish them.
- The operational level is the level at which campaigns and major operations are conducted and sustained to accomplish strategic objectives within theaters or AOs. It links the tactical employment of forces to strategic objectives.
- The tactical level is the employment of units in combat. It includes the ordered arrangement and maneuver of units in relation to each other, the terrain, and the enemy to translate potential combat power into victorious battles and engagements.

2-4. Understanding the interdependent relationship of all three levels of war helps commanders visualize a logical flow of operations, allocate resources, and assign tasks. Actions within the three levels are not associated with a particular command level, unit size, equipment type, or force or component type. The concept of strategic, operational, and tactical intelligence operations aids JFCs and their J2s in visualizing the flow of intelligence from one level to the next. The concept facilitates allocating required collection, analysis, production, and dissemination resources; and facilitates the assignment of appropriate intelligence tasks to national, theater, component, and supporting intelligence elements.

STRATEGIC

2-5. The President and the Secretary of Defense use strategic intelligence to develop national strategy and policy, monitor the international situation, prepare military plans, determine major weapon systems and force structure requirements, and conduct strategic operations.

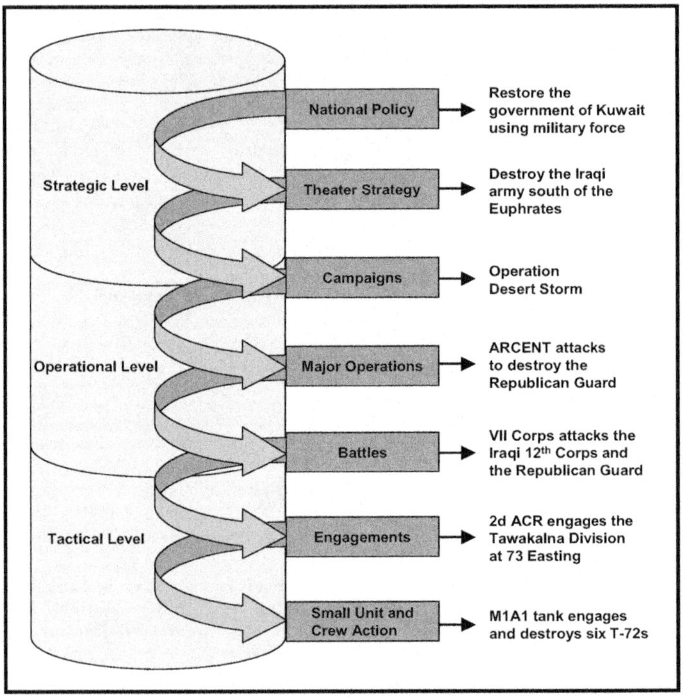

Figure 2-1. The Levels of War.

2-6. Intelligence supports joint operations across the full spectrum of military operations. It determines the current capabilities, and forecasts future developments, of adversaries or potential adversaries that could affect the national security and interests. A strategic operation also produces the intelligence required by combatant commanders to prepare strategic estimates, strategies, and plans to accomplish missions assigned by higher authorities. Theater intelligence includes determining when, where, and in what strength the adversary will stage and conduct theater level campaigns and strategic unified operations. The intelligence staff should also focus predictive analysis efforts on identifying strategic threat events and how these events will impact US actions at the strategic, operational, and tactical levels. Intelligence operations support IO as well. Intelligence operations support strategic planning by—

- Developing strategic intelligence policy.
- Preparing the strategic collection plan.
- Allocating national intelligence resources.

2-7. Intelligence requirements are those requirements generated from the staff's IRs regarding the enemy and environment that are not a part of the CCIRs (PIRs and FFIRs). Intelligence requirements require collection and can provide answers in order to identify indicators of enemy actions or intent, which reduce the uncertainties associated with an operation. Significant changes (that is, branches and sequels) with an operation usually lead to changes in intelligence requirements. Of particular importance is information relating to enemy or threat strategic vulnerabilities, strategic forces, strategic centers of gravity, and any capabilities relating to the development and employment of CBRNE.

2-8. Global and regional issues and threats are identified and reported to the President and the Secretary of Defense, as well as to the senior military leadership and the combatant commanders. Intelligence requirements include any foreign developments that could threaten the US, its citizens abroad, or multinational military, political, or economic interests. Intelligence also includes identifying hostile reactions to US reconnaissance activities and indications of impending terrorist attacks (I&W). For a complete listing of strategic tasks, refer to CJCSM 3500.04C.

OPERATIONAL

2-9. Combatant commanders and subordinate JFCs and their component commanders are the primary users of intelligence. At the operational echelons, intelligence—

- Focuses on the military capabilities and intentions of enemies and threats.
- Provides analysis of events within the AOI and helps commanders determine when, where, and in what strength the adversary might stage and conduct campaigns and major operations.
- Supports all phases of military operations, from mobilization all the way through redeployment of US forces, and continues during sustainment.
- Supports all aspects of the joint campaign.
- Identifies adversary centers of gravity and HVTs.
- Provides critical support to friendly IO.

2-10. The JFC and staff allocate intelligence resources and request support from national agencies, other theaters, and multinational partners. During stability operations and support operations, operational intelligence includes training and assisting multinational partners in conducting intelligence operations.

2-11. Combatant commanders use intelligence concerning the nature and characteristics of the battlespace to determine the type and scale of operations. Intelligence also aids in determining the impact of significant regional features and hazards on the conduct of both friendly and adversary operations. Significant regional factors include the natural environment, political, informational, economic, industrial, infrastructure, geographic, demographic, topographic, hydrographic, climatic, populace, cultural, medical, lingual, historical, and psychological aspects of the AOI. Intelligence analysis also assists in determining the ROE and other restrictions which will affect operations in the JFC's AO.

2-12. Intelligence relating to the adversary's military and nonmilitary capabilities assists in determining the adversary's ability to conduct military operations. Factors that operational intelligence addresses include mobilization potential, force structure (including alliance forces), force dispositions, equipment, military doctrine, C2 structure, and their MDMP. Intelligence includes the continuous refinement of the OBs for the entire array of the enemy's forces in the AOI. For a complete listing of operational tasks, refer to CJCSM 3500.04C.

US Army Intelligence and Security Command (INSCOM)

2-13. The U.S. Army has vested its intelligence at the operational level with INSCOM, the major Army command (MACOM) responsible for the Army's intelligence forces above Corps. INSCOM's mission is to conduct and support dominant intelligence, security, and information operations for military commanders and national decisionmakers. The INSCOM strategy is to provide superior information and information capabilities to Army commanders, while denying the same to adversaries. Headquarters (HQ), INSCOM, in coordination with its major subordinate commands (MSCs), provides a myriad of general intelligence support operations. INSCOM is providing a globally focused, rapidly deployable, knowledge based, adaptively force packaged capability, supporting commanders and leaders with actionable intelligence at the point of decision. INSCOM serves as the national to tactical Intelligence bridge.

Army Space Program Office (ASPO)

2-14. ASPO executes the Army's Tactical Exploitation of National Capabilities Program (TENCAP). The program focuses on exploiting current and future tactical potential of national systems and integrating the capabilities into the Army's tactical decisionmaking process. Army TENCAP systems enable the tactical commander maximum flexibility to satisfy intelligence needs under a wide range of operational scenarios. ASPO is the point of contact (POC) for all tactical activities between MACOMs or users and the National Reconnaissance Office (NRO).

TACTICAL

2-15. Tactical commanders use intelligence for planning and conducting battles and engagements. Relevant, accurate, predictive, and timely intelligence allows tactical units to achieve an advantage over their adversaries. Precise and predictive intelligence, on the threat and targets, is essential for mission success. Predictive intelligence also enables the staff to better identify or develop ECOAs. Tactical intelligence—

- Identifies and assesses the enemy's capabilities, COAs, and vulnerabilities, as well as describes the battlespace.
- Seeks to identify when, where, and in what strength the enemy will conduct tactical level operations.
- Provides the commander with information on imminent threats to the force including those from terrorists, saboteurs, insurgents, and foreign intelligence collection.
- Provides critical support to friendly IO.
- Develops and disseminates targeting information and intelligence.

2-16. Intelligence provides the tactical commander with the information and intelligence he requires to successfully employ combat forces against enemy forces. Thus, intelligence tasks support the execution of battles and engagements. These intelligence tasks are different from those at other levels due to their ability to immediately influence the outcome of the tactical commander's mission. They include information gathered from tactical sources, such as combat information, interrogations, debriefings, and eliciting information from captured or misplaced personnel. For a complete listing of tactical tasks, refer to FM 7-15.

CATEGORIES OF INTELLIGENCE

2-17. Directly or indirectly, the unit receives intelligence from throughout the US intelligence community. The intelligence operations and products of national, joint, and service organizations that make up the intelligence community fall into one of six categories:

- I&W.
- Current intelligence.
- GMI.
- Target intelligence.
- S&T intelligence.
- CI.

2-18. The categories of intelligence are distinguishable from each other primarily by the purpose of the intelligence product. The categories can overlap and some of the same intelligence is useful in more than one category. Depending upon the echelon, intelligence organizations use specialized procedures to develop each category of intelligence. The following information describes each category.

INDICATIONS AND WARNINGS

2-19. Indications and warnings (I&W) are those intelligence activities intended to detect and report time-sensitive intelligence information on foreign developments that could involve a threat to the US or allied and/or coalition military, political, or economic interests or to US citizens abroad. I&W includes forewarning of enemy actions or intentions; the imminence of hostilities; insurgency; nuclear or non-nuclear attack on the US, its overseas forces, or alliled and/or coalition nations; hostile reactions to US reconnaissance activities; terrorist attacks; and other similar events. (JP 3-13). I&W comes from time-sensitive information and analysis of developments that could involve a threat to the US and multinational military forces, US political or economic interests, or to US citizens. While the G2/S2 is primarily responsible for producing I&W intelligence, each element, such as the MPs conducting PIO, within every unit contributes to I&W through awareness of the CCIRs and reporting related information.

CURRENT INTELLIGENCE

2-20. Current intelligence involves the integration of time-sensitive, all-source intelligence and information into concise, accurate, and objective reporting on the battlespace and current enemy situation. One of the most important forms of current intelligence is the enemy situation portion of the COP. The G2/S2 is responsible for producing current intelligence for the unit. Current intelligence supports ongoing operations across full spectrum operations. In addition to the current situation, current intelligence should provide projections of the enemy's anticipated situations (estimates) and their implications on the friendly operation.

GENERAL MILITARY INTELLIGENCE

2-21. GMI is intelligence concerning military capabilities of foreign countries or organizations or topics affecting potential US or multinational military operations relating to armed forces capabilities, including OB, organization, training, tactics, doctrine, strategy, and other factors bearing on military strength and effectiveness and area and terrain intelligence. This broad category of intelligence is normally associated with long-term planning at the national level. However, GMI is also an essential tool for the intelligence staff and should be in place long before the start of preparations for a particular military operation. An up-to-date, comprehensive intelligence database is critical to the unit's ability to plan and prepare rapidly for the range of operations and global environments in which it may operate. GMI supports the requirement to quickly respond to differing crisis situations with corresponding intelligence spanning the globe. One of the many places to get information for GMI is the medical intelligence database. The G2/S2 planner develops his initial IPB from GMI products.

2-22. The G2/S2 develops and maintains the unit's GMI database on potential threat forces and environments based on the commander's guidance. As an essential component of intelligence readiness, this database supports the unit's planning, preparation, execution, and assessment of operations. The G2/S2 applies and updates the database as it executes its intelligence production tasks.

TARGET INTELLIGENCE

2-23. Target intelligence is the analysis of enemy units, dispositions, facilities, and systems to identify and nominate specific assets or vulnerabilities for attack, re-attack, or exploitation (for intelligence). It consists of two mutually supporting production tasks: target development and combat assessment.

- **Target development** is the systematic evaluation and analysis of target systems, system components, and component elements to determine HVTs for potential attack through maneuver, fires, or non-lethal means.
- Once attacked, **combat assessment** provides a timely and accurate estimate of the affects of the application of military force (lethal or non-lethal) an IO on targets and target systems based on predetermined objectives.

SCIENTIFIC AND TECHNICAL INTELLIGENCE

2-24. Scientific and technical intelligence (S&TI) is the product resulting from the collection, evaluation, analysis, and interpretation of foreign S&T information which covers foreign developments in basic and applied research and in applied engineering techniques and S&T characteristics, capabilities, and limitations of all foreign military systems, weapons, weapon systems, and materiel, the research and development (R&D) related thereto, and the production methods employed for their manufacture.

2-25. S&T intelligence concerns foreign developments in basic and applied sciences and technologies with warfare potential. It includes characteristics, capabilities, vulnerabilities, and limitations of all weapon systems, subsystems, and associated materiel, as well as related R&D. S&T intelligence also addresses overall weapon systems and equipment effectiveness. Specialized organizations—such as the Defense Intelligence Agency (DIA) Missile and Space Intelligence Center (MSIC), INSCOM, Air Missile Defense (AMD), Area Air and Missile Defense Command (AAMDC), and National Ground Intelligence Center (NGIC)—produce this category of intelligence. The G2/S2 establishes instructions within SOPs, orders, and plans for handling and evacuating captured enemy materiel (CEM) for S&T intelligence (S&TI) exploitation.

COUNTERINTELLIGENCE

2-26. As previously defined at the end of Chapter 1, CI attempts to identify and recommend countermeasures to the threat posed by foreign intelligence security services (FISS) and the ISR activities of non-state entities such as organized crime, terrorist groups, and drug traffickers. CI analysis incorporates information from all sources as well as the results of CI collection, investigations, and operations to analyze the multidiscipline threat posed by foreign intelligence services and activities.

INTELLIGENCE COMMUNITY

2-27. There are many organizations in the intelligence community that support military operations by providing specific intelligence products and services. The J2/G2/S2 and his staff must be familiar with these organizations and the methods of obtaining information from them as necessary. Figure 2-2 shows organizations that compose the intelligence community.

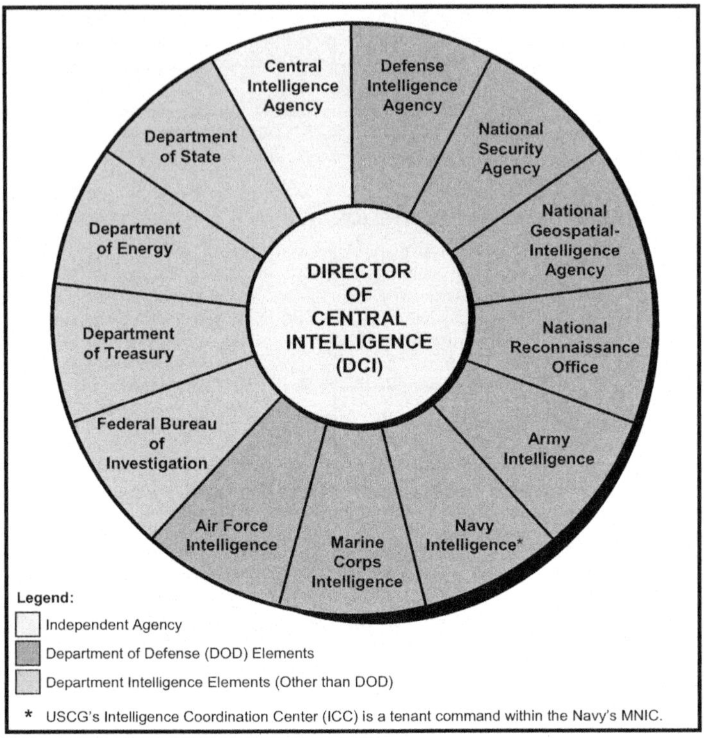

Figure 2-2. Intelligence Community Membership.

DOD AGENCIES

Defense Intelligence Agency (DIA)

2-28. DIA is a combat support agency and a major collector and producer in the defense intelligence community. The DIA supports the full spectrum of operations to include basic MI, counterterrorism, counterdrugs, medical intelligence, WMD and proliferation, United Nations (UN) peacekeeping and multinational support, missile and space intelligence, noncombatant evacuation operations (NEOs), targeting, combat assessment, and battle damage assessment (BDA).

National Security Agency (NSA)

2-29. NSA ensures cryptologic planning and support for joint operations. Working with the tactical cryptologic units of a command, the NSA provides SIGINT and information security (INFOSEC), encompassing communications security (COMSEC) and computer security, as well as telecommunications support and operations security (OPSEC). The people and equipment providing SIGINT, INFOSEC, and OPSEC constitute the United States Cryptologic System (USCS). The NSA, through the USCS, fulfills cryptologic command and/or management, readiness, and operational responsibilities in support of military operations according to the Secretary of Defense tasking, priorities, and standards of timeliness.

National Geospatial-Intelligence Agency (NGA)

2-30. The NGA (formerly National Imagery and Mapping Agency [(NIMA)] mission is to provide timely, relevant, and accurate intelligence and geospatial information in support of national security objectives of the United States. The Director of NGA advises the Secretary of Defense, Director Central Intelligence (DCI), Chairman of the Joint Chiefs of Staff (JCS), and the combatant commanders on imagery, IMINT, and geospatial information. The Operations Directorate, Customer Support Office, is the focal point for interface with external customers, including the JCS, combatant commands, services, and national and defense agencies.

National Reconnaissance Office (NRO)

2-31. The NRO develops and operates unique and innovative space reconnaissance systems and conducts intelligence-related activities essential for US security. The role of the NRO is to enhance US government and military information superiority across full spectrum operations. NRO responsibilities include supporting I&W, monitoring arms control agreements, and performing crisis support to the planning and conduct of military operations. The NRO accomplishes its mission by building and operating IMINT and SIGINT reconnaissance satellites and associated communications systems

US Navy (USN)

2-32. Naval intelligence products and services support the operating naval forces, the Department of the Navy, and the maritime intelligence requirements of national level agencies. Naval intelligence operates the National Maritime Intelligence Center (NMIC). Naval intelligence responsibilities include maritime intelligence on global merchant affairs, counter-narcotics, fishing issues, ocean dumping of radioactive waste, technology transfer, counter-proliferation, cryptologic related functions, CI, I&W support, management of Coast Guard collection, and development of new weapons systems and countermeasures.

US Marine Corps (USMC)

2-33. USMC intelligence provides pre-deployment training and force contingency planning for requirements that are not satisfied by theater, other service, or national capabilities. The Marine Corps Intelligence Agency (MCIA) handles the integration, development, and application of GMI, technical information, all-source production, and open-source information.

US Air Force (USAF)

2-34. USAF ISR assets fill a variety of roles to meet US national security requirements. The USAF operates worldwide ground sites and an array of airborne ISR platforms to meet national level intelligence requirements. To support day-to-day USAF operations and to meet specific USAF requirements, intelligence professionals at the wing and squadron levels use suites of interoperable analysis tools and dissemination systems to tailor information received from all levels and agencies in the intelligence community. USAF responsibilities include all-source information on aerospace systems and potential adversaries' capabilities and intentions, cryptologic operations, I&W, IO, and criminal investigative and CI services.

NON-DOD AGENCIES

2-35. Although the primary focus of non-DOD members of the intelligence community is strategic intelligence and support to the President and the Secretary of Defense, these agencies also produce intelligence and intelligence products that support operational and tactical ARFOR. This responsibility includes assessing potential issues and situations that could impact US national security interests and objectives. These agencies identify global and regional issues and threats. Some of the intelligence products and services these agencies provide are essential to accurate assessment of the threat and battlespace, particularly during stability operations and support operations.

Central Intelligence Agency (CIA)

2-36. The CIA's primary areas of expertise are in HUMINT collection, imagery, all-source analysis, and the production of political and economic intelligence. CIA and military personnel staff the CIA's Office of Military Affairs (OMA). As the CIA's single POC for military support, OMA negotiates, coordinates, manages, and monitors all aspects of agency support for military operations. This support is a continuous process, which the agency enhances or modifies to respond to a crisis or developing operation. Interaction between OMA and the DCI representatives to the Office of the Secretary of Defense (OSD), the Joint Staff, and the combatant commands facilitates providing national level intelligence in support of joint operations, contingency and operations planning, and exercises.

Department of State, Bureau of Intelligence and Research

2-37. The Bureau of Intelligence and Research coordinates programs for intelligence, analysis, and research; it produces intelligence studies and current intelligence analyses essential to foreign policy determination and execution. Its subordinate Bureau of International Narcotics Matters develops, coordinates, and implements international narcotics control assistance activities. It is the principal POC and provides policy advice on international narcotics control matters for the Office of Management and Budget (OMB), the NSC, and the White House Office of National Drug Control Policy (ONDCP). The Bureau also oversees and coordinates the international narcotics control policies, programs, and activities of US agencies.

Department of Energy, Office of Nonproliferation and National Security

2-38. The Department of Energy's Office of Nonproliferation and National Security assists in the development of the State Department's policy, plans, and procedures relating to arms control, nonproliferation, export controls, and safeguard activities. Additionally, this office is responsible for—

- Managing the department's R&D program.
- Verifying and monitoring arms implementation and compliance activities.
- Providing threat assessments and support to headquarters and field offices.

Department of the Treasury

2-39. The US Treasury Department's intelligence-related missions include producing and disseminating foreign intelligence relating to US economic policy and participating with the Department of State in the overt collection of general foreign economic information.

Federal Bureau of Investigation (FBI)

2-40. The FBI is the principal investigative arm of the Department of Justice (DOJ) and has primary responsibility for CI and counterterrorism operations conducted in the US. CI operations contemplated by any other organizations in the US must be coordinated with the FBI. Any overseas CI operation conducted by the FBI must be coordinated with the CIA.

Department of Homeland Security

2-41. The mission of the Department of Homeland Security is to develop and coordinate the implementation of a comprehensive national strategy to secure the US from terrorist threats or attacks. The organizational construct of homeland security mission is a framework of prepare, deter, preempt, defend, and respond. Component agencies will analyze threats and intelligence, guard our borders and airports, protect our critical infrastructure, and coordinate the response of our nation for future emergencies. The component agencies of the department include—

- US Coast Guard (USCG).
- US Customs Service.
- US Border Patrol.
- Federal Emergency Management Agency (FEMA).
- Immigration and Naturalization Service (INS).
- Transportation Security Administration.
- Federal Protective Service.
- Office of Domestic Preparedness.

United States Coast Guard (USCG)

2-42. The USCG, subordinate to the Department of Homeland Security, has unique missions and responsibilities as both an armed force and a law enforcement agency (LEA), which makes it a significant player in several national security issues. The USCG intelligence program supports counter-drug operations, mass seaborne migration operations, alien migration interdiction operations, living marine resource enforcement, maritime intercept operations, port status and/or safety, counterterrorism, coastal and harbor defense operations, and marine safety and/or environmental protection.

Other Agencies

2-43. There are a number of US Government agencies and organizations, not members of the intelligence community, that are responsible for collecting and maintaining information and statistics related to foreign governments and international affairs. Organizations such as the Library of Congress, the Departments of Agriculture and Commerce, the National Technical Information Center, US Information Agency, US Information Service, and the US Patent Office are potential sources of specialized information on political, economic, and military-related topics. The intelligence community may draw on these organizations to support and enhance research and analysis and to provide relevant information and intelligence for planners and decisionmakers. Many other US Government agencies may become directly involved in supporting DOD especially during stability operations and support operations. (See JP 2-02 for a description of agency support to joint operations and intelligence.) These organizations include—

- Department of Transportation.
- Disaster Assistance Response Team within the Office of Foreign Disaster.
- US Agency for International Development.
- NGA.

UNIFIED ACTION INTELLIGENCE OPERATIONS

2-44. In unified action, ARFOR synchronize their actions with those of other participants to achieve unity of effort and to accomplish the combatant commander's objectives. Unified action links subordinates to the combatant commander under combatant command (command authority) (COCOM). Multinational, interagency, and nonmilitary forces work with the combatant commander through cooperation and coordination. Combatant commanders form theater strategies and campaigns, organize joint forces, designate operational areas, and provide strategic guidance and operational focus to subordinates. The aim is to achieve unity of effort among many diverse agencies in a complex environment. Subordinate JFCs synchronize joint operations in time and space, direct the action of foreign military forces (multinational operations), and coordinate with governmental and NGOs (interagency coordination) to achieve the same goal.

2-45. The J2 staff provides intelligence promptly, in an appropriate form, and by any suitable means to those who need it. Intelligence personnel ensure that the consumers understand the intelligence and assist them as they apply the intelligence to their operations.

2-46. Dissemination requires establishing appropriate communications systems and procedures. The J2 and other intelligence personnel must fully participate in all operation planning and execution, and develop close working relationships with the JFC and other staff elements.

2-47. The commander and staff assess intelligence operations to determine their effectiveness and to make any necessary improvements. The intelligence process functions presented in Chapter 4 provide the criteria for evaluating intelligence operations.

JOINT INTELLIGENCE OPERATIONS

2-48. The JTF commanders and his intelligence staff must—

- Understand the intelligence requirements of superior, subordinate, and component commands.
- Identify organic intelligence capabilities and shortfalls.
- Access theater and national systems to ensure appropriate intelligence and CI products are available to the JTF.

2-49. The JTF's intelligence effort focuses on integrating multi-source information and multi-echelon intelligence into all-source intelligence products that provide clear, relevant, and timely knowledge of the enemy and battlespace. These products must be in formats that are readily understood and directly usable by the recipient in a timely manner. They must neither overload the user nor the communications architecture.

2-50. The J2 is the JTF commander's focal point for intelligence. The J2 directly supports the JFC's responsibilities for determining objectives, directing operations, and evaluating the effects of those operations. The J2 supports the execution of the plan with the intelligence needed to sustain the operation, attain joint force objectives, provide support to subordinate commands, and continually support FP efforts. The J2 analyzes the potential threat situation and provides assessments to support friendly opportunities. The J2 then supports the execution of the plan with the operational intelligence needed to sustain the operations, attain joint force objectives, and support FP. To maintain the initiative, the JFC will seek to get inside the adversary's decisionmaking cycle; that is, the JFC will seek to develop procedures and an organization in order to receive new and accurate intelligence and respond to the new situation faster than the adversary. The J2 must help in identifying the adversary's decisionmaking cycle and identifying vulnerabilities that may be exploited. The J2 also—

- Ensures the provision of the required ISR support to the JTF and its subordinate functional and service components.
- Assists the JTF commander in defining intelligence responsibilities and PIRs.
- Actively participates in joint staff planning and in planning, preparing, executing, and assessing ISR efforts. This includes leading the JTF's joint collection board and providing representation at the joint targeting board.

2-51. Figure 2-3 shows a typical JTF J2 organization. The overall organization of the JTF and operations will dictate actual composition of the J2. At a minimum, a core element of analytical, ISR management, and administrative capabilities is required.

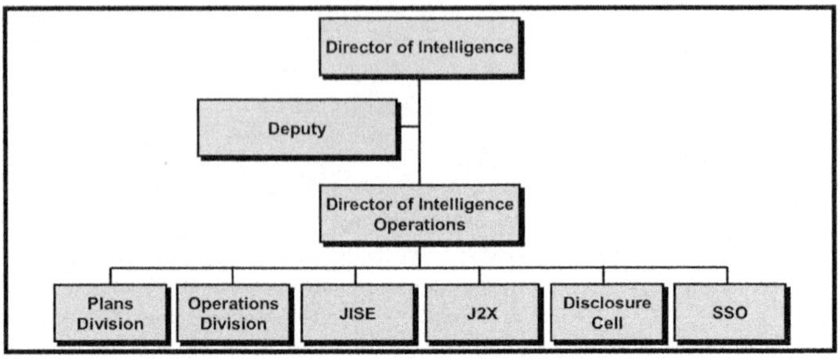

Figure 2-3. Typical Joint Task Force J2 Organization.

Considerations In Joint Intelligence Operations

2-52. The Army, Air Force, Navy, Marine Corps, and SOF must provide globally responsive assets to support the combatant commanders' theater strategies and the national security strategy. The capabilities of the other armed forces complement those of ARFOR. During joint operations, they provide support consistent with JFC-directed missions. When conducting joint intelligence operations, there are a number of unique problems that can arise due to the complexity of integrating the efforts of the different services and commands. Elements affecting joint intelligence operations among the different services include the following:

- Intelligence liaison is critical to the success of intelligence operations and requires early establishment, particularly between units that have not routinely trained together and possess differing capabilities. As a minimum, organizations exchange liaison teams with the higher echelon organization. Additional liaison may be necessary to facilitate joint force collection, production, or dissemination requirements. Liaison teams—
  - Support planning and C2 of intelligence operations.
  - Ensure timely two-way flow of intelligence between commands.
  - Manage intelligence and resource requirements of the subordinate command.
  - Advise the commander on service ISR capabilities, limitations, and employment.
- Commanders and staffs use IPB to understand the battlespace and develop or refine plans and orders. IPB products exchanged between echelons ensure a common picture of the battlespace and estimate of the situation.
- Communications considerations for joint operations include—
  - Planning for intelligence communications transition to facilitate execution of branches or sequels to the plan or to accommodate shifting of the main effort from one force to another.

    - Identifying the initial communications architecture to include establishing procedures and protocols for information exchanges (databases, text, imagery, voice, and video).
    - Balancing the availability of service-unique intelligence systems between echelons or services. This may require each service providing additional resources. The senior commander is responsible for allocating resources.
    - Disseminating intelligence between commands and services. Additional communications equipment, intelligence terminals, and personnel may be required to balance capabilities between services.
- Identifying the databases each service possesses or has access to; determining which databases will support the operation and, if necessary, merge them into a single database; and ensure access by the entire force.
- Providing a focal point for subordinate command access to national or joint intelligence is essential. The senior commander will request and allocate resources required to support this access.

2-53. The JFC's intelligence requirements, concept of operation, and intent drive the ISR effort. The different organizations and services participating in joint intelligence operations must continuously share information, intelligence, and products to satisfy requirements. See FM 2-01 for details on intelligence requirements and RM. Activities required to facilitate an effecttive joint collection effort include—

- Coordinating intelligence and ISR operations to optimize capabilities of collection assets and reduce duplication of effort.
- Integrating supporting national and theater intelligence collection assets into the intelligence synchronization plan.
- Establishing procedures for tracking and handing off HPTs between services and echelons.
- Establishing procedures for cueing Army and other service collection assets.
- Coordinating airspace for ISR assets.
- Maximizing available linguist capabilities. Shortages of military linguists trained in target languages may require cross-leveling Army and other service linguists.

2-54. Reporting and intelligence production considerations in joint operations include—

- Establishing production criteria and thresholds that produce timely and relevant intelligence keyed to the commander's intelligence and targeting requirements. The ISM, attack guidance matrix (AGM), and HPT list are examples of tools used to support joint intelligence and targeting efforts. The sharing of these products to all echelons and services is crucial.
- Establishing common methodology and criteria for producing the BDA and supporting the combat assessment function.
- Establishing set standards for the number and frequency of periodic reports such as intelligence summaries (INTSUMs).

Joint Intelligence Architecture

2-55. In addition to the J2 staffs at every joint level of command, the key organizations in the joint intelligence architecture are the National Military Joint Intelligence Center (NMJIC) (see Figure 2-4), the Joint Intelligence Centers (JICs) or Joint Analysis Centers (JACs) of the unified commands, and, when formed, the JTF's joint intelligence support element (JISE). Working together, these organizations play the primary role in managing and controlling joint intelligence operations. The formal relationships which link these organizations facilitate information management and optimize complementary intelligence functions by echelon without obstructing the timely flow of intelligence up, down, or laterally.

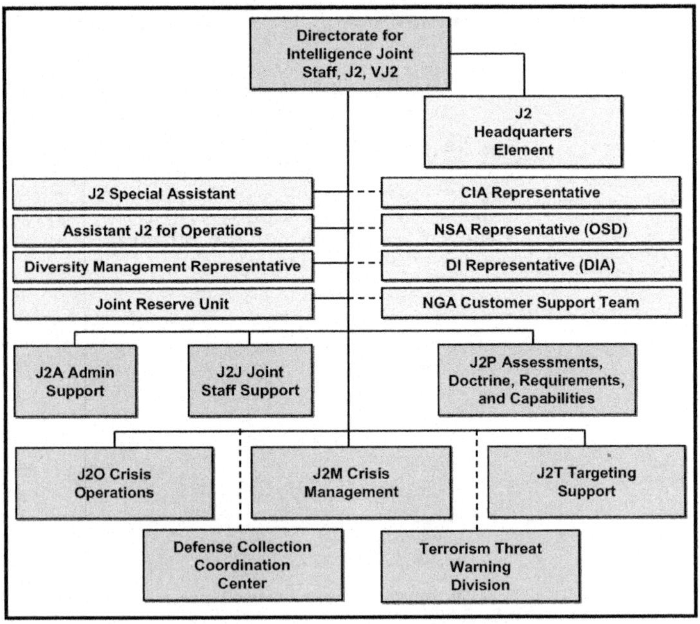

Figure 2-4. National Military Joint Intelligence Center.

2-56. The NMJIC is the focal point for intelligence activities in support of joint operations. The NMJIC is the CJCS J2 intelligence watch within the National Military Command Center (NMCC) in the Pentagon. When the assigned or attached assets cannot satisfy the combatant commander and the JTF commander's crisis-related and time-sensitive intelligence requirements, the NMJIC, as part of the J2's staff, sends tasks to appropriate agencies to fill the requirements. During crises, the NMJIC expands as necessary to establish a working group, intelligence task force, or an expanded intelligence task force. The NMJIC participates in targeting by developing national level target lists. When requested, the NMJIC supports the theater in performing BDA. Besides supporting the combatant commands and JTFs, the NMJIC supports any multinational partners and prescribed international organizations.

2-57. The JIC is organized in accordance with the combatant commander's prerogatives, but normally performs the general functions described in JP 2-0 and specific unified command intelligence TTP. The JIC is responsible for providing intelligence to JTF and its subordinates during military operations. If the JIC cannot meet the combatant commander's requirements, the JIC forwards requests to the NMJIC or to subordinate command levels through established channels, using standard command procedures. In some cases, the JIC may also ensure timely support by approving a direct communication path, in advance, between requesters such as the JTF JISE and outside producers. In this case, the parties must inform the JIC of all requests as they occur. This method is most appropriate when the parties require products and services, which the JIC does not routinely produce.

2-58. At the discretion of the JTF commander, the J2 can establish a JISE during the crisis or preparation stage for operations to augment the J2 staff. The JISE is a tailored subset of the JIC or the intelligence organization of the service component designated as the JTF headquarters.

2-59. Under the direction of the J2, the JISE normally manages the intelligence collection, production, and dissemination for the JTF. The JISE provides intelligence to JTF operational forces and performs common intelligence functions. By design, the JISE is scaleable and can expand to meet the needs of the JTF and the operating environment. It is composed of analytical experts, analysis teams, and ISR managers that provide services and products, which the JTF, JTF staff, and subordinate components require. These experts, mentioned above, focus on solving the operational intelligence problems of concern to the JTF commander. The JISE's capability to perform all-source analysis and ISR synchronization is key to producing operational intelligence that is timely, relevant, and complete. Figure 2-5 illustrates the features of a typical JISE.

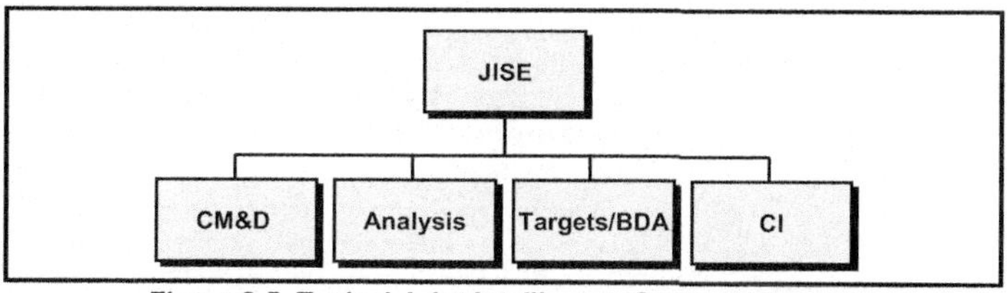

Figure 2-5. Typical Joint Intelligence Support Element.

Joint Task Force Intelligence Organizations

2-60. In addition to the JISE, the JTF commander and J2 may require other supporting JICs or teams based on projected operations. The JTF commander may make a request to the NMJIC for specific national intelligence agency capabilities. The NMJIC evaluates and coordinates the JTF commander's requirements with the J3, J5, and national intelligence agencies and tailors the composition of the deployment packages to meet those needs. The deployment packages, such as the National Intelligence Support Team (NIST), provide access to the entire range of capabilities resident in the national intelligence agencies and can focus those capabilities on the JTF commander's intelligence requirements. The J2X manages and coordinates the HUMINT and CI activities of national, theater, and service components operating within the JTF's JOA. The Joint Captured Materiel Exploitation Center (JCMEC) assists in management of recovery, exploitation, and disposal of captured enemy equipment (CEE). The JTF commander's requirements dictate the composition and tailoring of such deployment packages.

Augmentation Considerations

2-61. Depending on the scale of the operations, the intelligence organizations described above and those of the JTF's subordinate command may require personnel augmentation. Optimum use of available intelligence assets is essential to ensure quality support in meeting the JTF commander's requirements.

2-62. The demand for additional intelligence increases significantly during crisis and wartime operations. As the need arises for more intelligence personnel, the intelligence presence also increases at all command levels. Locating additional intelligence personnel and knowing how to integrate those personnel into the operation is vital. The JTF J2 should identify intelligence personnel augmentation requirements in accordance with the Chairman, Joint Chiefs of Staff (CJCS) Instruction 1301.01. The combatant commander's joint table of mobilization and distribution (JTMD) also contains the standing augmentation requirements. The JTMD should reflect the need for either IMA or Individual Ready Reserve (IRR) personnel. The combatant commander and the JTF refine personnel requirements and initiate requests when they anticipate or start an operation.

2-63. A consideration for the JTF when requesting support or augmentation is that these national level teams and individual augmentees are not totally self-contained elements; rather they require logistic, information, and other support from the supported command. Each deployment is unique based on mission, duration, team composition, and capabilities required. A full NIST, for example, requires a private, access-controlled area within a sensitive compartmented information facility (SCIF) work environment and dedicated secure communications.

2-64. For more information on intelligence operations as they apply to other armed services, see the individual service intelligence doctrine. See also JP 2-0 series, JP 3-0, JP 3-55, and JP 5-0 for more details on joint intelligence operations and considerations.

Multinational Intelligence

2-65. Multinational intelligence operations take place within the structure of an alliance or coalition. Some multinational military organizations, such as the North Atlantic Treaty Organization (NATO) and the UN Command in the Republic of Korea (ROK), are highly structured and enduring. Others, such as the coalition formed during the Gulf War, are less formal and temporary.

2-66. In multinational operations, the multinational force commander (MNFC) exercises command authority over a military force composed of elements from two or more nations. Therefore, in most multinational operations, the JTF must share intelligence, as necessary, for mission accomplishment with foreign military forces and coordinate exchange of intelligence with those forces.

2-67. In some circumstances, the JTF may need to seek authority to go outside the usual political-military channels to provide information to NGOs. The JTF must tailor intelligence policy and dissemination criteria to each multinational operation.

INTELLIGENCE REACH

2-68. Military forces use intelligence reach to rapidly access information, receive support, and conduct collaboration and information sharing with other units and organizations unconstrained by geographic proximity, echelon, or command.

2-69. Intelligence reach supports minimizing the deployed footprint of ISR assets. By providing enhanced information and tailored intelligence products, intelligence reach can greatly enhance the intelligence capabilities of the unit and play a significant role in improving the commander's ability to make decisions on the battlefield. It allows the commander to harness national, joint, foreign, and other military organization resources. Table 2-1 shows examples of partners and sources for intelligence reach.

2-70. Detailed planning and training are critical to the success of intelligence reach operations. The following are steps that the staff can take to ensure optimal use, operability, and effectiveness of intelligence reach:

- Establish data exchange methods and procedures.
- Establish electronic message transfer procedures.
- Establish homepages for identified forces.
- Establish POCs for I&W centers, production centers, Theater JICs, DIA, INSCOM, and their major subordinate commands such as NGIC and the theater intelligence brigades and groups.
- Ensure the intelligence staff has the necessary personnel, training, automated systems, bandwidth, and resources to conduct intelligence reach.
- Determine IRs through staff planning. Develop production requirements for identified intelligence gaps.
- Order geospatial products for the projected joint AOI.
- Establish and maintain a comprehensive directory of intelligence reach resources before deployment and throughout operations. The value of intelligence reach will greatly increase as the staff develops and maintains ready access to rich information resources. These resources are numerous and may include, for example, Army, Joint, DOD, non-DOD, national, commercial, foreign, and university research programs. Know what types of information the resources can provide. Continuously expand the resource directory through identification of new resources.

- Use intelligence reach first to fill intelligence gaps and requirements and answer RFIs. This technique can preclude unnecessary tasking or risk to limited ISR assets.
- Maintain continuous situational understanding and anticipate intelligence requirements. Use intelligence reach to fulfill these requirements and provide the results to the commander and staff for planning and decisionmaking.
- Exchange intelligence reach strategies with other units.
- Present the information retrieved through intelligence reach in a usable form. Share the information derived from intelligence reach with subordinate, lateral, and higher echelons. Ensure follow-on forces have all information as well.

Table 2-1. Examples of Intelligence and Reach Partners and Sources.

ARMY	SERVICES	JOINT	DOD
ACE ISE TSE (TIB/TIG) 902d NGIC ATCAE ARISCs	ONI NMIC AIA NAIC MCIA	USEUCOM JAC USSOUTHCOM JIC USSOCOM JIC USSPACECOM CIC USJFCOM AIC USCENTCOM JIC USTRANSCOM JIC USSTRATCOM USPACOM JIC USNORTHCOM JIC	DIA CMO MSIC AFMIC DHS NGA NSA RSOCs
NON-DOD	**NATIONAL**	**COMMERICAL**	**FOREIGN**
DOE FBI DOS DEA FEMA INS	DCI NIC CIA NRO INR DOT, Office of Intel Support	RAND Jane's Defence Weekly Economic Intelligence Unit CNN Reuters Associated Press United Press International	DIS NFHQ DIO

INTELLIGENCE REACH COMPONENTS

2-71. Intelligence reach requires the G2/S2 to develop a strategy on how best to support the unit's mission with intelligence reach capabilities. There are six basic components of the strategy:

- Push
- Database Access.
- Pull.
- Broadcast Services.
- Collaborative Tools.
- Requirements Management.

Push

2-72. Push occurs when the producers of intelligence or information are knowledgeable of the customer's requirements and are able to send the desired intelligence to the customer without further requests. Push is accomplished through the most appropriate or efficient dissemination means. Unless coordinated as an acceptable dissemination method, push is not accomplished solely by posting intelligence products on a web page. The entity that posts the document must ensure the intended recipient is aware of the product's location or has received the product.

2-73. Push begins with the statement of intelligence interest (SII). The SII establishes the unit's intelligence and IRs. The SII is prepared by intelligence staff organizations to register their interest in receiving recurring hardcopy and softcopy reports, studies, and publications covering a wide variety of intelligence subjects. The J2/G2/S2 works with the Department of the Army Production and Dissemination Management (DAPDM) to establish a profile for information transfer. Intelligence reports or information meeting the unit's requirements are then automatically sent directly to the unit's classified or unclassified networks or communication systems.

2-74. Organizations that push information and data down to subordinate units must be careful of the amount, detail, and focus of the information they are sending. Too much information will overwhelm the subordinate unit. Crucial information may be lost in the midst of an overabundant flow of information, much of which is of little use. Intelligence reach often runs more efficiently if the user pulls the information or at least focuses the intelligence producers to send the appropriate information with the correct level of detail at the appropriate time.

Database Access

2-75. Access to local, theater, DOD, non-DOD, and commercial databases allows analysts to leverage stored knowledge on topics ranging from basic demographics to OB information. A validated DIA Customer Number (acquired by the J2/G2/S2) in combination with Secret Internet Protocol Network (SIPRNET) and Joint Worldwide Intelligence Communications System (JWICS) connectivity establishes access to most of the databases online. The challenge for an analyst is to gain an understanding of the structure, contents, strengths, and weaknesses of the database regardless of database type. Additionally, the procedures are often difficult for extracting portions or downloading and transferring the database to unit level automated information systems.

2-76. Each intelligence discipline has unique databases established and maintained by a variety of agencies. Database access is accomplished through unit or agency homepages via SIPRNET (Intelink-S) and JWICS (Intelink). The DAPDM office, upon approving the unit's SII, validates the requirement for access to the majority of these databases. Units coordinate with the DA dissemination and program manager (DPM) and/or the agency for access to those databases requiring passwords and permissions beyond normal Intelink-S and Intelink access.

Pull

2-77. Pull occurs when the requestor is familiar enough with existing databases and products to be able to anticipate the location of the desired information. Knowledge of both the types and locations of intelligence databases can greatly increase the efficiency of the Intelligence BOS by saving time and effort on the part of analysts at every echelon.

2-78. The G2/S2 must also ensure that intelligence reach capability extends to multinational forces, augmentees, and other services or organizations working as part of the JTF. The G2/S2 must forward relevant intelligence to units which do not possess the necessary automation to conduct intelligence reach.

Broadcast Services

2-79. Broadcast services are an integrated, interactive dissemination system, focusing on tactical user's IRs using joint message formats. The theater broadcast's data prioritization and processing occur through information management. Selected tactical, theater, DOD, and national collectors broadcast their messages using prescribed message formats. The analyst decides which types of messages are required. Broadcast service data is responsive to the operational commander's information needs and allows tactical users to construct successively detailed pictures of the battlespace.

2-80. Tactical users require intelligence and information to be available at the lowest classification level possible (for example, collateral SECRET) and releasable, as necessary, to multinational partners. Broadcast service information management will adhere to national policy as it relates to the provision of data to NATO or multinational partners.

Collaborative Tools

2-81. Collaboration is the sharing of knowledge, expertise, and information, normally online, and may take numerous forms. Collaborative tools are computer-based tools (groupware) that help individuals work together and share information. They allow for virtual online meetings and data sharing. As an example of the use of collaborative tools, the President or the Secretary of Defense and DOD during a crisis situation will establish a number of crisis action teams (CATs) or joint interagency working groups (JIAWGs). These groups or teams are formed to address specified subjects or topics in support of the warfighter or decisionmaker. The groups will focus on the crisis and normally publish their products on a homepage. Analysts with online access can participate in a number of ways. This includes passively accessing the homepages to study the products, sending queries to the identified POCs, or by having their organization join the CAT or JIAWG. Table 2-2 shows some examples of collaborative tools.

Requirements Management

2-82. The RM system provides a mechanism for submitting RFIs, tasking, and managing ISR assets. Analysts who are trained and familiar with the RM process and the various tasking procedures can leverage its systems for refined information. Each intelligence discipline has established procedures for requesting specific information. It is therefore advantageous to have someone familiar with each discipline's procedures to participate in synchronizing the ISR effort. For example, an analyst receives a HUMINT-based intelligence information report, which provides information on an event or subject. If he requires more refined data or clarification, he should submit a time-sensitive requirement or a source-directed requirement. This creates a two-way communication from the field to the collectors without creating additional standing requirements. The more familiar the analyst is with the RM process, the better he can leverage it.

2-83. The intelligence reach component of RM includes the ability of an intelligence officer at any level to request information, which is beyond what is available at his location. The normal procedure for obtaining intelligence or information not obtainable through the use of available ISR assets is to submit an RFI to the next higher echelon. Users enter RFIs into the RM system. See ST 2-33.5 for more information on intelligence reach operations and FM 2-01 for more information on RM.

Table 2-2. Examples of Collaborative Tools.

Tool	Description
Chat (Audio and Text)	Used to conduct conversations online.
Whiteboard	Permits real-time display of drawings, pictures, or documents for group discussion and comment. Participants can annotate in real time as well.
Bulletin Board	Used to post notices and facilitate discussions on any topic.
Video	Used at a desktop computer or a video teleconferencing (VTC) center to allow the person or group to see with whom they are communicating.
Discussion Groups (Newsgroups)	Topics are posted to a website for discussion and comment where participants can follow a line of discussion on a topic.
File-Sharing Tools	Virtual file cabinets allow information to be stored on web servers, and are available to anyone having access to the site and electronic permission to use the files.
Presentation Tools	Used in a virtual auditorium to allow lectures and briefings to be given to an audience.
Application Sharing	An entire team can use an application running on one computer to revise documents.
Text Tools	Allows live text input and editing by group members. Once complete, the text document can be copied into word processing software.
E-mail	Electronic mail.
Persistent Capability	The ability to preserve files, briefings, or other team or project material for future reference. Properly organized, it becomes an information management device and is invaluable to a long-term effort.
Instant Messaging	This allows real-time exchange of notes and messages.

PART TWO

Intelligence in Full Spectrum Operations

Part Two provides a primer of what and who constitutes both MI and the Intelligence BOS and the process that these entities use in order to provide warfighters and decisionmakers with the intelligence they require in order to accomplish their respective missions.

Chapter 3 discusses the role of MI and the Intelligence BOS within full spectrum operations. It provides an overview of intelligence readiness, particularly the intelligence requirements associated with force projection. The concept of MI asset technical control is discussed as a complement to, not a replacement of, the Army's command and support relationships.

Chapter 4 presents the Intelligence Process—the Intelligence BOS methodology that accomplishes the primary focus of intelligence in full spectrum operations, which provides the warfighter with effective intelligence.

Chapter 3

Fundamentals in Full Spectrum Operations

FULL SPECTRUM OPERATIONS

3-1. Army commanders at all echelons may combine different types of operations simultaneously and sequentially to accomplish missions. Full spectrum operations include offensive, defensive, stability, and support operations. Missions in any environment require ARFOR that are prepared to conduct any combination of these operations:

- **Offensive operations** aim to destroy or defeat an enemy. Their purpose is to impose US will on the enemy and achieve decisive victory.
- **Defensive operations** defeat an enemy attack, buy time, economize forces, or develop conditions favorable for offensive operations. Defensive operations alone normally cannot achieve a decision. Their purpose is to create conditions for a counteroffensive that allow ARFOR to regain the initiative.
- **Stability operations** promote and protect US national interests by influencing the threat, political, and information dimensions of the OE through a combination of peacetime developmental, cooperative activities, and coercive actions in response to crises. Regional security is supported by a balanced approach that enhances regional stability

and economic prosperity simultaneously. ARFOR presence promotes a stable environment.

- **Support operations** employ ARFOR to assist civil authorities, foreign or domestic, as they prepare for or respond to crises and relieve suffering. Domestically, ARFOR respond only when the President or Secretary of Defense direct. ARFOR operate under the lead federal agency and comply with provisions of US law, to include the Posse Comitatus and Stafford Acts. (See FM 3-0.)

3-2. Intelligence supports the commander across full spectrum operations. It helps the commander decide when and where to concentrate sufficient combat power to overwhelm the enemy. ISR is essential for the commander to preclude surprise from the enemy, maintain the initiative on the battlefield, and win battles. Commanders at all levels synchronize intelligence with the other BOSs to maximize their ability to see and strike the enemy simultaneously throughout the AO.

3-3. Every soldier in the command is responsible for detecting and reporting enemy activities, dispositions, and capabilities. This task is critical because the environment we operate in is characterized by violence, uncertainty, complexity, and asymmetric methods by the threat. The increased situational awareness that soldiers develop through personal contact and observation is a critical element of that unit's ability to more fully understand the OE. However, soldiers collect information, they are not intelligence collectors. While medical personnel cannot be assigned ISR tasks due to their Geneva Convention category status, medical personnel who gain information through casual observation of activities in plain view while discharging their humanitarian duties will report the information to their supporting intelligence element.

OFFENSIVE OPERATIONS

3-4. Offensive operations at all levels require effective intelligence to help the commander avoid the enemy's main strength and to deceive and surprise the enemy. During offensive operations, intelligence must provide the commander with the composition, disposition, limitations, employment characteristics, and anticipated enemy actions in a timely enough manner for the commander to significantly affect the enemy commander's decision cycle. It ensures commanders have the intelligence they need to conduct offensive operations with minimum risk of surprise.

DEFENSIVE OPERATIONS

3-5. The immediate purpose of defensive operations is to defeat an enemy attack. Commanders defend to buy time, hold key terrain, hold the enemy in one place while attacking in another, or destroy enemy combat power while reinforcing friendly forces. Intelligence should determine the enemy's strength, COAs, and location of enemy follow-on forces. The defending commander can then decide where to arrange his forces in an economy-of-force role to defend and shape the battlefield. Intelligence support affords him the time necessary to commit the striking force precisely.

3-6. Intelligence in area defensive operations identifies, locates, and tracks the enemy's main attack and provides the commander time to allocate sufficient combat power to strengthen the defense at the point of the enemy's main effort. Intelligence should also identify where and when the commander can most decisively counterattack the enemy's main effort or exploit enemy vulnerabilities.

3-7. Although the battlefield is normally organized as decisive, shaping, and sustaining operations, commanders conducting defensive operations may delineate intelligence in battlefield organizational terms of deep area, close area, and rear area.

STABILITY OPERATIONS

3-8. The environment is often much more complex during stability operations and as a result intelligence is often more complex. In fact, intelligence is even more important a factor (or operational multiplier) during stability operations. As a result, the commander must be even more involved in and knowledgeable of ISR (to include ISR operations the commander controls and other higher level ISR operations that may be occurring within his AO) during stability operations. For example, the commanders must understand the complex details of HUMINT and special access program (SAP) operations.

3-9. The commander requires the appropriate intelligence and IPB products in order to determine how best to influence the threat, political and information dimensions of the operational environment, and enhance regional stability. The identification and analysis of characteristics of the terrain and weather, politics, infrastructure, health status, civilian press, attitudes, and culture of the local populace and all possible threats are important in conducting stability operations. A lack of knowledge concerning local politics, customs, and culture could lead to US actions which attack inappropriate targets or which may offend or cause mistrust among the local population.

SUPPORT OPERATIONS

3-10. Support operations are usually nonlinear and noncontiguous. Commanders designate the decisive, shaping, and sustaining operations necessary for mission success. However, determining the intelligence requirements that drive the ISR effort intending to identify any potential threat's centers of gravity and decisive points may require a more complex and unorthodox thought process than that used in offensive and defensive operations. The G2/S2 may have to define the enemy differently. In support operations, the adversary is often the effects of disease, hunger, or disaster on a civilian population. US forces conducting support operations must also fully understand the organization and identity of key figures or groups within the region where they are operating, as these figures may influence greatly the actions of the population—both civilian and military.

ELEMENTS OF COMBAT POWER

3-11. Combat power is the ability to fight. It is the total means of destructive or disruptive force, or both, that a military unit or formation can apply against the adversary at a given time. Commanders combine the elements of combat power to meet constantly changing requirements and defeat an enemy. The elements of combat power are—

- **Manuever** is the employment of forces, through movement combined with fire or fire potential, to achieve a position of advantage with respect to the enemy to accomplish the mission. Maneuver is the means by which commanders concentrate combat power to achieve surprise, shock, momentum, and dominance.
- **Firepower** provides the destructive force essential to overcoming the enemy's ability and will to fight. Firepower magnifies the effects of maneuver by destroying enemy forces and restricting his ability to counter friendly actions; maneuver creates the conditions for the effective use of firepower.
- **Leadership** is the most dynamic element of combat power. Confident, audacious, and competent leadership focuses on the other elements of combat power and serves as the catalyst that creates conditions for success. Leadership provides purpose, direction, and motivation in all operations.
- **Protection** is the preservation of fighting potential of a force so the commander can apply maximum force at the decisive time and place. Protection is neither timidity nor risk avoidance.
- **Information** enhances leadership and magnifies the effects of maneuver, firepower, and protection. Today, ARFOR use information collected to increase their situational understanding before engaging the enemy. Information from the COP and running estimates, transformed into situational understanding, allows commanders to combine the elements of combat power in new ways.

THE FOUNDATIONS OF ARMY OPERATIONS

3-12. Understanding the principles of war and tenets of Army operations is fundamental to operating successfully across the range of military operations. The principles of war and tenets of Army operations form the foundation of Army operational doctrine. (Refer to FM 3-0 for a full description of the Principles of War and Tenets of Army Operations.)

- **The Principles of War** provide general guidance for conducting war and military operations other than war at the strategic, operational, and tactical levels. The principles are the enduring bedrock of Army doctrine.
- **The Tenets of Army Operations** build upon the principles of war. They further describe the characteristics of successful operations. These tenets are essential to victory.

THE OPERATIONAL FRAMEWORK

3-13. The operational framework consists of the arrangement of friendly forces and resources in time, space, and purpose with respect to each other and the enemy or situation. It consists of the **AO, battlespace**, and the **battlefield organization**. The framework establishes an area of geographic and operational responsibility and provides a way for commanders to visualize how to employ forces against the enemy. Commanders design an operational framework to accomplish their mission by defining and arranging its three components. They use the framework to focus combat power. The operational framework provides a mechanism through which the commander can focus the Intelligence BOS effort. Understanding the operational framework, AOIR, and intelligence coordination line (ICL) and their relationship to each other is key to planning and executing ISR operations.

- **AO** is an operational area defined by the JFC for land and naval forces. AOs do not typically encompass the entire operational area of the JFC but should be large enough for component commanders to accomplish their missions and protect their forces. AOs should also allow component commanders to employ their available systems to the limits of their capabilities. The AO is the basic control measure for all operations that defines the geographical area for which a particular unit is responsible. The commander—
  - Assumes responsibility for intelligence, maneuver, fires, terrain management, security, and movement within his AO.
  - Establishes control measures within his AO to assign responsibilities, coordinate intelligence, fires, and maneuver, and to control other activities.
- **Battlespace** is the environment, factors, and conditions that must be understood to successfully apply combat power, protect the force, or complete the mission. This includes air, land, sea, space, and the included enemy and friendly forces; facilities; weather; terrain; the EMS; and the information environment within the operational areas and AOIs. The G2/S2 performs IPB and synchronizes ISR activities throughout the battlespace as determined by the commander's METT-TC considerations.
  - Area of influence is the geographical area in which a commander can directly influence operations by maneuver of FS systems normally under the commander's command or control. Areas of influence surround and include the associated AO.
  - AOI is that area of concern to the commander, including the area of influence, areas adjacent thereto, and extending into enemy territory to the objectives of current or planned operations. This area also includes areas occupied by enemy forces who could jeopardize the accomplishment of this mission. (See JP 1-02.)
  - Information environment refers to a commander's battlespace that encompasses information activity affecting the operation. To envision that part of the information environment that is within their battlespace, commanders determine the information activities that affect their own operational capabilities and opposing C2 and information systems.
  - Force Projection Bases are the intermediate staging bases and power projection platforms.
  - Home Stations are the permanent locations of active and RC units. To a significant degree, events occurring at home station affect the morale and performance of deployed forces. Thus, the commander's battlespace encompasses all home station functions.

- **Battlefield Organization** is the allocation of forces in the AO by purpose. It consists of three all-encompassing categories of operations:
  - Decisive Operations are those that directly accomplish the task assigned by higher headquarters. Decisive operations conclusively determine the outcome of major operations, battles, and engagements.
  - Shaping Operations are operations at any echelon that create and preserve conditions for the success of decisive operations.
  - Sustaining Operations are operations at any echelon that enable shaping and decisive operations by providing CSS, rear area and base security, movement control, terrain management, and infrastructure development.

3-14. AOIs extend into enemy territory, to the objectives of current or planned operations. They include areas occupied by enemy forces that could jeopardize the accomplishment of the mission. AOIs also serve to focus intelligence development and IO directed at factors outside the AO that may affect the operation. (FM 3-0, para 4-79) The scheme of support includes the coordination of reconnaissance and surveillance missions and AOIs with a joint force or higher headquarters and lateral units to answer the intelligence requirements within the AOI. The G2/S2 monitors enemy activities within the AOI and provides intelligence on enemy activities that may affect the unit.

3-15. The AOIR is an area allocated to a commander in which the commander is responsible for the provision of intelligence within the means at the commander's disposal. It is an area allocated to the commander in which the commander is responsible for the collection of information concerning the threat and the environment and the analysis of that information in order to produce intelligence. Higher headquarters also ensure through intelligence handovers, collection management, and deconfliction that problems with duplication, confliction, and command and control do not occur in the AOIR. (FM 2-19.402/FM 34-80-2) They include the available ISR assets, capability of the G2/S2 section, available intelligence architecture, and METT-TC considerations. The AOIR cannot extend beyond a unit's AOI; however, it can be smaller than its AO as well as vary (expand or contract) during an operation. An example of when a unit's AOIR is smaller than its AO is when a higher headquarters ISR effort covers an area within the unit's AO.

3-16. ICLs designate the boundary between AOIRs. The G2/S2 establishes ICLs to facilitate coordination between higher, lateral, and subordinate units; coordinates with the G3/S3 to direct subordinates to track enemy units and HPTs in their areas; and hands over intelligence responsibility for areas of the battlefield. The establishment of ICLs ensures that there are no gaps in the collection effort; that all echelons are aware of the location, mission, and capabilities of other assets. The G2/S2 keeps abreast of collection activities in progress (all echelons) and battlefield developments through the ICLs. ICLs help—

- Facilitate coordination between higher, lateral, and subordinate units.

- Ensure higher, lateral, and subordinate units share information and intelligence as enemy entities and HPTs move into, within, or transition between AOIRs.
- Specify intelligence responsibility for areas of the battlefield.

ARMY CAPABILITIES

3-17. Commanders combine AC and RC ARFOR—consisting of different types of units with varying degrees of modernization—with multinational forces and civilian agencies to achieve effective and efficient unified action. A broad range of organizations makes up the institutional Army that supports the field Army. Institutional Army organizations design, staff, train, and equip the force. The institutional Army assists in effectively integrating Army capabilities. It does this through leadership and guidance regarding force structure, doctrine, modernization, and budget. (See FM 100-11.)

TASK ORGANIZATION

3-18. The Army supports JFCs by providing tailored force packages to accomplish joint missions and dominate enemies and situations on land. Trained and equipped AC and RC units comprise these force packages. Within these force packages, Army commanders organize groups of units for specific missions. They reorganize for subsequent missions when necessary. This process of allocating available assets to subordinate commanders and establishing their command and support relationships is called task organizing. A temporary grouping of forces designed to accomplish a particular mission is a task organization. The ability of ARFOR to tailor (select forces based upon a mission) and task organize (temporarily organize units to accomplish a tactical mission) gives them extraordinary agility. It allows operational and tactical level commanders to organize their units to make the best use of available resources. The ability to task organize means ARFOR can shift rapidly among offensive, defensive, stability, and support operations.

COMBINED ARMS

3-19. The fundamental basis for the organization and operations of ARFOR is combined arms. Combined arms is the synchronized or simultaneous application of several arms—such as infantry, armor, field artillery, engineers, air defense, and aviation—to achieve an effect on the enemy that is greater than if each arm was used against the enemy separately or sequentially. The ultimate goal of Army organization for operations remains success in joint and combined arms warfare. Its combined arms capability allows commanders to form Army combat, combat support (CS), and CSS forces into cohesive teams focused on common goals.

COMMAND AND SUPPORT RELATIONSHIPS

3-20. Establishing clean command and support relationships is fundamental in organizing for all operations. These relationships can achieve clear responsibilities and authorities among subordinate and supporting units. The commander designates command and support relationships within his authority to weigh the decisive operation and support his scheme of maneuver. Some forces available to a commander are given command or support relationships that limit his authority to prescribe additional relationships. Command and support relationships carry with them varying responsibilities to the subordinate unit by parent and gaining units. By knowing the inherent responsibilities of each command and support relationship, a commander may organize his force to establish clear relationships.

3-21. Command relationships establish the degree of control and responsibility commanders have for forces operating under their tactical control (TACON). When commanders establish command relationships, they determine if the command relationship includes administrative control (ADCON). Figure 3-1 shows command and support relationships.

3-22. Support relationships define the purpose, scope, and effect desired when one capability supports another. Support relationships establish specific responsibilities between supporting and supported unit. Army support relationships are direct support (DS), general support (GS), general support-reinforcing (GSR), and reinforcing.

3-23. While not an actual C2 function, technical control often affects certain intelligence operations. Technical control ensures adherence to existing policies or regulations and provides technical guidance for MI activities, particularly HUMINT, SIGINT, and CI operations. Commanders direct operations but often rely on technical expertise to plan, prepare, execute, and assess portions of the unit's collection effort. Technical control also involves translating ISR tasks into the specific parameters used to focus highly technical or legally sensitive aspects of the ISR effort. Technical control includes, but is not limited to—

- Defining, managing, or guiding the employment of specific ISR assets.
- Identifying critical technical collection criteria such as technical indicators.
- Recommending collection techniques, procedures, or assets.
- Conducting operational reviews.
- Conducting operational coordination.
- Conducting specialized training for specific MI personnel or units.

3-24. An example of technical control is the Prophet control team converting the PIR and SOR sets from the MDMP process and assigning times and anticipated enemy frequencies for subordinate Prophets to collect.

	IF RELATIONSHIP IS:	INHERENT RESPONSIBILITIES ARE:							
		Has Command Relationship with:	May Be Task Organized by:	Receives CSS from:	Assigned Position or AO By:	Provides Liaison To:	Establishes/ Maintains Communications with:	Has Priorities Established by:	Gaining Unit Can Impose Further Command or Support Relationship of:
COMMAND	Attached	Gaining unit	Gaining unit	Gaining unit	Gaining unit	As required by gaining unit	Unit to which attached	Gaining unit	Attached; OPCON; TACON; GS; GSR; R; DS
COMMAND	OPCON	Gaining unit	Parent unit and gaining unit; gaining unit may pass OPCON to lower HQ. Note 1	Parent unit	Gaining unit	As required by gaining unit	As required by gaining unit and parent unit	Gaining unit	OPCON; TACON; GS; GSR; R; DS
COMMAND	TACON	Gaining unit	Parent unit	Parent unit	Gaining unit	As required by gaining unit	As required by gaining unit and parent unit	Gaining unit	GS; GSR; R; DS
COMMAND	Assigned	Parent unit	Parent unit	Parent unit	Gaining unit	As required by parent unit	As required by parent unit	Parent unit	Not Applicable
SUPPORT	Direct Support (DS)	Parent unit	Parent unit	Parent unit	Supported unit	Supported unit	Parent unit; Supported unit	Supported unit	Note 2
SUPPORT	Reinforcing (R)	Parent unit	Parent unit	Parent unit	Reinforced unit	Reinforced unit	Parent unit; reinforced unit	Reinforced unit: then parent unit	Not Applicable
SUPPORT	General Support Reinforcing (GSR)	Parent unit	Parent unit	Parent unit	Parent unit	Reinforced unit and as required by parent unit	Reinforced unit and as required by parent unit	Parent unit; then reinforced unit	Not Applicable
SUPPORT	General Support (GS)	Parent unit	Parent unit	Parent unit	Parent unit	As required by parent unit	As required by parent unit	Parent unit	Not Applicable

NOTE 1. In NATO, the gaining unit may not task organize a multinational unit (see TACON).

NOTE 2. Commanders of units in DS may further assign support relationships between their subordinate units and elements of the supported unit after coordination with the supported commander.

Figure 3-1. Army Command and Support Relationships and Inherent Responsibilities.

Chapter 4

Intelligence Process in Full Spectrum Operations

THE INTELLIGENCE PROCESS

4-1. Commanders use the operations process of plan, prepare, execute, and assess to continuously design and conduct operations (see Figure 4-1). The commander cannot successfully accomplish the activities involved in the operations process without information and intelligence. The design and structure of intelligence operations support the commander's operations process by providing him with intelligence regarding the enemy, the battlefield environment, and the situation.

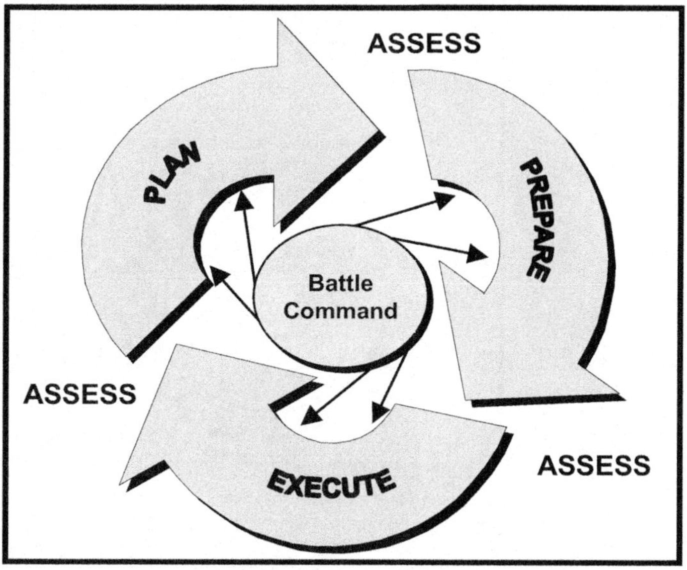

Figure 4-1. The Operations Process.

4-2. Intelligence operations consist of the functions that constitute the intelligence process: **plan, prepare, collect, process, produce**, and the three common tasks of **analyze, disseminate,** and **assess**. Just as the activities of the operations process overlap and recur as circumstances demand, so do the functions of the intelligence process. Additionally, the analyze, disseminate, and assess functions of the intelligence process occur continuously throughout the intelligence process.

4-3. The operations process and the intelligence process are mutually dependent. The commander, through the operations process, provides the guidance and focus through CCIRs and PIRs that drives the intelligence process; the intelligence process provides the continuous intelligence essential to the operations process. Intelligence about the enemy, the battlefield environment, and the situation allows the commander and staff to develop a plan, seize and retain the initiative, build and maintain momentum, and exploit success (see Figure 4-2). The intelligence process is just one of the mechanisms that provides input to build the COP and facilitate the commander's situational understanding.

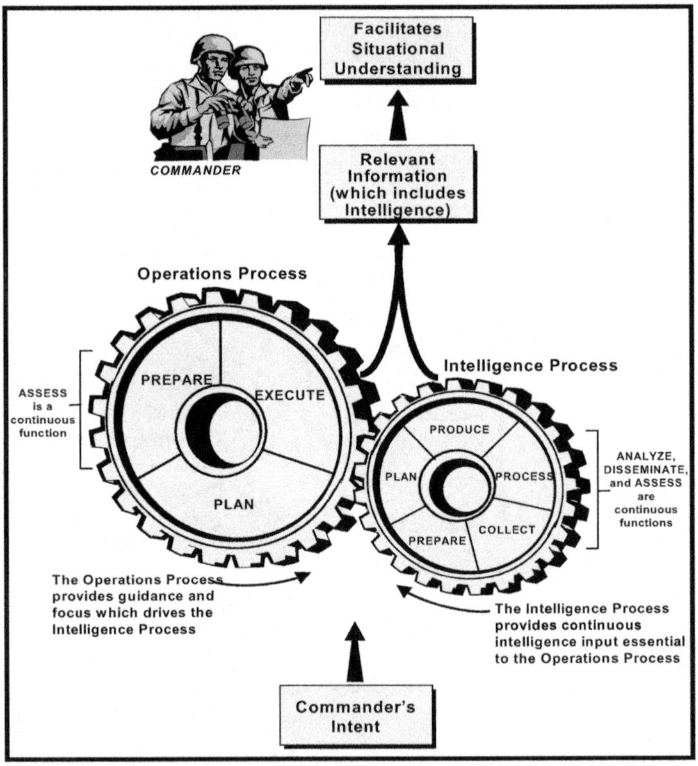

Figure 4-2. The Relationship Between the Operations and Intelligence Processes.

PLAN

4-4. The planning step of the intelligence process consists of activities that include assessing the situation, envisioning a desired outcome (also known as setting the vision), identifying pertinent information and intelligence requirements, developing a strategy for ISR operations to satisfy those requirements, directing intelligence operations, and synchronizing the ISR effort. The commander's intent, planning guidance, and CCIRs (PIRs and FIRs) drive the planning of intelligence operations. Planning, managing, and coordinating these operations are continuous activities necessary to obtain information and produce intelligence essential to decisionmaking.

COORDINATE

4-5. Staff and leaders coordinate with various elements, units, and organizations to ensure the necessary resources, linguist support (see Appendix B), information, intelligence, training, and procedures are in place to facilitate effective intelligence operations.

- Coordination for Movement of ISR Assets. All ISR assets at one time or another will move through or near another unit's AO. To avoid fratricide, ISR elements must coordinate with units, G3/S3, G2/S2, and each other, as well as coordinate with the fire support officer (FSO) to establish no-fire areas and/or other control measures around ISR assets and the air defense officer (ADO) in reference to aerial ISR assets in order to establish the appropriate weapons control status.
- Coordination for Information and Intelligence. The intelligence staff must prepare and practice coordination with personnel from all MI units, non-MI units, other service components, and multinational organizations that may contribute to or facilitate the collection effort. This coordination enables the G2/S2 to share and update databases, information, and intelligence and ensures connectivity with those organizations. All units are sources of relevant information regarding the enemy and the operational environment.
- Liaison. In order to accomplish the mission, exchange information and intelligence, move through certain areas and ensure FP, it may be necessary to coordinate with many different elements, organizations, and local nationals of the country in which friendly forces are conducting operations. Local nationals include police, town officials, foreign military forces, and political and other key figures within the AO. Operations may also necessitate coordination with other US and multinational forces; for example, the International Police Task Force (IPTF), Joint Commission Observers (JCO), Organization for Security and Cooperation in Europe (OSCE), Allied Military Intelligence Battalion (AMIB), and Defense HUMINT Service (DHS).
- Movement. Coordination with the G3/S3 ensures ease of movement and safe passage of friendly forces through an area. Coordinating movement also helps avoid fratricide.

INTELLIGENCE BOS CONSIDERATIONS

4-6. The Intelligence BOS is a unified system that anticipates and satisfies intelligence needs. Commanders ensure its proper employment by clearly articulating intent, designating CCIRs (PIRs and FFIRs), and prioritizing targets. Commanders must, however, understand the limitations of the Intelligence BOS to preclude unrealistic expectations of the system. The following are Intelligence BOS limitations:

- Intelligence only reduces uncertainty on the battlefield; it does not eliminate it entirely. The commander will always have to determine the presence and degree of risk involved in conducting a particular mission.
- The Intelligence BOS is composed of finite resources and capabilities. Intelligence systems and soldiers trained in specific ISR skills are limited in any unit. Once lost to action or accident, these soldiers and systems are not easily replaceable; for some, it may not be possible to replace them during the course of the current operation. The loss of soldiers and equipment can result in the inability to detect or analyze enemy actions. The loss of qualified language-trained soldiers, especially soldiers trained in low-density languages or skills, could adversely affect intelligence operations as well.
- The Intelligence BOS cannot effectively and efficiently provide intelligence without adequate communications equipment, capacity, and connectivity. Commanders and G2/S2s must ensure communications support to intelligence has the appropriate priority.
- Commanders and G2/S2s cannot expect that higher echelons will automatically send them everything they need. While intelligence reach is a valuable tool, the push of intelligence products from higher echelons does not relieve subordinate staffs from conducting detailed analysis and focusing the efforts of higher headquarters. Nor can they expect products pushed to them to be always at the level of detail they require. Commanders and G2/S2s must focus higher echelons by clearly articulating and actively pursuing intelligence requirements. By providing higher echelons with a clear picture of the required intelligence products, commanders can also narrow the flow of intelligence and information and preclude being overwhelmed by too much information.

4-7. Commanders should be aware that intelligence collection is enabled by, and subject to, laws, regulations, and policies to ensure proper conduct of intelligence operations. While there are too many to list here specifically, categories of these legal considerations include United States Codes (USCs), Executive Orders, National Security Council Intelligence Directives (NCSIDs), Army Regulations, United States Signal Intelligence Directives (USSIDs), SOFAs, ROE, and other international laws and directives.

PREPARE

4-8. The prepare step includes those staff and leader activities which take place upon receiving the OPORD, OPLAN, WARNO, or commander's intent to improve the unit's ability to execute tasks or missions and survive on the battlefield. These activities include—

- Effecting necessary coordination in accordance with the OPORD, METT-TC, unit SOP.
- Establishing and testing the intelligence architecture. This activity includes complex and technical issues like hardware, software, communications, COMSEC materials, network classification, technicians, database access, liaison officers (LNOs), training, funding, and TTP.
- Establishing an intelligence team attitude. This activity includes knowing different unit's and organization's capabilities, training the necessary collective skills, establishing effective relationships with different units and organizations, developing mutual battle rhythms and TTP, and leveraging the right architectures and collaboration tools.
- Coordinating effective analytic collaboration. Effective analytic collaboration is necessary to maximize the complementary analytic capabilities of different units and organizations that produce intelligence within the same theater of operations. Coordinating this collaboration is an effort-intensive activity that requires careful mutual planning, division of labor, defined responsibilities, and procedures for adapting to changing circumstances as they develop.
- Establishing reporting procedures.
- Conducting IPB.
- Producing Intelligence Estimates.
- Presenting briefings.
- Ensuring staff and personnel are trained. If personnel are not adequately trained at this point, they must be trained or the leader must evaluate the risk they bring to the operation.
- Planning refinement, brief-backs, SOP reviews, rehearsals, and coordinating with various elements and organizations.
- Establishing other troop-leading procedures (TLPs) or coordination, as necessary, in accordance with METT-TC factors.

G2/S2 PREPARATION ACTIVITIES

4-9. The G2/S2 takes numerous steps before mission execution to ensure intelligence operations run smoothly and effectively. These steps include, but are not limited to, the following:

- Conduct rehearsals.
- Conduct communication rehearsals.
- Review and update available databases and IPB products.
- Review applicable SOPs, Army Regulations, DA Pamphlets, Field Manuals, and ROE for guidance in conducting intelligence operations.
- Plan and practice actions supporting likely contingencies, or the branches or sequels to an operation.
- Ensure coordination measures are still in effect.
- Ensure training (individual and collective).

- Verify communications protocols with theater and higher headquarters and subordinate and lateral units.
- Update intelligence databases.
- Update the forces with the most recent intelligence on the AO immediately before mission execution.

VERIFICATION

4-10. Coordination for or requesting provisions or services is only the first step in acquiring them. It is crucial that staff and leaders check to verify that procedures, personnel, equipment, and services are in place and ready for mission execution.

REHEARSALS

4-11. Rehearsals help units prepare for operations by either verifying that provisions and procedures are in place and functioning or identifying inadequacies, which staff and leaders must remedy. They allow participants in an operation to become familiar with and to translate the plan into specific actions that orient them to their environment and other units when executing the mission. They also imprint a mental picture of the sequence of key actions within the operation and provide a forum for subordinate and supporting leaders and units to coordinate. (FM 6-0)

REPORTING PROCEDURES

4-12. The timely and accurate reporting of CCIRs (PIRs and FFIRs) and IRs is key to successful operations. All assets should know when, how often, and what format to use when reporting. The G2/S2 must verify the frequencies, alternate frequencies, and reactions during jamming, as well as the LTIOV for specific information to be reported. Unit SOPs provide the proper reporting procedures.

4-13. The G2/S2 coordinates with the unit staff, subordinate and lateral commands, and higher echelon units to ensure that specific reporting assets, personnel, equipment (especially communications), and procedures are in place. The G2/S2 requests or establishes the appropriate message addresses, routing indicators, mailing addresses, and special security office (SSO) security accreditation for units.

COMMUNICATIONS

4-14. Staff and leaders must work closely with the G6/S6 or signal officer (SIGO) to coordinate for the required communication links. The unit may require classified and unclassified network connections for their equipment. If elements of the unit will be working outside the range of the unit's communications systems, then it is necessary to coordinate for global or extended range communications. Leaders must obtain the required type and amount of communications equipment and related components as well as the latest fills and frequencies. They must possess and be familiar with all the instructions, passwords, policies, regulations, and directives conducive to OPSEC. They must also ensure soldiers are trained in the use and procedures involved in operating communications equipment.

SITUATION UPDATES

4-15. Each staff section and element conducts activities to maximize the operational effectiveness of the force. Coordination and preparation are just as important, if not more important, as developing the plan. Staff preparation includes assembling and continuously updating estimates. For example, continuous IPB provides accurate situational updates for commanders.

INTELLIGENCE HANDOFF

4-16. A well-executed intelligence hand-off will ensure a smooth and seamless transition between units. It is important that the incoming unit becomes familiar with the operation as soon as possible to avoid compromising the intelligence production and flow of the mission. The following are points to consider during a mission hand-off:

- Briefings and reports (learn what briefings are required and when as well as report formats and requirements).
- Past, present, and planned activities within the AOI.
- Established SOPs (know procedures for reporting; intelligence contingency funds [ICFs] and incentive use if applicable; emplacement and use of ISR equipment).
- Key personalities (introductions are required; establish rapport and a good working relationship with all key personalities).
  - Key personnel on the base or camp (their responsibilities; how to contact them).
  - Key personnel in other US and multinational service components (coordinate for exchange of information and intelligence).
  - Key personalities from surrounding towns (local figures).
  - Key national level political and military figures.
- Supporting units (know where to go for provisions, information, or assistance and POCs within those organizations).
- Current attitudes (understand current attitudes and perspectives of the local populace).
- Equipment operation and idiosyncrasies (equipment may run on different applications; personnel may need to train on specific equipment and procedures).
- Area familiarization (identify NAIs, key terrain, minefields, and boundaries; know camp locations, routes and route names, checkpoints, towns, and troubled resettlement areas).

RULES OF ENGAGEMENT

4-17. Although ROE training was presented during the plan function of the intelligence process, leaders at all levels can take the opportunity during the prepare function to ensure their subordinates completely understand the ROE. It is also during this function that commanders may need to consider exceptions to, or modifications of, the ROE to facilitate HUMINT and CI collection or to enable the placement of ISR assets.

COLLECT

4-18. Recent ISR doctrine necessitates that the entire staff, especially the G3/S3 and G2/S2, must change their reconnaissance and surveillance mind-set to conducting ISR. The staff must carefully focus ISR on the CCIR (PIR and FFIR) but also enable the quick retasking of units and assets as the situation changes. This doctrinal requirement ensures that the enemy situation not just our OPLAN "drives" ISR operations. Well-developed procedures and carefully planned flexibility to support emerging targets, changing requirements, and the need to support combat assessment is critical. The G2/S2 and G3/S3 play a critical role in this challenging task that is sometimes referred to as "fighting ISR" because it is so staff intensive during planning and execution (it is an operation within the operation). Elements of all units on the battlefield obtain information and data about enemy forces, activities, facilities, and resources as well as information concerning the environmental and geographical characteristics of a particular area.

ISR TASKS

4-19. ISR tasks are the actions of the intelligence collection effort. ISR tasks consists of three categories:

- Intelligence.
- Surveillance.
- Reconnaissance.

4-20. Intelligence tasks are included in Annex B of the OPORD under Scheme of Intelligence. They include the following:

- **Intelligence Production**. Intelligence production includes analyzing information and intelligence and presenting intelligence products, conclusions, or projections regarding the OE and enemy forces in a format that enables the commander to achieve situational understanding.
- **Request for Information**. Submitting an RFI to the next higher or lateral echelon is the normal procedure for obtaining intelligence information not available through the use of available ISR assets. Users enter RFIs into an RFI management system where every other user of that system can see it. Hence, an echelon several echelons above the actual requester becomes aware of the request and may be able to answer it. A G2/S2 who receives an RFI from a subordinate element may use intelligence reach to answer RFIs.
- **Intelligence Reach**. Intelligence reach allows the commander to access the resources of national, joint, foreign, and other military organizations and units. Requestors can acquire information through push and pull of information, databases, homepages, collaborative tools, and broadcast services. (See Chapter 2 for more information on intelligence reach.)

4-21. For information on reconnaissance or surveillance tasks, refer to FM 7-15 and FM 3-55.

SPECIAL RECONNAISSANCE

4-22. Special reconnaissance (SR) is the complementing of national and theater intelligence collection assets and systems by obtaining specific, well-defined, and time-sensitive information of strategic or operational significance. It may complement other collection methods where there are constraints of weather, terrain, hostile countermeasures, and/or other systems availability. SR places US or US-controlled personnel conducting direct observation in hostile, denied, or politically sensitive territory when authorized. SOF may conduct these missions unilaterally or in support of conventional operations. (See JP 3-05 and FM 101-5-1.)

4-23. Army Special Operations Forces (ARSOF) elements conduct SR missions to obtain information not available through other means. SR operations encompass a broad range of collection activities to include reconnaissance, surveillance, and TA. SR complements national and theater collection systems that are more vulnerable to weather, terrain masking, and hostile countermeasures. SR missions provide intelligence or information that is often not available through other means. Typical SR missions include—

- TA and surveillance of hostile C2 systems, troop concentrations, deep-strike weapons, lines of communication (LOCs), WMD systems, and other targets.
- Location and surveillance of hostage, prisoner of war, or political prisoner detention facilities.
- Post-strike reconnaissance for BDA.
- Meteorologic, geographic, or hydrographic reconnaissance to support specific air, land, or sea operations.

4-24. For more information on special reconnaissance, see FM 3-05.102.

REPORTING

4-25. The most critical information collected is worthless if not reported in a timely manner. Collectors may report information via verbal, written, graphic, or electronic means. Unit SOPs must clearly state the transmission means of different types of reports (for example, sent by voice frequency modulated [FM] radios or by automated means). In general, the transmission of reports for enemy contact and actions, CCIRs, exceptional information, and NBC reports is by voice FM, and then followed up with automated reports. Commanders and staffs must remember that timely reporting, especially of enemy activity, is critical in fast-moving operations. Collectors must report accurate information as quickly as possible. Commanders and staff must not delay reports for the sole purpose of editing and ensuring the correct format. This is particularly true for reporting information or intelligence that answers the PIR.

TIME-SENSITIVE REPORTING

4-26. Intelligence and time-sensitive combat information that affects the current operation is disseminated immediately upon recognition. Combat information is unevaluated data, gathered by or provided directly to the tactical commander which, due to its highly perishable nature or the criticality of the situation, cannot be processed into tactical intelligence in time to satisfy the user's tactical intelligence requirements. Thus combat information is provided directly to the tactical commander (see JP 1-02). The routing of combat information proceeds immediately in two directions: directly to the commander and through routine reporting channels, which include intelligence analysis and production elements.

4-27. Time-sensitive information usually includes reports concerning enemy contact and actions and CCIRs.

PROCESS

4-28. The process function converts relevant information into a form suitable for analysis, production, or immediate use by the commander. Processing also includes sorting through large amounts of collected information and intelligence (multidiscipline reports from the unit's ISR assets, lateral and higher echelon units and organizations, and non-MI elements in the battlespace). Processing identifies and exploits that information which is pertinent to the commander's intelligence requirements and facilitates situational understanding. Examples of processing include developing film, enhancing imagery, translating a document from a foreign language, converting electronic data into a standardized report that can be analyzed by a system operator, and correlating dissimilar or jumbled information by assembling like elements before the information is forwarded for analysis.

4-29. Often collection assets must collect and process their data prior to disseminating it. MI systems have their own reporting and processing systems, the details of which are in the appropriate MI system manuals and technical manuals. Some collection assets, particularly air reconnaissance and ground scouts, can report relevant information that is immediately usable by the tactical commander (for example, for targeting purposes). However, the personnel in the reporting chain still process these reports by evaluating their relevancy and accuracy. In many cases, the output of a collection asset is data, or information of limited immediate use to a commander. Also, in certain situations ROE dictate a requirement for target confirmation by other sources.

4-30. The intelligence staff processes information collected by the unit's assets as well as that received from higher echelons. The intelligence staff processes many types of information and data from intelligence reach, unmanned aerial vehicle (UAV) imagery, radar imagery, mobile target indicators (MTIs), and HUMINT and SIGINT reports.

PRODUCE

4-31. In the production step, the G2/S2 integrates evaluated, analyzed, and interpreted information from single or multiple sources and disciplines into finished intelligence products. Like collection operations, the G2/S2 must ensure the unit's information processing and intelligence production are prioritized and synchronized to support answering the CCIRs (PIRs and FFIRs).

4-32. Intelligence products must be timely, relevant, accurate, predictive, and usable. The accuracy and detail of every intelligence product has a direct effect on how well the unit plans and prepares for operations. However, the G2/S2 and unit must use intelligence (no matter what form the intelligence is in) that meets the requirements but might not be as detailed or refined as possible or in a better form. A good answer on time is better than a more refined answer that is late.

4-33. The G2/S2 produces intelligence for the commander as part of a collaborative process. The commander drives the G2/S2's intelligence production effort by establishing intelligence and IRs with clearly defined goals and criteria. Differing unit missions, environments, and situations impose numerous and varied production requirements on the G2/S2 and his staff.

4-34. The G2/S2 must employ collaborative analysis techniques and procedures that leverage intelligence production capability of higher and subordinate echelons to meet these requirements. Proficiency in these techniques and procedures enables the G2/S2 to answer the commander's and staff's requirements regardless of the mission, environment, and situation. The G2/S2 and staff intelligence products enable the commander to—

- Plan operations and employ maneuver forces effectively.
- Recognize potential COAs.
- Employ effective tactics and techniques.
- Take appropriate security measures.
- Focus ISR.

COMMON INTELLIGENCE PROCESS TASKS

ANALYZE

4-35. Analysis occurs at various stages throughout the intelligence process. Personnel conducting intelligence operations at all levels analyze intelligence, information, and problems to produce intelligence, solve problems and, most importantly, answer the PIRs. Leaders at all levels conduct analysis to assist in making many types of decisions. An example is a HUMINT collector analyzing an intelligence requirement in order to determine the best possible collection strategy to use against a specific source.

4-36. Analysis in RM is critical to ensuring the IRs receive the appropriate priority for collection. The intelligence staff analyzes each requirement to determine its feasibility, whether or not it supports the commander's intent, and to determine the best method of satisfying the IRs. The staff also analyzes collected information to determine if it satisfies requirements.

4-37. During the produce function, the intelligence staff analyzes information from multiple sources to develop all-source intelligence products. The intelligence staff analyzes information and intelligence to ensure the focus, prioritization, and synchronization of the unit's intelligence production is in accordance with the PIRs.

4-38. In situation development, the intelligence staff analyzes information to determine its significance relative to predicted ECOAs and the CCIRs (PIRs and FFIRs). Through predictive analysis, the staff attempts to identify enemy activity or trends that represent opportunities or risks to the friendly force.

They use the indicators developed for each ECOA and CCIRs (PIRs and FFIRs) during the MDMP as the basis for their analysis and conclusions.

DISSEMINATE

4-39. Successful operations at the tactical and operational levels require an increased ability to synchronize fires, have faster access to intelligence, and enhance situational understanding and effective FP. Timely and accurate dissemination of intelligence is key to the success of these and other operations. Commanders must receive combat information and intelligence products in time and in an appropriate format to support decisionmaking. Additionally, sharing the most current all-source information and intelligence at all echelons is essential for commanders to maintain situational understanding.

4-40. To achieve this, it is imperative that the commander and staff establish and support a seamless intelligence architecture—including an effective dissemination plan—across all echelons to ensure information and intelligence flow in a timely manner to all those who need them. Intelligence and communications systems continue to evolve in their sophistication, application of technology, and accessibility to the commander. Their increasing capabilities also create an unprecedented volume of information available to commanders at all echelons. Finally, the commander and staff must have a basic understanding of these systems and how they contribute to the Intelligence BOS. A dissemination plan can be a separate product, or integrated into existing products such as the ISR synchronization plan or ISM, the decision support template (DST), or decision support matrix (DSM).

Dissemination Procedures

4-41. The G2/S2 and intelligence personnel at all levels assess the dissemination of intelligence and intelligence products.

4-42. Disseminating intelligence simultaneously to multiple recipients is one of the most effective, efficient, and timely methods. This can be accomplished through various means; for example, push, broadcast. However, within the current tactical intelligence architecture, reports and other intelligence products move along specific channels. The staff helps streamline information distribution within these channels by ensuring dissemination of the right information in a timely manner to the right person or element. There are three channels through which commanders and their staffs communicate: command, staff, and technical.

4-43. **Command Channel**. The command channel is the direct chain-of-command link that commanders, or authorized staff officers, use for command-related activities. Command channels include command radio nets (CRNs), video teleconferences (VTC), and the Maneuver Control System (MCS).

4-44. **Staff Channel**. The staff channel is the staff-to-staff link within and between headquarters. The staff uses the staff channel for control-related activities. Through the staff channel, the staff coordinates and transmits intelligence, controlling instructions, planning information, provides early warning information, and other information to support C2. Examples of staff channels include the operations and intelligence radio net, telephone, the staff huddle, VTC, and the BOS-specific components of the Army Battle Command System (ABCS).

4-45. **Technical Channel.** Staffs typically use technical channels to control specific combat, CS, and CSS activities. These activities include fire direction and the technical support and sensitive compartmented information (SCI) reporting channels of intelligence and ISR operations. The SIGINT tasking and reporting radio net, intelligence broadcast communications, and the wide area networks (WANs) supporting single intelligence discipline collection, processing, and production are examples of technical channels.

Presentation Techniques and Procedures

4-46. The staff's objective in presenting information is to provide the commander with relevant information. Table 4-1 lists the three general methods that the staff uses to present information and meet its information objective. Systems within the ABCS contain standard report formats, maps, and mapping tools that assist the staff in presenting information in written, verbal, and graphic form. Audio and video systems such as large format displays and teleconferencing systems enable the staff to use a combination of the methods in multimedia presentations.

Table 4-1. Presentation Methods and Products.

Method	Products
Written Narrative	Reports, Estimates, and Studies
Verbal Narrative	Briefing (information, decision, mission, and staff)
Graphic	Charts, Overlays, and Electronic Displays

Intelligence Communications Architecture

4-47. The intelligence communications architecture transmits intelligence and information to and from various ISR elements, units, and agencies by means of automation and communication systems. With the continued development of sensors, processors, and communications systems, it is increasingly important to understand the requirements of establishing an effective communications architecture. The G2/S2 must identify the Intelligence BOS specific requirements of the unit's overall communications architecture. Refer to FM 2-33.5 for more information on intelligence communications reach. The following are some (but not all) of the questions which the staff must answer in order to establish the intelligence communications architecture:

- Where are the unit's collectors?
- What and where are the unit's processors?
- Where are the unit's intelligence production elements?
- Where are the unit's decisionmakers?
- How does the unit disseminate information from its producers to its decisionmakers and/or consumers?

- Are the systems which the unit's collectors, producers, processors, and consumers use compatible with each other? If not, what is the plan to overcome this challenge?
- How can the unit access databases and information from higher and other agencies?

ASSESS

4-48. Assessment is the continuous monitoring—throughout planning, preparation, and execution—of the current situation and progress of an operation, and the evaluation of it against criteria of success to make decisions and adjustments. Assessment plays an integral role in all aspects of the intelligence process. Assessing the situation and available information begins upon receipt of the mission and continues throughout the intelligence process. The continual assessment of intelligence operations and ISR assets, available information and intelligence, the various aspects of the battlefield environment, and the situation are critical to—

- Ensure the CCIRs (PIRs and FFIRs) are answered.
- Ensure intelligence requirements are met.
- Redirect collection assets to support changing requirements.
- Ensure operations run effectively and efficiently.
- Ensure proper use of information and intelligence.
- Identify enemy efforts at deception and denial.

4-49. During planning, the intelligence staff conducts a quick initial assessment of the unit's intelligence posture and holdings, status of intelligence estimates, and any other available intelligence products. From this assessment the commander issues his initial guidance and a WARNO.

4-50. While the majority of the unit is engaged in preparation, the ISR effort should already have begun. It is during this period when the prepare and execute activities of the operations process overlap, that the G2/S2 assesses the current situation as well as the progress of ISR operations.

4-51. During execution the intelligence staff continues assessing the effectiveness of the ISR effort while at the same time assessing the results and products derived from the collection effort. The critical aspects of assessment at this point include determining whether the PIRs have been answered, will be answered with the current ISR operations, or which ISR operations to adjust in order to answer the CCIRs (PIRs and FFIRs). This type of assessment requires sound judgment and a thorough knowledge of friendly military operations, characteristics of the AO and AOI, and the threat situation, doctrine, patterns, and projected future COAs.

PART THREE

Military Intelligence Disciplines

Part Three provides a more detailed explanation of the intelligence disciplines introduced in Part One of this manual.

Chapter 5 defines and discusses the roles and fundamentals of the all-source intelligence discipline.

Chapter 6 defines and discusses the roles, fundamentals, and generic organization of the HUMINT discipline.

Chapter 7 defines and discusses the roles, fundamentals, and generic organization of the IMINT discipline.

Chapter 8 defines and discusses the roles, fundamentals, and generic organization of the SIGINT discipline.

Chapter 9 defines and discusses the roles, fundamentals, and generic organization of the MASINT intelligence discipline.

Chapter 10 defines and discusses the roles, fundamentals, and generic organization of the TECHINT discipline.

Chapter 11 defines and discusses the roles, fundamentals, and generic organization of the CI discipline.

Chapter 5

All-Source Intelligence

DEFINITION

5-1. All-source intelligence is defined as the intelligence products, organizations, and activities that incorporate all sources of information and intelligence, including open-source information, in the production of intelligence. All-source intelligence is a separate intelligence discipline, as well as the name of the function used to produce intelligence from multiple intelligence or information sources.

ROLE

5-2. The operational environment provides an ever-growing volume of data and information available from numerous sources, from which the commander can use to achieve situational understanding. His situational understanding enables him to make decisions in order to influence the outcome of the operation; prioritize and allocate resources; assess and take risks; and understand the needs of the higher and subordinate commanders. The commander depends upon a skilled G2/S2 working within his intent to support his ISR effort and provide all-source intelligence analysis conclusions and projections of future conditions or events.

FUNDAMENTALS

5-3. All-source intelligence production satisfies intelligence requirements. It provides an overall picture of the adversary and the battlespace. It reduces the possibility of error, bias, and misinformation through the use of multiple sources of information and intelligence.

PLAN

5-4. The utilization of all-source intelligence facilitates the development of accurate and concise contingency-specific plans, orders, and intelligence products. Additionally, the G2/S2 must retrieve, update, or develop any required intelligence databases. The most important all-source products are the modified combined obstacle overlay (MCOO), event templates, ECOA sketches, and the HPT list provided during initial IPB in support of the plan function. Through all-source intelligence, the commander can make informed decisions about COAs and adversary capabilities.

5-5. The intelligence staff coordinates its efforts with other elements such as the engineer terrain team, the unit surgeon, the Air Force weather team, and other assets or elements that can support the analytical effort.

5-6. IPB plays a significant role in the planning phase of the intelligence process. It is a systematic, continuous process of analyzing the threat and environment and is designed to support staff estimates and the MDMP. The IPB is led by the G2/S2, with participation by the entire staff. IPB allows the commander and staff to make informed decisions, develop COAs, and focus ISR efforts where they are most needed. IPB continuously assists the commander and staff in focusing ISR assets on the appropriate targets at critical points in time and space.

5-7. The time available for the IPB process may not permit the luxury of conducting each step in detail. Overcoming time limitations requires the commander to identify those parts of IPB that are most important to the commander in planning and executing his mission. Applying the specific steps or degree of detail performed varies according to METT-TC. There are four steps in the IPB process: define the battlefield environment, describe the battlefield's effects, evaluate the threat, and determine ECOAs.

Define the Battlefield Environment

5-8. In this step, the G2/S2—

- Identifies characteristics of the battlefield that will influence friendly and threat operations including terrain (mobility), weather, hydrological data, infrastructure, and civilian demographics. Produces accurate, timely, and predictive IPB products that depict the aspects of the battlefield.

- Identifies the limits of the command's AO and AOI.
- Identifies gaps in current intelligence holdings, identifies IRs, and recommends CCIRs (PIRs and FFIRs).

5-9. Defining the battlefield environment includes identifying enemy forces (their location, mobility, general capabilities, and weapon ranges) and all other aspects of the environment that could have an effect on the unit's ability to accomplish the mission. Depending on the situation, these considerations may also include—

- Geography, terrain, and weather of the area.
- Information environment to include but not limited to computer and communications systems and capabilities, data acquisition systems and capabilities, media access and distribution, areas prone to electromagnetic interferences, and systems to generate electromagnetic interference.
- Population demographics (ethnic groups, religious sects, age distribution, health status, income groups).
- Medical threat to include but not limited to endemic and epidemic diseases, occupational and environmental health hazards, and poisonous and toxic plants and animals.
- Political or socio-economic factors, including factions, clans, and gangs.
- Infrastructures, such as transportation or telecommunications, and critical decisionmaking infrastructures and supporting information systems.
- ROE or legal restrictions, such as international treaties or agreements.

Describe the Battlefield's Effects

5-10. This IPB step deals with the effects of the battlefield environment on the current operations and potential enemy and friendly COAs. It begins with the assessment of existing and projected conditions of the battlefield environment, which the staff accomplishes through terrain analysis, weather analysis, and analysis of other characteristics of the battlefield. To conduct terrain analysis, the staff uses maps, reconnaissance, and other specialized terrain products (maps, overlays, databases, software). These products address such factors as wet or dry cross-country mobility, transportation systems (road and bridge information), vegetation type and distribution, surface drainage and configuration, surface materials (soils), ground water, manmade structures, and obstacles. The results of evaluating the terrain's effects should be expressed by identifying areas of the battlefield that influence each COA.

Evaluate the Threat

5-11. Evaluating the threat involves determining the threat force capabilities and the doctrinal principles and TTP that threat forces prefer to employ. The result of this evaluation produces a threat model, which portrays how threat forces normally execute operations and how they have reacted to similar situations in the past. The threat model includes an evaluation of the threat's strengths, weaknesses, and vulnerabilities, including an evaluation of typical HVTs.

Determine ECOAs

5-12. Determining ECOAs involves the identification and development of likely ECOAs that will influence accomplishing the friendly mission. Developing ECOAs is a form of predictive intelligence analysis and production. The procedures for this step include—

- Identifying the enemy's likely objectives and predicting the desired end-state.
- Identifying the full set of COAs available to the enemy.
- Predicting the enemy's most likely and most dangerous COAs.
- Evaluating and prioritizing each COA.
- Developing each COA in the amount of detail that time permits.
- Identifying initial ISR and collection requirements.

5-13. After conducting the initial IPB, the staff, primarily the G2/S2, identifies gaps in the available intelligence, develops the initial PIRs and IRs, and develops the initial input to the ISR plan based on the commander's guidance.

5-14. Indicators are the basis for situation development. Indicators are activities that will confirm or deny the event specified in an intelligence requirement. They are any positive or negative evidence of enemy activity or characteristic of the AO that points toward enemy capabilities, vulnerabilities, or intentions. Individual indicators cannot stand alone. Each indicator is integrated with other factors and indicators before patterns are detected and enemy intentions established. Indicators are developed by the analysts in the G2/S2 section. The event matrix shows the threat activities, or indicators, to look for in each NAI, and the timelines during which each NAI should be active. All indicators are developed to answer the commander's PIRs and IRs. The analyst uses indicators to correlate particular events or activities with probable ECOAs and to determine what events or activities must occur for an enemy to follow a particular COA. The ability to read indicators (including recognition of enemy deception indicators) contributes to the success of friendly operations. The analyst integrates information from all sources to confirm indicators of enemy activities. As indicators are detected and confirmed, PIRs are answered.

PREPARE

Activities

5-15. All-source activities during the prepare function include—

- Conducting rehearsals.
- Conducting communications rehearsals and verifiying communications protocols with higher, lateral, and subordinate units.
- Planning and practicing actions that support likely contingencies, branches, or sequels.
- Reviewing and updating available databases and IPB products.
- Ensuring control and coordination measures are still in effect.

- Updating the force with the most recent intelligence before mission execution.

Intelligence Estimates

5-16. The intelligence estimate is a continuous process that is the product of all actions the intelligence staff performs throughout the MDMP. The intelligence estimate provides a timely and accurate evaluation of the enemy and the AO (and often the AOIR) at a given time. It provides the background the G2/S2 uses to portray enemy actions during COA analysis.

5-17. The G2/S2 must clearly understand the weather and terrain effects and the ability to visualize the battlespace before producing the intelligence estimate. This understanding facilitates accurate assessments and projections regarding the enemy: enemy situation (including strengths and weaknesses), enemy capabilities and an analysis of those capabilities (COAs available to the enemy), and conclusions drawn from that analysis. The intelligence estimate's conclusion identifies the enemy's most likely COA and most dangerous COA. ECOAs must include a sketch of the COA with a task and purpose for the enemy's actions. Any gaps in the intelligence estimate are identified as information requirements.

5-18. The intelligence estimate may be written or oral. At the tactical level, especially during operations and exercises, intelligence estimates are usually delivered orally, supported by graphic displays and other decision support tools. During contingency planning, especially at corps level and above, estimates are usually written. During deliberate planning at joint headquarters, estimates are always written (see JP 5-00.2). However, the intelligence estimate should always be prepared as thoroughly as time and circumstances permit. A comprehensive intelligence estimate considers both the tangible (quantifiable) and the intangible aspects of the enemy's operations. It translates enemy strengths, weapon systems, training, morale, and leadership into combat capabilities and projections of future enemy actions.

5-19. Different sections of the intelligence estimate receive more emphasis during different activities within the operations process.

- During planning, the most important decision the commander makes is selecting a COA on which to base the plan. Thus, the sections of the intelligence estimate that focus on the commander's selecting the most appropriate COA are the most important at that time.
- During preparation, the intelligence estimate must focus on the commander's decisions that affect the ability of the unit to execute the upcoming operation. The intelligence estimate—and functions—that supports these decisions must focus on answering the CCIRs (PIRs and FFIRs) and guiding the ISR effort.
- During execution, the intelligence estimate focus is to support the anticipated command decisions. The most important action of the intelligence estimate—and the ISR effort— is to answer the PIR. However, it is also during the execution phase that the intelligence estimate must look ahead: anticipating branches or sequels to the

current operation, transition to other operations, changes in the ECOAs, and required adjustments to the ISR effort.

5-20. Refer to FM 2-01 for examples of intelligence estimates.

COLLECT

5-21. The collect intelligence process function within the all-source intelligence discipline is limited to gathering the necessary information, intelligence, and intelligence products required to perform analysis. Thus, all-source intelligence depends upon the other intelligence disciplines to perform collection. This is also the primary reason why OSINT is defined as a source of information and intelligence and not a separate intelligence discipline.

PROCESS

5-22. The process function converts relevant information into a form suitable for analysis, production, or immediate use by the commander. Processing also includes sorting through large amounts of collected information and intelligence to identify and exploit the information which is pertinent to the commander's intelligence requirements and facilitates situational understanding.

PRODUCE

5-23. All-source intelligence analysis and production—

- Drives collection to answer the PIR.
- Provides the enemy situation.
- Provides INTSUMs and other intelligence reports.
- Supports situational understanding.
- Provides predictive estimates of enemy actions; specifically, ECOAs.
- Provides continuously updated IPB.
- Provides all-source target packages (or folders).

5-24. As previously stated, the intelligence estimate is a continuous process. The commander and staff constantly collect, process, store, display, and disseminate information. Staff members update their estimates as they receive new information, such as—

- When they recognize new facts.
- When they replace assumptions with facts or find their assumptions invalid.
- When they receive changes to the mission or when changes are indicated.

5-25. Technological advances and NRT information allow estimates to be continuously updated; the running estimate and its intelligence component, the intelligence running estimate, exemplify this.

Running Estimate

5-26. Running estimates are continuously updated estimates based on new information as the operation proceeds. They serve as a staff technique to support the commander's visualization and decisionmaking, as well as the staff's tool for assessing during preparation and execution. In the running estimate, staff officers continuously update their conclusions and recommendations as they evaluate the impact of new facts. Current doctrine emphasizes the COP as the primary tool that provides the current situation and, when merged with the running estimates, facilitates commanders achieving situational understanding. All staff sections (BOSs) provide their respective input to the COP. The COP as used today, combined with the running estimate, is predictive and enhances our current ability to collect, process, store, display, and disseminate information.

5-27. The portion of the COP that depicts the enemy situation is currently limited to displaying the locations and dispositions of enemy and threat forces in a relatively static manner, sometimes referred to as snapshots in time. The enemy situation portion of the COP requires analysis to provide the required level of detail. The COP and running estimates sufficiently provide effective support for battle command of knowledge-based organizations.

5-28. The main differentiation between the running estimate and the old staff estimates is the emphasis on not only continuously updating the facts of the estimate but also continuously updating the conclusions and recommendations while including projections of future conditions of the entire battlespace. While the COP is primarily a display of current intelligence and information, the running estimate requires the merging of the staff's cognitive processes with automation applications. The primary focus of the staff's cognitive process is to present predictive or anticipatory intelligence in support of the commander's decisionmaking or situational understanding. See Figure 5-1.

5-29. The running estimate is a product of the entire battle staff. Just as all BOSs contribute to the COP, they will also contribute their portion of the running estimate. The running estimate integrates the running estimates from each BOS. The Intelligence BOS input to the running estimate is the intelligence running estimate.

Intelligence Running Estimate

5-30. The intelligence running estimate is a continuous flow and presentation of relevant information and predictive intelligence that, when combined with the other staff running estimates, enables the decisionmaker's visualization and situational understanding of the AOI in order to achieve information superiority. The intelligence running estimate requires constant verification to support situational understanding of the current situation as well as predictive assessments for future operations.

5-31. While it is possible to employ the concept of the intelligence running estimate today, the true seamless and continuous update feature will be achieved with technological enablers that are not yet present in the force. In the future, technology should allow the information from the running estimate to become implanted as a part of the COP. However, there will always be a place for some separate context, orally or written, separate from the COP display to add fidelity and assist in the commander's visualization and decisionmaking. Additional training must be implemented not just within MI but also throughout the force in order to institute the running estimate concept.

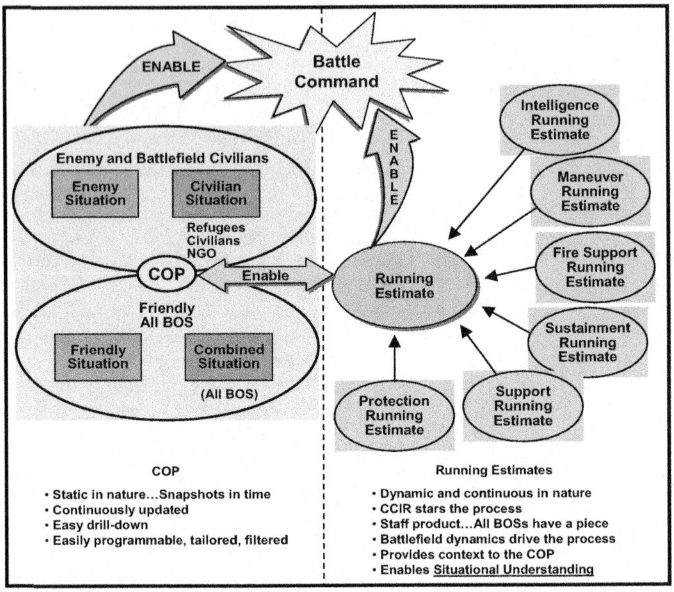

Figure 5-1. Example of the COP and a Running Estimate.

ANALYZE

5-32. Analysis is key in converting combat information and all intelligence from each discipline into all-source intelligence products and targeting information. From the multiple sources of information received, the staff analyzes and identifies critical information; determines the relevancy and accuracy of the information; and reaches conclusions about the information. These conclusions are either immediately disseminated or are used to form the basis of intelligence analysis products.

DISSEMINATE

5-33. The intelligence staff extracts and disseminates pertinent intelligence from products developed as a result of all-source analysis. The intelligence staff—

- Rapidly disseminates time-sensitive, all-source analysis intelligence and intelligence products to higher, lateral, and subordinate commands in order to keep all commanders abreast of current developments in the situation and battlespace in accordance with unit SOPs.

- Disseminates all-source analysis intelligence to other G2/S2 intelligence activities for additional processing or detailed analysis and exploitation in accordance with unit SOPs.

5-34. For intelligence reach operations all-source intelligence products are available and disseminated in a variety of forms. It is incumbent on the requestor to ensure that the all-source product can be transmitted over the available communications systems. This includes verifying the appropriate security level of the communication system.

ASSESS

5-35. The commander, the intelligence staff, and the intelligence consumers assess the analysis, production, and dissemination of all-source intelligence. The staff identifies intelligence gaps, emerging operational requirements, duplication of effort, or new targets or threats and redirects all-source activities as appropriate to meet intelligence requirements. Throughout the assessment process, the G2/S2, the intelligence staff, and intelligence users should be alert for evidence of possible enemy denial and deception efforts. The staff assesses intelligence and information for—

- Consistency with the current situation and threat trends.
- Accuracy and confirmation by other intelligence sources or disciplines.
- Source reliability and credibility.
- Pertinence to PIRs, intelligence requirements, RFIs, and other intelligence tasks or requests.

Chapter 6

Human Intelligence

DEFINITION

6-1. HUMINT is the collection by a trained HUMINT Collector of foreign information from people and multimedia to identify elements, intentions, composition, strength, dispositions, tactics, equipment, personnel, and capabilities. It uses human sources as a tool and a variety of collection methods, both passively and actively, to gather information to satisfy the commander's intelligence requirements and cross-cue other intelligence disciplines.

ROLE

6-2. The role of HUMINT Collectors is to gather foreign information from people and multiple media sources to identify adversary elements, intentions, composition, strength, dispositions, tactics, equipment, personnel, and capabilities. It uses human sources and a variety of collection methods to gather information to satisfy the commander's intelligence requirements, and cross-cue other intelligence disciplines.

HUMINT FUNCTIONS

6-3. HUMINT functions are interrelated, mutually supporting, and can be derived from one another. No single function or technical capability can provide a full understanding of our adversaries. HUMINT functions are defined below and shown in Figure 6-1.

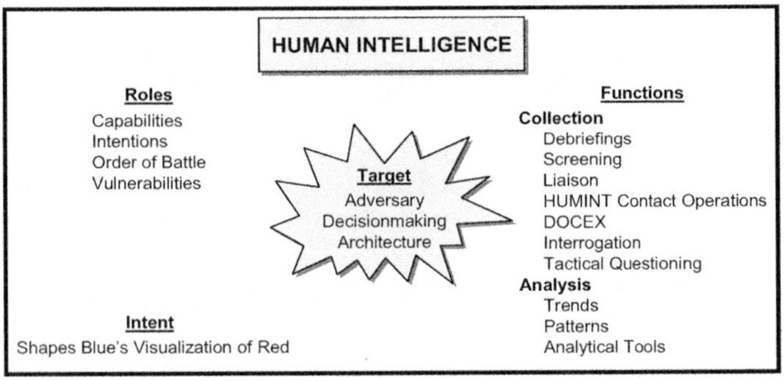

Figure 6-1. HUMINT Roles and Functions.

COLLECTION

6-4. HUMINT collection methods and operations include—

- **Debriefings.** Debriefings are the systematic questioning of individuals to procure information to answer specific collection requirements by direct and indirect questioning techniques. Sources for debriefings are categorized as friendly forces, US and non-US civilians to include refugees, displaced persons, and local inhabitants.
- **Screening.** Screening is the process of evaluating and selecting human and media sources for the prioritized collection of information in support of commander's PIRs and IRs. Screening categorizes and prioritizes sources based on the probability of a particular source having priority information and the degree of difficulty of extracting that information from the source. While screening is not in itself an information collection technique, it is vital to the rapid collection of information.
- **Liaison.** HUMINT elements conduct liaison with US, multinational, and HN military and civilian agencies, to include NGOs. Liaison is conducted to obtain information of interest and to coordinate or deconflict HUMINT activities.
- **HUMINT Contact Operations.** HUMINT contact operations are tactically oriented, overt collection activities that use human sources to identify attitude, intentions, composition, strength, dispositions, tactics, equipment, target development, personnel, and capabilities of those elements that pose a potential or actual threat to US and multinational forces. These forces provide early warning of imminent danger to US and multinational forces and contribute to the MDMP.
- **Document Exploitation (DOCEX).** DOCEX is the systematic extraction of information from all media formats in response to collection requirements.
- **Interrogation.** Interrogation is the systematic effort to procure information to answer specific collection requirements by direct and indirect questioning techniques of persons who are in the custody of the forces conducting the questioning.

UNIT SUPPORT TO HUMINT COLLECTION

6-5. Small units contribute to HUMINT collection through a number of different ways.

Tactical Questioning

6-6. Tactical questioning can provide critical information for situational understanding. Tactical questioning is the expedient initial questioning for information of immediate tactical value. Soldiers conduct tactical questioning based on the unit's SOP, ROE, and the order for that mission. Unit leaders must include specific guidance for tactical questioning in the order for appropriate missions. The unit S3 and S2 must also provide specific guidance down to the unit level to help guide tactical questioning. For a more detailed discussion of tactical questioning, see ST 2-91.6.

Combat Patrolling

6-7. Through observing and interacting with the local environment during the conduct of missions, handling EPWs/detainees, and handling captured documents, soldiers serve as the commander's "eyes and ears" whether—

- Performing traditional offensive or defensive operations.
- Performing a patrol in a stability operation.
- Manning a TCP or a roadblock in a support operation.
- Occupying an observation post.
- Passing through an area in a convoy.
- Performing any operation that involves observing and reporting on elements of the environment and activities in the AO.

ISR Operations

6-8. The information that the soldier reports as a result of completed missions up the chain of command forms a vital part of planning and operations. Careful and quick handling of EPWs/detainees and documents also helps the ISR effort. Unit headquarters must ensure that the reports they receive and forward contain all the subtle detail provided by the small unit. For tactical operations, there are four levels of reporting:

- Immediate reporting of information of critical tactical value, based either on predetermined criteria or common sense.
- Normal reporting, submitted before the unit S2 section performs the debriefing.
- Debriefing performed by the S2 section.
- Follow-up reporting, submitted after the unit S2 section performs the debriefing.

6-9. The four levels of reporting facilitate the unit S2 section capturing all the fine details of the small unit's activities for all-source analysis and future planning. The unit S2 must ensure that information of HUMINT and CI value is reported to the G2X.

ANALYSIS

6-10. IPB, all-source, and single-source analysis are used to template adversarial OB. HUMINT elements analyze operational taskings to determine the best collection methods to employ in satisfying CCIRs (PIRs and FFIRs) and IRs. As information is collected, it is analyzed for completeness and accuracy to identify significant facts for subsequent interpretation and inclusion in intelligence products, and passing to other collectors. This analysis also helps to identify collection gaps and focus or refocus collection efforts. Analysis provides the commander with situational understanding of the battlespace to execute operations. Raw information, open-source material, and finished intelligence products are analyzed in response to local and national requirements. Analysis occurs at all levels from tactical to strategic.

- At the tactical level, HUMINT teams focus their efforts on supporting mission requirements and contributing to the all-source COP.

- Operational analysis is used to assess adversary intentions, capabilities, dispositions, and their regional impacts.
- Strategic analysis supports national and Army programs through compilation of local and regional analysis into global assessments; identifies technology development trends and patterns in military activities and capabilities.

6-11. HUMINT products consist of, but are not limited to, target nomination, input to threat and vulnerability assessments, intelligence estimates, and intelligence information reports. Finalized intelligence derived from HUMINT activities is incorporated into joint and national intelligence databases, assessments, and analysis products. HUMINT products are also incorporated into the COP to support situational awareness. HUMINT production takes place at all levels.

- Operational and tactical production includes tactical alerts, spot reports, and current intelligence; input to threat and/or vulnerability assessments tailored to specific activities, units, installations, programs or geographic areas, and target studies to support contingency planning and major exercises; studies of military activities and capabilities.
- Strategic products include assessments supporting national and Army information requirements on foreign technology development; worldwide assessments of the organization, location, funding, training, operations capabilities and intentions of terrorist organizations; analyses of the capabilities of international narcotics trafficking organizations.

OPERATIONAL EMPLOYMENT

6-12. Army HUMINT must support the full spectrum of military operations. Commanders require a well-trained HUMINT force consisting of AC, RC, civilian government, and contractor personnel. The HUMINT force is focused on and dedicated to the collection of data and information relevant to the commander's PIRs and IRs. Effective employment of Army HUMINT elements in all phases of operations and at all levels from tactical to strategic will remain paramount to ensuring that commanders have the best pictures possible of their adversaries. HUMINT Collectors use the following collection methods.

LEVELS OF EMPLOYMENT

6-13. **Strategic and Departmental.** Strategic and departmental operations will be conducted by HUMINT elements supporting national, DOD, and DA required missions (for example, support to NATO and special operations and missions). The Army strategic HUMINT force resides in DHS. Trained Army HUMINT professionals must be available to augment DOD HUMINT efforts and coordinate between Army and Joint or DOD elements during crisis and war. Strategic HUMINT activities are generally carried out by departmental level HUMINT elements or Army HUMINT assets directly assigned to DOD and Joint positions. The Army overt HUMINT capability at the strategic level will be deployed and under operational control (OPCON) of the Army Service Component Commander (ASCC).

6-14. **Operational.** HUMINT operational missions must support combatant commanders, generally in geographic AORs. Operational HUMINT assets provide capabilities to support theater requirements for HUMINT collection. Operational HUMINT elements will focus on threat identification and capabilities. As with all levels of employment, HUMINT activities and functions will include DOCEX, collection, operations, analysis, and production. HUMINT elements must be capable of quickly transitioning from a peacetime mission to crisis operations to support combatant commander requirements. Operational elements may also be deployed to support or reinforce tactical forces in CONOPS.

6-15. **Tactical.** Army HUMINT play a crucial role in supporting tactical forces. HUMINT teams conduct operations throughout the battlespace during CONOPS. CONOPS support activities include debriefings, screenings, contact operations, and interrogations. HUMINT activities in CONOPS focus on the threat and assisting the senior intelligence officer (SIO) and commander in understanding the threat's decisionmaking process. During peacetime, organic tactical HUMINT teams conduct activities pursuant with approved regulations and command guidance.

SUPPORT TO CONTINGENCY OPERATIONS

6-16. **Initial Phase.** The initial phase of operations from peacetime military engagement (PME) to major theater war (MTW) lays the foundation of future team operations. In general, the priority of effort focuses inward on security of operating bases, areas of troop concentration, and C2 nodes to identify the collection threat to US forces that could be used by adversarial elements to plan hostile acts against US activities and locations.

6-17. **Continuation Phase.** Once security of the operating bases has been established, the operational focus of HUMINT teams shifts outside the operating base to continue to detect, identify, and neutralize the collection threat to US forces as well as to provide I&W of hostile acts targeting US activities. The HUMINT team uses several collection methods, to include HUMINT contact operations, elicitation, and liaison, to answer the supported commander's requirements.

6-18. **HUMINT Team.** A key element to the HUMINT team's success is the opportunity to spot, assess, and develop relationships with potential sources of information. Operating as independent teams, without being tied to combat assets, enables the HUMINT team's maximum interaction with the local population, thereby maximizing the pool of potential sources of information. A second key element of a HUMINT team's success is its approachability to the local population. A soft posture enables a HUMINT team to appear as non-threatening as possible. Experience has shown that the local population in general is apprehensive of fully and openly armed patrols and soldiers moving around population centers. During some operations civilian attire or non-tactical vehicles may be used to lower the

HUMINT team profile. NOTE: In some special situations, these measures are taken to make the operation less visible to the casual observer. Also, in some cultures, sharing food and beverages among friends is expected; exceptions to restrictions or general orders should be considered to facilitate successful HUMINT team operations, many of which are geared towards developing relationships with potential sources of information.

SUPPORT TO INSTALLATIONS AND OPERATING BASES

6-19. Commanders may restrict personnel to base camps and installations during the initial stages of operations (PME to MTW), when the operational environment is being assessed, or as a temporary expedient when the threat level exceeds the ability to provide reasonable FP. Operational restrictions minimize the risk to the HUMINT team, but minimizing its collection potential may increase the risk to the force as a whole. While confined to an installation or a base camp, the HUMINT team can maintain a limited level of information collection by—

- Screening locally employed personnel.
- Debriefing combat and ISR patrols.
- Debriefing friendly force personnel who are in contact with the local population.
- Conducting limited local open-source information collection.
- Contributing to the threat and vulnerability assessments of the base camp.

TACTICS, TECHNIQUES, AND PROCEDURES

6-20. At the HUMINT team level, team members conduct mission analysis and planning specific to their AO. Backwards planning and source profiling are used extensively to choose HUMINT targets. To verify adequate area coverage, the HUMINT team may periodically develop and use HUMINT target overlays and other HUMINT analytical tools that illustrate the HUMINT situation, identify gaps, and help refocus the collection effort.

6-21. The HUMINT team is also in constant contact with the supported S2 and the other ISR assets (Scouts, PSYOP, CA, and MPs) in order to coordinate and deconflict operations and to cross-check collected information. The supported unit S2, with the help of the HUMINT team, regularly and systematically debriefs all ISR assets.

6-22. The HUMINT team must be integrated into the supported unit's ISR plan. The HUMINT operational management team (OMT) chief will advise the supported unit on the specific capabilities and requirements of the team to maximize mission success.

OPERATIONAL RISK MITIGATION

6-23. The employment of HUMINT teams includes varying degrees of contact with the local population. As the degree of contact with the population increases, both the quantity and quality of collection increases. In many instances, however, there is a risk to the team inherent with increased exposure to the local population. The decision at what level to employ a team is METT-TC dependent. The risk to HUMINT assets must be balanced with the need to collect priority information and to protect the force as a whole. ROE, SOFA, direction from higher headquarters, and the overall threat level may also restrict the deployment and use of HUMINT teams. The commander should consider exceptions to the ROE to facilitate HUMINT collection.

6-24. Risks are minimized through the situational awareness of HUMINT team members. They plan and rehearse to readily react to any situation and carry the necessary firepower to disengage from difficult situations. If it becomes necessary to call for assistance, adequate and redundant communications equipment is critical. These scenarios and actions should be trained prior to deployment into a contingency area and rehearsed continuously throughout the deployment.

6-25. A supported unit commander is often tempted to keep the HUMINT team "inside the wire" when the threat condition (THREATCON) level increases. The supported commander must weigh the risk versus potential information gain when establishing operational parameters of supporting HUMINT teams. This is necessary especially during high THREATCON levels when the supported unit commander needs as complete a picture as possible of the threat arrayed against US and multinational forces.

6-26. When it is not expedient to deploy the HUMINT team independently due to threat levels or other restrictions, the team can be integrated into other ongoing operations. The HUMINT team may be employed as part of a combat, ISR, or MP patrol or used to support CA, PSYOP, engineer, or other operations. This method reduces the risk to the team while allowing a limited ability to collect information. It has the advantage of placing the team in contact with the local population and allowing it to spot, assess, and interact with potential sources of information. However, this deployment method restricts collection by subordinating the team's efforts to the requirements, locations, and timetables of the unit or operation into which it is integrated and does not allow for the conduct of sensitive source operations. **This method of employment should be considered a last resort.**

HUMINT EQUIPMENT

6-27. Basic C2, transportation, and weapons requirements do not differ significantly from most soldier requirements and are available as unit issue items. However, HUMINT teams have unique communications, collection, processing, and mission-specific requirements.

COMMUNICATIONS

6-28. Dedicated and secure long-range communications are key to the success of the HUMINT team mission. HUMINT team operations require a secure, three-tiered communications architecture consisting of inter/intra-team radios, vehicle-based communications, and a CI and HUMINT base station.

6-29. The HUMINT team must have access to existing communications networks such as the tactical local area network (LAN). The HUMINT team must also be equipped with its own COMSEC devices. It is imperative that the HUMINT team acquire access to the public communication system of the HN. This can be in the form of either landlines or cellular telephones. Such access enables the HUMINT team to develop leads which can provide early indicators to US forces.

Interoperability

6-30. Communications systems must be equipped with an open-ended architecture to allow for expansion and compatibility with other service elements, government organizations, NGOs, and multinational elements to effectively communicate during CONOPS. All ISR systems must be vertically and horizontally integrated to be compatible across all BOSs and with Legacy and Interim Force elements.

Satellite Communications On-The-Move (SOTM)

6-31. To provide real time or NRT information reporting, HUMINT elements must have the capability to transmit voice, data, imagery, and video while on the move. HUMINT teams must be able to transmit while geographically separated from their parent unit while operating remotely. This broadband requirement can only be achieved through a SATCOM capability and must be achievable while mobile.

HUMINT COLLECTION AND PROCESSING SYSTEMS

6-32. The HUMINT team must rely on automation to achieve and maintain information dominance in a given operation. With time, effective collection planning and management at all echelons, the HUMINT team can collect a wealth of information. The sorting and analysis of this information in a timely and efficient manner is crucial to operations. Automation helps the HUMINT team to report, database, analyze, and evaluate the collected information quickly and to provide the supported unit with accurate data in the form of timely, relevant, accurate, and predictive intelligence.

6-33. Automation hardware and software must be user friendly as well as interoperable among different echelons and services. They must interface with the communications equipment of the HUMINT team as well as facilitate the interface of audiovisual devices. Technical support for hardware and software must be available and responsive.

6-34. The demand for accurate and timely HUMINT reporting, DOCEX, and open-source information has grown tremendously. Biometric (physiological, neurological, thermal analysis, facial and fingerprint recognition) technologies will allow rapid identification; coding and tracking of adversaries and human sources; and cataloging of information concerning enemy prisoners of war (EPWs), detainees, and civilians of HUMINT interest on the battlefield. Biometrics will also provide secure authentication of individuals seeking network or facility access.

6-35. HUMINT teams work with multinational forces, and other foreign nationals, and require the ability to communicate in their respective languages. Often HUMINT personnel have little or no training in the target language, and lack of skilled interpreters can hinder HUMINT activities. HUMINT teams require textual and voice translation devices, source verification, and deception detection machines (biometrics) to improve collection capability and accuracy.

6-36. HUMINT teams require dynamic machine language translation (MLT) tools that provide both non-linguists and those with limited linguist skills a comprehensive, accurate means to conduct initial screenings and basic interviews in a variety of situations. HUMINT elements will focus on in-depth interviews and communications with persons of higher priority. MLT tools will minimize reliance on contract linguists and will allow soldiers to concentrate on mission accomplishment.

MISSION SPECIFIC

6-37. The HUMINT team may conduct night operations and must be equipped with NVDs for its members and photographic and weapons systems. The HUMINT team also may operate in urban and rural areas, where the threat level can vary from semi-hostile to hostile. The safety of the HUMINT team can be enhanced with equipment that can detect, locate, suppress, illuminate, and designate hostile optical and E-O devices. In addition, high power, gyro-stabilized binoculars, which can be used from a moving vehicle, also increase the survivability of the HUMINT team. It also gives the team another surveillance and collection device.

6-38. Some of the HUMINT team missions may require the documentation of incidents. The teams can use the following equipment in their open-source collection efforts.

- Small, rugged, battery-operated digital camcorders and cameras which are able to interface with the collection and processing systems as well as communication devices.
- GPSs that can be mounted and dismounted to move in the AO efficiently.
- Short-range multichannel RF scanning devices that can also identify frequencies which enhance their security.

6-39. In some cases HUMINT teams require a stand-off, high resolution optical surveillance and recording capability that can provide target identification at extended ranges to protect the intelligence collector while avoiding detection by the adversary target. An advanced optical capability provides intelligence collectors the ability to locate and track adversarial targets (passive and hostile) for identification, collection, and target exploitations.

INTEGRATION OF LINGUISTS

6-40. Integrating linguists into the HUMINT team should take place as soon as possible. Security clearances and contractual agreements will help the team determine the level of integration.

6-41. Along with the basic briefing of what is expected of the civilian linguists as interpreters, HUMINT teams should be informed about the civilians' chain of command and the scope of their duties beyond interpreting. The HUMINT team leader must ensure that linguists are trained and capable of completing all tasks expected of them.

BATTLE HAND-OFF

6-42. HUMINT teams are always engaged. A good battle hand-off is critical to smooth transition and mission success. The battle hand-off can directly contribute to mission success or failure of the outgoing team, but especially of the incoming team. The battle hand-off begins the first day the HUMINT team begins to operate in an AO. Regardless of how long the team believes it will operate within the AO, it must ensure there is a seamless transition to an incoming team, other US unit, or agency. The HUMINT team accomplishes this transition by establishing procedures for source administration, database maintenance, and report files.

6-43. Teams must plan and implement a logical and systematic sequence of tasks to enable an incoming team to assume the operations in the AO. Adequate time must be allotted for an effective battle hand-off. In some environments, a few weeks may be necessary to accomplish an effective battle hand-off. Introductions to sources of information, especially HUMINT contact operations sources, are critical and teams must prioritize their time. During this time the outgoing HUMINT team must familiarize the new HUMINT team with all aspects of the operation, which includes past, present, and planned activities within the AO. Area orientation is critical. These include major routes, population centers, potential hot spots, and other points of interest (such as police stations, political centers, and social centers).

ORGANIZATION

6-44. HUMINT activities require a complex C2 relationship to ensure that the requirements of the supported commander are fulfilled while balancing the need for strict integrity and legality of HUMINT operations. This complex relationship balances the role of the SIO as the requirements manager and the *2X as the mission manager with the MI commander as the asset manager. ("*2X" indicates 2X functions at all levels.)

COMMAND VERSUS CONTROL

6-45. ARFOR will normally deploy as part of a joint and/or multinational operation. In all cases, commanders at each echelon will exercise command over the forces assigned to their organization. Command includes the authority and responsibility for effectively using resources, planning for and employment of forces, and ensuring that forces accomplish assigned missions. Leaders and staffs exercise control to facilitate mission accomplishment.

6-46. While the MI commander supervises subordinates and produces reports, the *2X synchronizes activities between intelligence units and provides single-source processing and limited analysis. While the MI commander takes care of the operators executing missions, the *2X obtains the data and reports from higher echelons required to execute the missions.

STAFF RESPONSIBILITIES AND FUNCTIONS

*2X Staff

6-47. The *2X staff is responsible for the integration, correlation, and fusion of all Human Sensor information into the Intelligence BOS within the *2X

AOIR. The *2X is also responsible for analyzing adversary collection, terrorist and sabotage activities, developing countermeasures to defeat threat collection activities, identifying and submitting collection requirements to fill collection gaps, and providing input to the all-source picture regarding adversary intelligence activities.

*2X Staff Officer

6-48. The *2X Staff Officer provides CI and HUMINT collection expertise. The *2X—

- Is the single focal point for all matters associated with CI and HUMINT in the AOIR.
- Is the CI and HUMINT advisor to the G2 and commander.
- Is an extension of the collection manager and ensures that the best asset or combinations of assets are used to satisfy information requirements.
- Along with his subordinate elements (CICA, HOC, OSC, CI Analysis Cell [CIAC], and HAC), exercises technical control over his assigned Army CI and HUMINT elements in the designated AOIR.
- Is the principal representative of the G2 and the commander when coordinating and deconflicting CI and HUMINT activities with national or theater agencies operating in the AOIR.
- Supports specific RM efforts in conjunction with the requirements manager through—
  - Planning and coordinating CI and HUMINT operations.
  - Reviewing and validating HUMINT requirements.
  - Recommending assignment of tasks to specific collectors.
  - Conducting liaison with non-organic HUMINT collection. This liaison includes national level and multinational force assets for source deconfliction and special activities outside the *2X AOIR.
- Will provide OMTs with capability to reach back to current database information, technical information and guidance, and source deconfliction necessary to monitor the collection activities of the HUMINT teams.

The HUMINT Analysis Cell (HAC)

6-49. The HAC is the single-source fusion point for all HUMINT reporting and operational analysis. It determines gaps in reporting and coordinates with other analysis teams and technical controllers to cross-cue other collection sensor systems. The HAC—

- Uses analytical tools to develop long-term collection plans and provides reporting feedback that supports all HUMINT and CI entities in the supported command's AOIR.
- Produces and disseminates HUMINT products and provides input to intelligence summaries.
- Uses analytical tools found at the ACE or JISE to develop long-term analyses and provides reporting feedback that supports the J/G/S2X, HUMINT operations section, OMTs, and HUMINT teams.
- Produces country and regional studies tailored to HUMINT collection.

- Compiles target folders to assist J/G/S2X assets in focusing collection efforts.
- Analyzes and reports on trends and patterns found in HUMINT reporting.
- Analyzes source reliability and credibility as reflected in reporting and communicating that analysis to the collector.
- Develops and maintains databases specific to HUMINT collection activities that directly support the collection efforts of HUMINT teams and are directly accessible by HUMINT teams.
- Provides collection requirements input to the HOC.
- Supports RM through the development of HUMINT SIRs based on command PIRs.
- Answers HUMINT-related RFIs.

6-50. For intelligence reach operations, HUMINT products are available and disseminated in a variety of forms. It is incumbent on the requestor to ensure that the HUMINT product can be transmitted over the available communications systems. This includes verifying the appropriate security level of the communications systems.

HUMINT Operations Cell (HOC)

6-51. The HOC in the *2X coordinates and synchronizes all HUMINT activities in the AOIR. The HOC exercises technical control over all HUMINT entities in the designated AOIR and deconflicts HUMINT activities with higher, lower, and adjacent HUMINT elements. The HOC accomplishes all responsibilities through coordination with the operational units and the CICA and operations support cell (OSC). The HOC tracks all HUMINT activities in the AOR. The J/G2X uses this information to advise the SIO on all HUMINT activities conducted within the AO. The HOC—

- Exercises technical control of all HUMINT assets and coordinates and deconflicts HUMINT activities in the deployed AO.
- Establishes and maintains a HUMINT source database.
- Coordinates and supervises HUMINT FP source operations conducted by all services and components in the AO.
- Develops and manages collection requirements for HUMINT in coordination with the requirements manager.
- Develops and provides the HUMINT portion of the intelligence synchronization plan to the J/G/S2X and requirements manager for inclusion in the intelligence synchronization plan.
- Coordinates the activities of HUMINT collectors assigned or attached to interrogations and debriefing facilities.
- Expedites preparation of intelligence information reports and their distribution to consumers at all levels.
- Performs liaison with HN and US national HUMINT organizations.

Operations Support Cell

6-52. The OSC in the *2X staff maintains the source registry for all HUMINT activities in the designated AOIR. The OSC provides management of intelligence property book operations, source incentive programs, and ICFs.

6-53. Integrating linguists and DOD emergency-essential civilians, such as technical support contractors, into the HUMINT team should take place as soon as possible. Security clearances and contractual agreements will help the team determine the level of integration.

6-54. Along with the basic briefing of what is expected of the civilian linguists as interpreters and the emergency-essential civilians as support personnel, HUMINT teams should be informed about the civilians' chain of command and the scope of their duties beyond interpreting and technical support. The HUMINT team leader must ensure that linguists and emergency-essential civilians are trained and capable of completing all tasks expected of them.

Counterintelligence Coordinating Authority

6-55. The CICA coordinates all CI activities for a deployed force. There can be only one CICA in a theater of operations. When multiple echelons exist, the highest echelon has the CICA and subordinate G2X offices have a CICA. Depending on the size and scope of the operation, the CICA could be the unified command's CI staff officer; the CI or HUMINT staff officer from corps or division; or a senior warrant officer or branch CI officer designated by the unified command CI staff officer or task force commander. For more information about the CICA functions, see Chapter 11.

HUMINT TEAM STRUCTURE

Operational Management Team

6-56. The OMT is a four-person team consisting of a warrant officer (WO), two noncommissioned officers (NCOs), and a junior enlisted soldier. Civilians may be inserted into this structure when appropriate. Rank structure and standards of grade for OMTs vary depending upon the skill sets required and mission focus. HUMINT OMTs provide operational guidance for two to four HUMINT teams, depending on mission focus and operational tempo. When two or more HUMINT teams are deployed in DS of a maneuver element, an OMT also deploys to provide technical control. The OMT works closely with the supported S2 and analysis and control team (ACT) to furnish current threat information and answer the supported commander's PIRs and IRs. OMTs coordinate with the supported 2X and manage subordinate HUMINT team echelons to—

- Provide guidance and technical control of operational activity.
- Provide the collection and operational focus for HUMINT teams.
- Provide quality control and dissemination of reports for subordinate HUMINT teams.
- Conduct single-discipline HUMINT analysis, and assist in mission analysis for the supported commander.

- Act as a conduit between subordinate HUMINT teams, the HOC, and supported unit headquarters.
- Provide administrative support for subordinate HUMINT teams, to include reporting mission and equipment status to the HOC and the supported unit headquarters.
- Educate the supported commander on the capabilities of the HUMINT teams.
- Integrate the HUMINT teams directly into the maneuver commander's ISR planning.

HUMINT Team

6-57. The HUMINT team is a four-person team consisting of two NCOs and two junior enlisted personnel. Civilians may be inserted into this structure when appropriate. Rank structure and standards of grade for HUMINT teams vary depending upon the skill sets required and the mission focus. HUMINT teams are trained to execute the full range of HUMINT functions as defined in **HUMINT FUNCTIONS** section. However, they may be assigned to mission-focused elements such as DOCEX, interrogation, debriefing, or contact operations.

Chapter 7

Imagery Intelligence

DEFINITION

7-1. IMINT is intelligence derived from the exploitation of imagery collected by visual photography, infrared, lasers, multi-spectral sensors, and radar. These sensors produce images of objects optically, electronically, or digitally on film, electronic display devices, or other media.

ROLE

7-2. The role of imagery is to assist the commander in focusing and protecting his combat power. Imagery often enhances the commander's situational understanding of the battlespace. Other than direct human observation, imagery is the only intelligence discipline that allows the commander to see the battlefield in real time as the operation progresses. In those cases where maps are not available, digital imagery in hardcopy or softcopy can be used as a substitute. Imagery can also be used to update maps or produce grid-referenced graphics. Detailed mission planning often requires imagery, to include three-dimensional stereo images, in order to provide the degree of resolution necessary to support such specialized planning.

FUNDAMENTALS

7-3. Some imagery assets are very responsive to the individual commander's intelligence requirements. Some imagery systems can directly transmit imagery into the tactical operations center (TOC); examples include imagery from UAVs and the Joint Surveillance Target Attack Radar System (JSTARS). This direct downlink enables the G2/S2 to use the imagery as soon as possible instead of having to wait for finished imagery products. A note of caution is required, however, if not a trained imagery analyst because the G2/S2 could incorrectly interpret the imagery.

7-4. Imagery-related equipment has undergone a reduction in size as well as a reduction in the time it takes to provide products, particularly softcopy imagery. The modularity and size reduction of imagery analysis, processing, and display systems make transport easier; they also allow the commander to bring lesser amounts than in the past, while still retaining those systems (or subsystems) required to complete the mission. Additionally, data compression allows faster transmission of imagery products directly to the warfighter.

SOURCES OF IMAGERY

7-5. There are three general sources of imagery: national, civil, and commercial. National imagery traditionally refers to imagery collected by DOD imagery systems. However, there are other sources of imagery provided by non-national sources such as handheld photography (film, digital, and video), UAV imagery, and gun-camera images.

National

7-6. National systems are developed specifically for supporting the President of the United States, the Secretary of Defense, other national agencies, and US military forces. These systems respond to the needs of the nation and those of the combatant commands.

Civil

7-7. Civil imagery systems are usually government funded in terms of building, launching, and operating the system. In many, but not all, cases the agencies operating these civil imagery systems also process, distribute, and archive the imagery data or images.

Commercial

7-8. Commercial companies build, launch, and operate imagery systems for profit. In times of crises, license agreements with the US Government obligate US commercial satellite imaging systems to provide data only to the US Government at the market value. This protects information concerning US operations from adversarial exploitation from commercial systems such as the ACE Imagery Company. However, foreign commercial imagery systems are not bound to this arrangement, and thus may be used by our nation's adversaries. Commercial imagery has become increasingly valuable for many reasons:

- Due to its unclassified nature, civil and commercial imagery is useful in an open environment, may be released to multinational partners, and can be made available to the press. They are especially useful for geospatial products. Their use allows national systems more time to focus on other intelligence functions.
- Civil and commercial imagery offer radar and multi-spectral. Some offer large area collection useful for broad area coverage purposes.
- Commercial satellite imagery resolution varies from less than one meter to several kilometers.

7-9. The Central Imagery Tasking Office (CITO) is responsible for ordering commercial imagery. The Commercial Satellite Imagery Library is available to research DOD purchased commercial imagery. The G2/S2 should consult the CITO when forming commercial imagery requests. NGA will deliver the imagery primarily on CD-ROM media via courier or mail service. Limited digital or electronic delivery is available as well.

TYPES OF IMAGERY SENSORS

7-10. There are four types of imagery sensors: visible (optical), infrared, radar, and multi-spectral. Each sensor has a unique capability, with distinct advantages and disadvantages. The G2/S2 must understand each sensor's capability in order to select the best sensor for the mission and thus enable the user to better understand the intelligence received. Certain sensors are better suited for military operations than others. (See Table 7-1.)

Table 7-1. Sensor Characteristics Matrix.

SENSORS	ADVANTAGES	DISADVANTAGES
Visible (Optical) Best tool for daytime, clear weather, detailed analysis. Includes video and electro-optical.	• Affords a familiar view of a scene. • Offers system resolution that cannot be achieved in other optical systems or in thermal images and radars. • Preferred for detailed analysis and mensuration. • Offers stereoscopic viewing.	• Restricted by terrain and vegetation. • Limited to daytime use only. • Reduced picture size.
Infrared Best tool for nighttime, clear weather, detailed analysis. Includes Overhead Non-Imaging Infrared (ONIR).	• A passive sensor and is impossible to jam. • Offers camouflage penetration. • Provides good resolution. • Nighttime imaging capability.	• Not effective during thermal crossover periods. • Product not easily interpretable. • Requires skilled analysis. • Cannot penetrate clouds.
Radar Useful for detecting presence of objects at night and in bad weather. Includes synthetic aperture radar (SAR), coherent change detection (CCD), and MTI.	• All weather; can penetrate fog, haze, clouds, smoke. • Day or night use. • Does not rely on visible light nor thermal radiation. • Good standoff capability. • Large area coverage. • Allows moving target detection. • Foliage and ground penetration.	• Product not easily interpretable. • Requires skilled analysis. • Terrain masking inhibits use.
Multi-Spectral Imagery (MSI) Best tool for mapping purposes and terrain analysis.	• Large database available. • Band combinations can be manipulated to display desired requirements. • Images can be merged with other digital data to provide higher resolution.	• Product not easily interpretable. • Requires skilled analysis. • Computer manipulation requires large amounts of memory and storage; requires large processing capabilities.

PLAN

7-11. The first step in planning for IMINT is determining the need for IMINT products based on the PIRs and the initial IPB. The G2/S2 should research targets using online imagery databases early and request those imagery products that are not perishable for contingency planning. National and COCOM imagery databases may hold recently imaged targets that could meet the commander's immediate needs instead of requesting new imagery. The staff must clearly articulate their intelligence requirements to include communicating what the mission is and how the requested product will aid in mission accomplishment. The G2/S2 should submit the imagery requirement using established procedures such as those in the unit's SOP or as established by the COCOM.

7-12. The G2/S2 must also determine the specific imagery requirements so as not to burden the system with unnecessary requests. The desire for imagery products often exceeds the capabilities of the IMINT system. Therefore, it is imperative that the G2/S2 consider what type of analysis they require, and request only that which they require. The specifications of the request for IMINT products often affect the timeliness of the response. For example, determining if vehicles are tanks takes less time and requires less resolution than determining if a tank is a T-64 or a T-72.

7-13. Here are some of the roles IMINT can perform that the G2/S2 may consider when determining IMINT requirements:

- Imagery can detect and/or identify and locate specific unit types, equipment, obstacles, potential field fortifications, etc., from which intelligence analysts are able to analyze enemy capabilities and develop possible COAs.
- Imagery can also update maps and enhance the interpretation of information from maps. Detailed mission planning uses imagery to include stereo images for three-dimensional viewing of the terrain and many other geospatial uses.
- Imagery can be used as a substitute for maps when maps are not available. The most common application of this technique is by constructing imagery mosaics: a combination of two or more overlapping photographic prints that form a single picture.
- MTI displays or products can provide an NRT picture of an entity's movement by indicating its speed, location, and direction of travel. MTI systems do not differentiate friendly from enemy. Imagery assets, particularly MTI systems, are useful in cueing other ISR systems.
- Imagery can be used to support protection of the force by helping the commander visualize how his forces look—including their disposition, composition, and vulnerabilities—as exploited by enemy IMINT systems.
- Imagery analysts use combat assessment imagery to confirm destruction, determine the percentage of destruction, or whether the target was unaffected.

PREPARE

7-14. The G2/S2 IMINT-related actions during the prepare function of the intelligence process include establishing or verifying the portion of the intelligence communications architecture that supports IMINT display and analysis functions properly. Additionally, the G2/S2 must ensure that required IMINT analytical assets and resources are prepared to provide support or are available through intelligence reach. Lastly, the G2/S2 must also ensure IMINT reporting and dissemination channels and procedures are in place and rehearsals are conducted with all pertinent IMINT elements to ensure interoperability.

COLLECT

7-15. As previously mentioned, there are four types of imagery sensors. Depending on the type of sensor, it can record hardcopy or softcopy single frame or continuous (video). A given imagery target will not necessarily receive continuous coverage due to the possible conflict between the number and priority of targets and the number and availability of imagery assets. However, a commander may decide to have continuous surveillance of certain targets, for specified periods of time, usually using his own imagery assets (for example, UAV) even though this detracts from the commander's ability to use these assets for other imagery targets within his AOI.

PROCESS

7-16. The process function regarding IMINT involves converting imagery data into a form that is suitable for performing analysis and producing intelligence. Examples of IMINT processing include developing film, enhancing imagery, converting electronic data into visual displays or graphics, and constructing electronic images from IMINT data.

PRODUCE

7-17. The IMINT producer must ensure the IMINT product satisfies the associated intelligence requirements and that the product is in the required format. The quality and resolution of the product is highly dependent upon the type of sensor, the time of day, and the weather conditions, as well as the imagery analyst's ability to identify items, vehicles, equipment, and personnel within the images. Specific IMINT products are discussed in the **ANALYZE** section below.

ANALYZE

7-18. Timeliness is critical not only to IMINT collection but also to IMINT analysis and reporting. It is difficult to separate IMINT reporting from IMINT analysis in this discussion. This is demonstrated by the three phases of IMINT reporting presented below; all are dependent upon the timeliness requirements. Each phase represents a different degree of analysis and period of time available to accomplish the exploitation of the imagery.

- First Phase imagery analysis is the rapid exploitation of newly acquired imagery and reporting of imagery-derived information within a specified time from receipt of imagery. This phase satisfies priority requirements of immediate need and/or identifies changes or activity of immediate significance. First Phase imagery analysis results in an Initial Phase Imagery Report (IPIR).
- Second Phase imagery analysis is the detailed exploitation of newly acquired imagery and the reporting of imagery-derived intelligence and information while meeting the production and timeliness requirements. Other intelligence discipline source material may support phase two imagery as appropriate. Second phase imagery analysis results in a Supplemental Imagery Report (SUPR).
- Third Phase imagery analysis is the detailed analysis of all available imagery pertinent to an SIR, and the subsequent production and reporting resulting from this analysis within a specified time. This

phase provides an organized detailed analysis of an imagery target or topic, using imagery as the primary data source but incorporating data from other sources as appropriate.

7-19. The two types of imagery exploitation are national and direct support (DS).

- National exploitation is imagery exploitation that supports presidential requirements, National Security Council (NSC) requirements, congressional requirements, or requirements of a common concern to the intelligence community.
- DS exploitation is imagery exploitation that supports assigned missions of a single agency, department, or command (the warfighter).

7-20. IMINT assets will complete DS exploitation in order to satisfy (First Phase) requirements and report the results as soon as possible, but not later than 24 hours after receipt of the imagery. Collectors will complete national exploitation in order to satisfy (Second and Third Phases) requirements and report the results within the time specifications of each individual requirement.

DISSEMINATE

7-21. IMINT products are distributed or disseminated in hardcopy, softcopy, or direct viewing such as through remote terminals. The distribution of hardcopy products will be via couriers or other type of mail system. The dissemination of softcopy products will be either as hardcopy products (for example, CD-ROM and 3.5-inch disks) or electronically. The requestor must ensure that the requested product is transmittable over the available communications systems.

ASSESS

7-22. The requestor should immediately assess the imagery product upon receipt for accuracy and relevance to the original request. The requestor must then notify the producer and inform him of the extent to which the product answered the PIR. Providing feedback to the producer regarding the product helps ensure the producer will provide the required information in the correct format. The following are some of the questions which the requestor should consider when providing feedback to the producer.

- Is the format of the product acceptable?
- Is additional information needed on the product or future products?
- Is excess information included on the product?

Chapter 8

Signals Intelligence

DEFINITION

8-1. SIGINT is a category of intelligence comprising either individually or in combination all COMINT, ELINT, and FISINT, however transmitted; intelligence is derived from communications, electronics, and foreign instrumentation signals. SIGINT has three subcategories:

- COMINT – The intelligence derived from foreign communications by other than the intended recipients.
- ELINT – The technical and geo-location intelligence derived from foreign non-communications electromagnetic radiations emanating from other than nuclear detonations or radioactive sources.
- FISINT – Technical information and intelligence derived from the intercept of foreign electromagnetic emissions associated with the testing and operational deployment of non-US aerospace, surface, and subsurface systems. Foreign instrumentation signals include but are not limited to telemetry, beaconry, electronic interrogators, and video data links. (See JP 1-02)

8-2. SIGINT provides intelligence to the commander based upon intercepted communications and provides transmitter location data.

ROLE

8-3. SIGINT provides intelligence on threat capabilities, disposition, composition, and intentions. In addition, SIGINT provides targeting information for the delivery of lethal and non-lethal effects.

FUNDAMENTALS

8-4. It is important that the G2/S2 understand how SIGINT assets are organized not only within the Army but also throughout the DOD. The majority of SIGINT assets are located at EAC. SIGINT assets from all the armed services, combined with national SIGINT assets, work together to support commanders from the tactical to the strategic level. Only by understanding the SIGINT structure that transcends traditional service component boundaries can the G2/S2 understand how to use SIGINT effectively.

TECHNICAL CONTROL AND ANALYSIS ELEMENT (TCAE) ARCHITECTURE

8-5. The TCAE architecture supports Army SIGINT collectors and analysts by providing SIGINT-specific intelligence and guidance to SIGINT personnel. The SIGINT technical architecture complements existing C2 relationships; it does not replace the commander's authority or chain of command.

8-6. There is a TCAE at each Army operational echelon. The TCAE at each echelon provides a single POC and resource dedicated to supporting the commander at that level. TCAEs or SIGINT elements at varying echelons draw information from each other in order to provide a more complete and detailed intelligence picture for their respective commanders.

ARMY TCAE (ATCAE)

8-7. The ATCAE, established at the national level, plays a significant role in TCAE operations by providing technical support oversight and providing access to national databases—in coordination with the NSA and other intelligence organizations. Additionally, the ATCAE works closely with NSA's IO element assisting in operations, monitoring technical capabilities, and providing liaison with Army's 1st Information Operations Command (Land). (This unit was previously called Land Information Warfare Activity.)

8-8. The ATCAE supports the Army's Quick Reaction Capability (QRC) SIGINT and signals research and target development (SRTD) operations. The QRC systems are responsive to the ground force commander's requirements and allow the ability to conduct SIGINT operations against modern communications systems.

REGIONAL TCAE (RTCAE)

8-9. The Army established RTCAEs to allow supported units at all echelons access to SIGINT regarding their respective AOI. The Army placed an RTCAE within each Regional Security Operations Center (RSOC). An RSOC is a joint service facility established by the Director, NSA, to conduct continuous security operations on selected targets in support of national and warfighter intelligence requirements using remoting technologies. The RSOC embeds a focus on the tactical commander's SIGINT requirements at a national level. It also creates a framework for enhanced interoperability among service SIGINT activities, especially within the context of a deployed JTF.

8-10. In addition to each RSOC, the RTCAE works closely with Theater TCAEs as well as Corps and Division SIGINT elements with the same regional focus. It provides analytic support and tailored products and answers RFIs. During peacetime and real-world operations RTCAEs facilitate intelligence reach support. They provide tailored technical support packages, support for surge and survey operations, analytic expertise, and signal data access and digital file transfers of live collection for Army tactical units.

THEATER TCAE

8-11. The Theater TCAE performs SIGINT and EW technical control and analysis and management. It provides SIGINT technical support for assigned, OPCON, and lower echelon SIGINT resources deployed in the theater. This includes mission tasking, processing, analyzing, and reporting of SIGINT data, information, and intelligence. The TCAE provides direction for the Theater Collection and Exploitation Battalion's SIGINT mission and for other theater tactical SIGINT assets.

SIGINT THEATER INTELLIGENCE BRIGADE/GROUP (TIB/TIG)

8-12. The TIB/TIG conducts operational level multidiscipline SIGINT operations. It provides timely intelligence to the commander throughout the full spectrum of operations. The TIB/TIG is structured to provide ground and aerial SIGINT support. It executes the full range of SIGINT missions or tasks.

SIGINT AT ECHELONS CORPS AND BELOW

8-13. There is a varying mixture of SIGINT assets within each corps. A reference listing the types of SIGINT assets found at echelons corps and below is listed in ST 2-50.

PLAN

8-14. An important SIGINT planning consideration is that if at all possible, SIGINT collection should be employed in conjunction with another intelligence discipline collection system. SIGINT is often used to cue, and be cued by, other ISR assets.

8-15. During planning, the G2/S2 should retrieve, update, or develop any required SIGINT databases. This includes effecting coordination with other SIGINT assets or elements that can support the operation.

8-16. SIGINT ground collection assets are usually placed in proximity to enemy signal sources due to the limited height of collector antennas, the low power output of threat or enemy emitters, and the line-of-sight (LOS) constraints imposed by terrain. Ground-based SIGINT teams are most effective when positioned to—

- Maximize threat emitter interception. This allows teams to overcome the constraints of threat emitter characteristics usually allowed by proximity to the threat.
- Minimize system receiver interference. This increases the potential capability of the team to acquire threat emitters in a timely manner.
- Optimize overlapping areas of intercept coverage. This ensures coverage of the AOIR and allows targets to be handed off from team to team if necessary.

8-17. Aerial SIGINT assets have additional planning requirements. Coordination must be effected with Army Airspace Command and Control (A2C2) for ingress and egress routes, restricted operating zones (ROZs) where the aircraft can travel on an intercept track, as well as with the supporting air operations center to determine the availability of asset types and times.

PREPARE

8-18. The G2/S2 ensures the SIGINT unit and asset leaders have effected all necessary coordination and conducted rehearsals. This includes establishing or verifying the operation of the SIGINT technical architecture. The G2/S2 also ensures all required SIGINT assets and resources are available, SIGINT reporting and dissemination channels and procedures are in place, and connectivity and interoperability exist with all pertinent SIGINT elements.

8-19. SIGINT OPCON is the authoritative direction of SIGINT activities, including tasking and allocation of effort, and the authoritative prescription of those uniform techniques and standards by which SIGINT information is collected, processed, and reported.

8-20. SIGINT operational tasking is the authoritative operational direction of and direct levying of SIGINT information needs by a military commander on designated SIGINT resources. These requirements are directive, regardless of other priorities, and are conditioned only by the capability of those resources to produce such information. Operational tasking includes authority to deploy all or part of the SIGINT resources for which SIGINT operational tasking authority (SOTA) has been delegated.

8-21. SOTA is the military commander's authority to operationally direct and levy SIGINT requirements on designated SIGINT resources; it includes authority to deploy and redeploy all or part of the SIGINT resources for which SOTA has been delegated.

COLLECT

8-22. SIGINT performs two major collection activities: signals intercept and direction finding (DF).

Signals Intercept

8-23. Signals intercept are those SIGINT actions used to search for, intercept, and identify threat electromagnetic signals for the purpose of immediate threat recognition. Signals intercept provides information required to answer PIRs, and other intelligence requirements in support of the ISR effort.

Direction Finding

8-24. Even when threat radio operators use COMSEC procedures, SIGINT teams can often intercept and approximate the location of the threat's signals. Specifically, SIGINT teams can use DF to determine—

- Movement of threat personnel or equipment.
- Locations of emitters associated with weapon systems and units.
- New emitter locations and confirm known emitter locations.
- Possible friendly targets the enemy intends to attack (lethal and non-lethal).

8-25. In addition to finding threat forces, DF operations can assist the (radio-equipped) friendly force by—

- Locating and vectoring assets or units during limited visibility.
- Locating downed aircraft and personnel radio beacons.
- Conducting signal security assessments.
- Locating sources of communication interference and jamming.

PROCESS

8-26. SIGINT processing involves converting intercepts of SIGINT into written and verbal reports, automated message, graphic displays, recordings, and other forms suitable for analysis and intelligence production. Since US forces routinely conduct operations against adversaries who speak languages other than English, SIGINT processing often also includes translation of these intercepts.

8-27. Due to the complexity of many SIGINT systems, automated processing may occur several times before SIGINT data or information receives any human interaction.

PRODUCE

8-28. The SIGINT producer must ensure the SIGINT product satisfies the associated intelligence requirements and that the product is in the required format. The quality, fidelity, and timeliness of SIGINT products are highly dependent upon the type of intercept, the collection system, the system's position in relation to the threat emitter, the weather (including space-based weather), as well as the SIGINT operator's ability to identify the appropriate threat signal activity.

8-29. SIGINT production results in some of the reports and formats mentioned in the process section above; however, the objective for SIGINT is to be used in an all-source analytical approach.

ANALYZE

8-30. The intelligence staff analyzes intelligence and information about the enemy's communications capabilities to determine appropriate SIGINT collection strategies. Conversely, a corresponding analysis of the friendly forces' SIGINT capabilities must be conducted to ensure the continued effectiveness of, or to improve upon, SIGINT collection.

8-31. SIGINT analysts also sort through large amounts of SIGINT and information and intelligence to identify and use only that which is pertinent to the CCIRs (PIRs and FFIRs).

DISSEMINATE

8-32. SIGINT of critical importance to the force, including answers to the CCIRs (PIRs and FFIRs), is disseminated via the most expeditious means possible. Due to the usually highly perishable nature of SIGINT, the most expeditious reporting means is often immediately augmented with a follow-up report or augmented by a report transmitted through additional means, enhancing the probability of receipt. Sometimes the most expeditious means of reporting critical SIGINT information to the commander is face to face.

8-33. For intelligence reach operations, SIGINT products are available and disseminated in a variety of forms: hardcopy, softcopy, direct viewing, or listening (television or radio). It is incumbent on the requestor to ensure that the SIGINT product can be transmitted over the available communications systems. This includes verifying the appropriate security level of the communications system.

ASSESS

8-34. The primary goal of the assess function when applied to SIGINT is to determine whether the results of SIGINT collection meet the requirements of the unit's ISR effort. SIGINT producers must assess all facets of SIGINT operations, from receipt of the ISR task to the dissemination of SIGINT, in an effort to determine their effectiveness. This assessment is not only directed at each SIGINT asset individually but also throughout the supporting SIGINT architecture and the unit's entire ISR effort.

8-35. The G2/S2 immediately assesses SIGINT products upon receipt for accuracy and relevance. He must inform the SIGINT producer of the extent to which the product answered the PIR or intelligence requirement. Providing feedback to the SIGINT producer—and collector—helps improve the effectiveness and efficiency of SIGINT.

Chapter 9

Measurement and Signatures Intelligence

DEFINITION

9-1. MASINT is technically derived intelligence that detects, locates, tracks, identifies, and/or describes the specific characteristics of fixed and dynamic target objects and sources. It also includes the additional advanced processing and exploitation of data derived from IMINT and SIGINT collection.

9-2. MASINT collection systems include but are not limited to radar, spectroradiometric, E-O, acoustic, RF, nuclear detection, and seismic sensors, as well as techniques for gathering NBC and other material samples.

9-3. It requires the translation of technical data into recognizable and useful target features and performance characteristics. Computer, communication, data, and display processing technologies now provide MASINT in support of commanders throughout the full spectrum of operations.

9-4. The subdisciplines within MASINT include, but are not limited to, the following:

- **Radar Intelligence (RADINT).** The active or passive collection of energy reflected from a target or object by LOS, bistatic, or over-the-horizon radar systems. RADINT collection provides information on radar cross-sections, tracking, precise spatial measurements of components, motion and radar reflectance, and absorption characteristics for dynamic targets and objectives. A SAR system, coupled with advanced MASINT processing techniques—
  - Provides a high resolution, day and night collection capability.
  - Can produce a variety of intelligence products that identify or provide change detection, terrain mapping, underwater obstacles, dynamic sensing of targets in clutter, and radar cross-section signature measurements.
- **Frequency Intelligence.** The collection, processing, and exploitation of electromagnetic emissions from a radio frequency weapon (RFW), an RFW precursor, or an RFW simulator; collateral signals from other weapons, weapon precursors, or weapon simulators (for example, electromagnetic pulse signals associated with nuclear bursts); and spurious or unintentional signals.
  - *Electromagnetic Pulses.* Measurable bursts of energy that result from a rapid change in a material or medium, resulting in an explosive force, produces RF emissions. The RF pulse emissions associated with nuclear testing, advanced technology devices, power and propulsion systems, or other impulsive events can be used to detect, locate, identify, characterize, and target threats.

- *Unintentional Radiation Intelligence (RINT).* The integration and specialized application of MASINT techniques against unintentional radiation sources that are incidental to the RF propagation and operating characteristics of military and civil engines, power sources, weapons systems, electronic systems, machinery, equipment, or instruments. These techniques may be valuable in detecting, tracking, and monitoring a variety of activities of interest.

- **E-O Intelligence.** The collection, processing, exploitation, and analysis of emitted or reflected energy across the optical portion (ultraviolet, visible, and infrared) of the EMS. MASINT E-O provides detailed information on the radiant intensities, dynamic motion, spectral and spatial characteristics, and the materials composition of a target. E-O data collection has broad application to a variety of military, civil, economic, and environmental targets. E-O sensor devices include radiometers, spectrometers, non-literal imaging systems, lasers, or laser radar (LIDAR).
  - *Infrared Intelligence (IRINT).* A subcategory of E-O that includes data collection across the infrared portion of the EMS where spectral and thermal properties are measured.
  - *LASER Intelligence (LASINT).* Integration and specialized application of MASINT E-O and other collection to gather data on laser systems. The focus of the collection is on laser detection, laser threat warning, and precise measurement of the frequencies, power levels, wave propagation, determination of power source, and other technical and operating characteristics associated with laser systems—strategic and tactical weapons, range finders, and illuminators.
  - *Hyperspectral Imagery (HSI).* A subcategory of E-O intelligence produced from reflected or emitted energy in the visible and near infrared spectrum used to improve target detection, discrimination, and recognition. HSI can detect specific types of foliage—supporting drug-crop identification; disturbed soil—supporting the identification of mass graves, minefields, caches, underground facilities or cut foliage; and variances in soil, foliage, and hydrologic features—often supporting NBC contaminant detection.

- **Spectroradiometric Products.** Include E-O spectral (frequency) and radiometric (energy) measurements. A spectral plot represents radiant intensity versus wavelength at an instant in time. The number of spectral bands in a sensor system determines the amount of detail that can be obtained about the source of the object being viewed. Sensor systems range from multispectral (2 to 100 bands) to hyperspectral (100 to 1,000 bands) to ultraspectral (1,000+ bands). More bands provide more discrete information, or greater resolution. The characteristic emission and absorption spectra serve to fingerprint or define the makeup of the feature that was observed. A radiometric plot represents the radiant intensity versus time. An example is the radiant intensity plot of a missile exhaust plume as the missile is in flight. The intensity or brightness of the object is a function of several conditions including its temperature, surface properties or material, and how fast it is moving. For each point along a time-intensity radiometric plot, a spectral plot can be generated based on the number of spectral bands in the collector.

- **Geophysical Intelligence.** Geophysical MASINT involves phenomena transmitted through the earth (ground, water, atmosphere) and manmade structures including emitted or reflected sounds, pressure waves, vibrations, and magnetic field or ionosphere disturbances.
  - *Seismic Intelligence.* The passive collection and measurement of seismic waves or vibrations in the earth surface.
  - *Acoustic Intelligence.* The collection of passive or active emitted or reflected sounds, pressure waves or vibrations in the atmosphere (ACOUSTINT) or in the water (ACINT). ACINT systems detect, identify, and track ships and submarines operating in the ocean.
  - *Magnetic Intelligence.* The collection of detectable magnetic field anomalies in the earth's magnetic field (land and sea). An example is a Remotely Emplaced Battlefield Surveillance System (REMBASS) sensor detection indicating the presence and direction of travel of a ferrous object.
- **Nuclear Intelligence (NUCINT).** The information derived from nuclear radiation and other physical phenomena associated with nuclear weapons, reactors, processes, materials, devices, and facilities. Nuclear monitoring can be done remotely or during onsite inspections of nuclear facilities. Data exploitation results in characterization of nuclear weapons, reactors, and materials. A number of systems detect and monitor the world for nuclear explosions, as well as nuclear materials production.
- **Materials Intelligence.** The collection, processing, and analysis of gas, liquid, or solid samples. Materials intelligence is critical to collection against NBC warfare threats. It is also important to analyzing military and civil manufacturing activities, public health concerns, and environmental problems. Samples are both collected by automatic equipment, such as air samplers, and directly by humans. Samples, once collected, may be rapidly characterized or undergo extensive forensic laboratory analysis to determine the identity and characteristics of the sources of the samples.

ROLE

9-5. MASINT provides intelligence to the commander throughout the full spectrum of operations to facilitate situational understanding. MASINT can thwart many of the camouflage, concealment, and deception techniques currently used to deceive ISR systems.

9-6. MASINT is perceived as a "strategic" discipline with limited "tactical" support capabilities. But, by application of real-time analysis and dissemination, MASINT has a potential ability to provide real-time situation awareness and targeting not necessarily available to the classic disciplines. Specifically, MASINT "sensors" have unique capabilities to detect missile launch, detect and track aircraft, ships, and vehicles; do non-cooperative target identification (NCTI), combat assessment, and BDA; and detect and track fallout from nuclear detonations. Often, these contributions are the first indicators of hostile activities. For example, two EXOCET-equipped

Mirage F-1s were shot down during the Operation DESERT STORM due to MASINT collection and analysis. As evidenced by Operation IRAQI FREEDOM (OIF), MASINT will play a decisive role in the targeting of smart munitions with the signatures (fingerprint) of the targets they are seeking (for example, infrared signatures).

9-7. The MASINT systems most familiar on today's battlefield are employed by ground surveillance and NBC reconnaissance elements.

9-8. MASINT spans the entire EMS and its capabilities complement, rather than compete with, the other intelligence disciplines. MASINT provides, to varying degrees, the capability to—

- Use automatic target recognition (ATR) and aided target recognition (AiTR).
- Penetrate manmade and/or natural camouflage.
- Penetrate manmade and/or natural cover, including the ability to detect subterranean anomalies or targets.
- Counter stealth technology.
- Detect recently placed mines.
- Detect natural or manmade environmental disturbances in the earth's surface not discernible through other intelligence means.
- Provide signatures (target identification) to munitions and sensors.
- Enhance passive identification of friend or foe.
- Detect the presence of NBC agents to include prior to, during, or after employment.
- Detect signature anomalies that may affect target-sensing systems.

FUNDAMENTALS

9-9. Before discussing the functions of the intelligence process within a MASINT, the following paragraph provides an overview of organizational structure of MASINT.

9-10. Within DOD, the DIA provides central coordination for MASINT collection efforts through the Central MASINT Office. Each service, in turn, has a primary command or staff activity to develop requirements and coordinate MASINT effort. Army responsibility currently resides with INSCOM. Army weapons systems programs that require MASINT information to support system design or operations submit requests through the Army Reprogramming Analysis Team (ARAT) or INSCOM channels for data collection and processing. The S&TI community also performs MASINT collection and processing primarily to support R&D programs and signature development. Every S&TI center has some involvement in MASINT collection or production that reflects that center's overall mission (for example, NGIC does work on armored vehicles and artillery). Service R&D centers such as the Communications-Electronics Command (CECOM) Research, Development, and Engineering Center (RDEC), and Night Vision and Electronic Systems Laboratory are also involved in developing sensor systems for collecting and processing MASINT.

9-11. In addition to supporting the S&TI mission, INSCOM units also execute limited ground-based operational collection to support Theater and ASCC PIRs. This capability will expand upon the standup of INSCOM TIBs and TIGs.

PLAN

9-12. Some MASINT sensors can provide extremely specific information about detected targets, whereas other sensors may only be capable of providing an indication that an entity was detected. Additionally, there are varying capabilities of detection, identification, and classification among MASINT sensors. It is these varying capabilities that require synchronizing the employment of MASINT sensors both within the MASINT discipline and within the ISR effort as a whole. See FM 2-01 for more specific information on ISR synchronization.

9-13. As previously mentioned, there are many types of MASINT sensors. Depending on the type of sensor employed, a given MASINT collection target or NAI may not necessarily receive continuous coverage due to the possible conflict between the number and priority of targets and the number and availability of MASINT assets. However, a commander may decide to have continuous surveillance of certain targets by using his own MASINT assets (for example, REMBASS or Improved-REMBASS).

9-14. Another consideration when planning MASINT missions is whether to use active, passive, or a combination of both when planning MASINT coverage.

PREPARE

9-15. The G2/S2 MASINT related actions during the prepare function of the intelligence process include properly establishing or verifying the MASINT portion of the intelligence communications architecture functions. Additionally, the G2/S2 must ensure that required MASINT analytical assets and resources are prepared to provide support or are available through intelligence reach. Since the products of MASINT are not as well known as products from other intelligence disciplines, the G2/S2 must be aware of the types of MASINT products available to support the operation, and then educate the rest of his unit's staff on the use of these MASINT products. Lastly, the G2/S2 must also ensure MASINT reporting and dissemination channels and procedures are in place and rehearsals are conducted with all pertinent MASINT elements to ensure interoperability.

COLLECT

9-16. MASINT provides information required to answer PIRs and other intelligence requirements in support of the ISR effort. As stated earlier in this chapter, MASINT collection must not only be synchronized within its own discipline but also be synchronized and integrated into the unit's overall ISR effort in order to be effective.

9-17. MASINT sensors are employed throughout the full spectrum of operations from a variety of platforms—sub-surface, ground, marine, and aerospace.

PROCESS

9-18. Just as in the other intelligence disciplines, MASINT involves dealing with huge volumes of data that have to be processed before beginning analysis and production. The process function regarding MASINT involves converting esoteric data into a form that is suitable for performing analysis and producing intelligence. MASINT processing can include relatively simple actions such as converting a REMBASS sensor activation into a report, to a complex task such as processing HSI into a report identifying the composition and concentrations of carcinogenic chemicals present in the emissions from a factory upwind from a US forces encampment.

PRODUCE

9-19. Effective and timely MASINT requires personnel with diverse skill sets. The MASINT producer must ensure the MASINT product satisfies the associated intelligence requirements and that the product is in the required format. The quality, fidelity, and timeliness of MASINT products are highly dependent upon the type of target, the collection system, the system's position in relation to the target or NAI, and the weather, as well as the MASINT system operator's ability to identify the appropriate threat activity.

9-20. The objective of MASINT production is to be used in an all-source analytical approach.

ANALYZE

9-21. The intelligence staff analyzes intelligence and information about the enemy's equipment, doctrine, and TTP to determine appropriate MASINT collection strategies. Conversely, a corresponding analysis of the friendly force's MASINT capabilities must be conducted to ensure the continued effectiveness of, or to improve upon, MASINT collection.

DISSEMINATE

9-22. MASINT of critical importance to the force, including answers to the PIRs, is disseminated via the most expeditious means possible.

9-23. For intelligence reach operations, MASINT products are available and disseminated in a variety of forms. The requestor must ensure that the MASINT product can be transmitted over the available communications systems. This includes verifying the appropriate security level of the communications system.

ASSESS

9-24. The primary goal of the MASINT assess function is to determine whether the results of MASINT collection and production meet the requirements of the unit's ISR effort. MASINT producers must assess all facets of MASINT operations, from receipt of the ISR task to the dissemination of MASINT, in an effort to determine the effectiveness of MASINT. This assessment is not only directed at each MASINT asset individually but also throughout the supporting intelligence communications architecture, to include intelligence reach and the unit's entire ISR effort.

9-25. Also, the G2/S2 immediately assesses MASINT products upon receipt for accuracy and relevance. He must inform the MASINT producer of the extent to which the product answered the PIR or intelligence requirement. Providing feedback to the MASINT producer—and collector—helps improve the effectiveness and efficiency of MASINT.

Chapter 10

Technical Intelligence

DEFINITION

10-1. TECHINT is intelligence derived from the collection and analysis of threat and foreign military equipment and associated materiel.

ROLE

10-2. The strength of the US military lies, in part, to the diversity and extent of its technology base. While the US aspires to be the leader in integrating technology, the threat can achieve temporary technological advantage in certain areas by acquiring modern systems or capabilities. The world arms market is willing to provide these advanced systems to countries or individuals with the resources to pay for them. A concerted TECHINT program is vital to providing precise direction and purpose within the US R&D process to ensure quick and efficient neutralization of this advantage.

10-3. The role of TECHINT is to ensure that the warfighter understands the full technological capabilities of the threat. With this understanding, the warfighter can adopt appropriate countermeasures, operations, and tactics.

10-4. TECHINT has two goals within its role:

- To ensure the US armed forces maintain technological advantage against any adversary.
- To provide tailored, timely, and accurate TECHINT support to the warfighter throughout the entire range of military operations. This includes providing US forces intelligence, information, and training on foreign weapons systems to an extent that allows their use of CEE.

FUNDAMENTALS

10-5. The G2/S2 must understand how TECHINT assets are organized in order to properly apply the intelligence process, as the majority of TECHINT assets are located at EAC.

DEFENSE INTELLIGENCE AGENCY

10-6. DIA manages and reviews overall TECHINT activities. The S&TI Directorate within DIA is the action element for TECHINT. This directorate coordinates with external TECHINT agencies on non-policy matters concerning the production of S&TI. The following organizations provide TECHINT support under the control of DIA:

- **Armed Forces Medical Intelligence Center (AFMIC).** AFMIC, based at Fort Detrick, MD, is a DOD intelligence production center under DIA control. AFMIC is responsible for exploiting foreign medical materiel. The director supports the Army Foreign Materiel Exploi-tation

Program (FMEP) and Army medical R&D requirements. The director coordinates planning, programming, and budgeting with the Army Deputy Chief of Staff, Intelligence (DCS, G2).

- **Missile and Space Intelligence Center (MSIC).** MSIC, based at Redstone Arsenal, AL, is a DOD intelligence production center under DIA control and supports the FMEP. The MSIC acquires, produces, maintains, and disseminates S&TI pertaining to missile and space weapons systems, subsystems, components, and activities. The S&TI produced at MSIC also covers foreign state-of-the-art technology and research applicable to missiles.
- **Defense HUMINT Service.** DHS conducts worldwide HUMINT operations in support of foreign materiel acquisition (FMA) and foreign materiel exploitation (FME).

10-7. The organizations and agencies discussed below constitute the Army TECHINT structure.

ARMY DCS, G2

10-8. The Army DCS, G2 exercises general staff responsibility for all Army TECHINT activities. The Army DCS, G2 forms policies and procedures for S&TI activities, supervises and carries out the Army S&TI program, coordinates DA staff and MSC requirements for TECHINT, and is responsible for the Army Foreign Materiel Program (FMP).

INSCOM

10-9. Under the direction of Headquarters, Department of the Army (HQDA), INSCOM is responsible for peacetime TECHINT operations. HQ, INSCOM, fulfills its responsibilities through its TECHINT oversight function and manages the Army's Foreign Materiel for Training (FMT) Program and FMEP. It provides the interface with strategic S&TI agencies in support of FME and organizes, trains, and equips EAC TECHINT organizations during peacetime. TECHINT exploitation within INSCOM is performed by the following elements:

- **National Ground Intelligence Center.** HQ, INSCOM, exercises direct OPCON over the NGIC. NGIC produces and maintains intelligence on foreign scientific developments, ground force weapons systems, and associated technologies. NGIC analysis includes but is not limited to military communications electronics systems, types of aircraft used by foreign ground forces, NBC systems, and basic research in civilian technologies with possible military applications.
- **203d Military Intelligence Battalion.** The 203d MI Battalion is a multi-component unit headquartered at Aberdeen Proving Ground, MD, and is the Army's sole TECHINT battalion. It performs the following functions:
  - Conducts TECHINT collection and reporting in support of validated S&TI objectives.
  - Acts as the HQDA executive agent for foreign materiel used for training purposes.

- Conducts TECHINT training for DOD analysts and RC TECHINT personnel.
- Supports INSCOM's FMA and FME operations as directed.
- Analyzes and exploits foreign captured enemy documents (CEDs), equipment, weapon systems, and other war materiel.
- Reports on the capabilities and limitations of enemy combat materiel.
- Provides reports alerting the command to the tactical threat posed by technical advances in new or recently discovered foreign or enemy materiel.
- Provides countermeasures to any enemy technical advantage.
- Provides foreign or enemy equipment for troop familiarization and training.
- Provides recommendations on the reuse of CEM.
- Supervises evacuating items of TECHINT interest.
- Provides task-organized battlefield TECHINT teams to support a subordinate command's TECHINT effort.

US ARMY MATERIEL COMMAND (AMC)

10-10. AMC plays a significant support role in TECHINT. Among AMC elements are a series of RDECs, the Army Research Laboratory System, and the Test and Evaluation Command (TECOM). Each conducts highly technical evaluations of foreign equipment. In peacetime, the AMC conducts FME on equipment purchased by each laboratory and RDEC for the intelligence community and for DOD as part of the International Materiel Evaluation Program (IMEP). AMC elements include—

- The Foreign Ordnance Exploitation Team, which is located at the Fire Support Armaments Center (FSAC) in the Picatinny Arsenal. This team exploits foreign ground ordnance and develops render safe procedures (RSPs) for foreign ordnance. It also prepares detailed intelligence reports to support explosive ordnance disposal (EOD), MI, and US munitions developers.
- The Science and Technology Center Europe and the Science and Technology Center Far East, which have the responsibility of collecting information on foreign technical developments by attending arms shows and technology exhibitions.
- The Soldier Biological Chemical Defense Command, which is the headquarters for the US Army Technical Escort Unit (TEU). The mission of the TEU is to collect and escort (transport) chemical, biological, radiological, and nuclear (CBRN) samples for testing and evaluation.

10-11. There are many other agencies with TECHINT responsibilities within the DOD. Refer to FM 34-54 for more information on TECHINT.

PLAN

10-12. TECHINT collection usually begins when an organization or individual reports the recovery or acquisition of unusual, new, or newly employed threat materiel. However, there are often indications that the threat may be using materiel not yet associated with the threat among the myriad intelligence products available. Conversely, it may be known that the threat is using a particular item, the capabilities of which are unknown to US forces. It is in these cases that a unit may receive a TECHINT related ISR task. It is conceivable that such a task may be reflected in the PIR. An example of such a task is the identification or verification of suspected external modifications on a particular model of an enemy's main battle tank; the results of which are linked to a commander's decision.

10-13. TECHINT related ISR tasks should include a mission or target folder. At a minimum, this folder should include a description of the item, with its associated major combat systems, as well as handling instructions, reporting instructions, and a photograph or sketch of the item if available.

10-14. Additionally, the G2/S2 is responsible for ensuring the staff coordinates and establishes a plan for evacuating the desired materiel.

PREPARE

10-15. The G2/S2 must ensure that required TECHINT analytical assets, resources, and evacuation means are prepared to provide support. This includes verifying coordination effected with the task-organized battlefield TECHINT teams from the 203d MI Battalion. The G2/S2 must also ensure the means to report and disseminate TECHINT results to the unit and its soldiers are in place so that they can immediately adopt appropriate countermeasures, operations, or tactics in order to enhance their survival and mission accomplishment.

COLLECT

10-16. TECHINT collection includes capturing, reporting, and evacuating CEM. TECHINT collection begins when an organization or individual reports the recovery or acquisition of threat materiel. An item of materiel is exploited at each level, and continues on through succeeding higher levels until an appropriate countermeasure to neutralize the item's capabilities is identified or developed.

10-17. Army personnel (soldiers and civilian) and units will normally safeguard CEM and report it through intelligence channels to the first TECHINT element in the reporting chain. The location of this TECHINT element will be in accordance with the METT-TC factors; however, there will usually be TECHINT representation at the Corps G2 or the COCOM J2. The TECHINT representative or element will verify if the type of materiel is of intelligence value and determine its further disposition in conjunction with the unit's staff.

PROCESS

10-18. TECHINT processing starts (simultaneously with collection) with the capture of a piece of equipment of TECHINT value. This confirms that the enemy is indeed employing this materiel. In accordance with METT-TC factors, a TECHINT team may move to the location of the item at the capture site or wait until the item is evacuated before conducting a hasty exploitation. After hasty exploitation, the team decides if further processing is required. If it is, the items are sent to the first (or nearest) Captured Materiel Exploitation Center (CMEC). If the item is deemed to yield no immediate tactical intelligence value, it may still be evacuated to the S&TI centers in CONUS for further analysis if the systems represent a change in the technological posture of an enemy.

PRODUCE

10-19. Battlefield TECHINT teams normally report initial and secondary examinations of CEM using either a preliminary technical report or a complementary technical report.

- A preliminary technical report—
  - Includes a general description of the item reported and recommended RSP.
  - Alerts others to information that can be used immediately by tactical units.
- A complementary technical report is more in-depth and—
  - Follows a secondary or an in-depth initial examination.
  - Allows the CMEC to compare new information with intelligence holdings.

10-20. At each successive echelon of exploitation, TECHINT analysts add to the overall body of information on an item by either adding to previous reports or by preparing new reports. The CMEC or other national level S&TI activities prepare more advanced technical reports and analyses. These reports include—

- Detailed technical reports.
- Translation reports.
- Special technical reports.

10-21. Other TECHINT products include—

- CMEC publications such as operator manuals, maintenance manuals, TECHINT bulletins, and tactical user bulletins.
- S&TI analysis bulletins.
- Foreign materiel exploitation reports.

ANALYZE

10-22. TECHINT analysts use checklists established by S&TI agencies and the CMECs to analyze each type of the adversary's equipment for which requirements exist. Analysis always begins with what is, and what is not, known about the piece of equipment. TECHINT units maintain procedures and plans for sampling, analyzing, and handling materiel.

DISSEMINATE

10-23. TECHINT of critical importance to the force, including answers to the PIR, is disseminated via the most expeditious means possible.

10-24. Routine TECHINT reports and products are usually transmitted through the unit's existing intelligence communications architecture in the format of an intelligence information report (IIR) format. For intelligence reach operations, TECHINT products are available and disseminated in a variety of forms. The requestor must ensure that the TECHINT product can be transmitted over the available communications systems. This includes verifying the appropriate security level of the communications systems.

ASSESS

10-25. The primary goal of the TECHINT assess function is to determine whether the results of TECHINT production meet the unit's PIR or intelligence requirements. The G2/S2 immediately assesses TECHINT products upon receipt for accuracy and relevance. He must inform the TECHINT producer of the extent to which the product answered the PIR or intelligence requirement. Providing feedback to TECHINT analysts helps improve the effectiveness and efficiency of TECHINT.

10-26. The G2/S2 also assesses the success of the unit's ISR effort in accomplishing any TECHINT associated ISR task and shares his assessment with the staff and the pertinent units or personnel.

Chapter 11

Counterintelligence

DEFINITION

11-1. CI counters or neutralizes intelligence collection efforts through collection, CI investigations, operations, analysis and production, and functional and technical services. CI includes all actions taken to detect, identify, exploit, and neutralize the multidiscipline intelligence activities of friends, competitors, opponents, adversaries, and enemies. It is the key intelligence community contributor to protect US interests and equities. Figure 11-1 shows the CI overview.

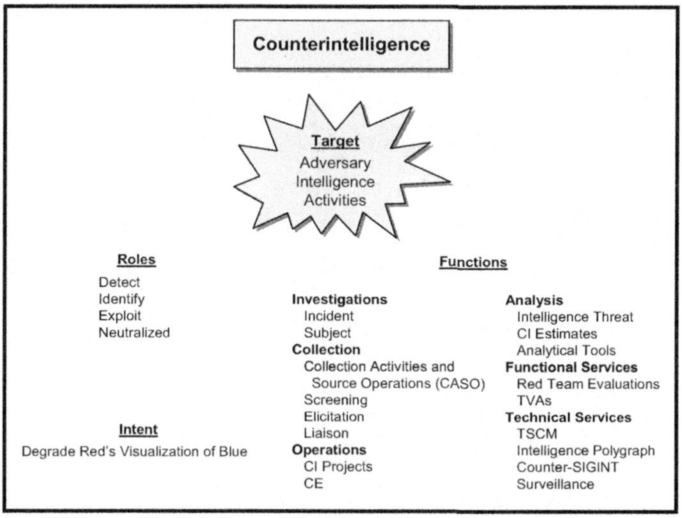

Figure 11-1. Counterintelligence Overview.

ROLE

11-2. The role of CI is to detect, identify, exploit, and neutralize all adversary intelligence entities targeting US and multinational interests. CI will focus on countering adversary intelligence collection activities targeting information or material concerning US personnel, activities, operations, plans, equipment, facilities, publications, technology, or documents—either classified or unclassified—without official consent of designated US release authorities, for any purpose that could cause damage or otherwise adversely impact the interests of national security of the US ability to fulfill national policy and objectives. Adversary intelligence threats include but are not limited to any (US, multinational, friendly, competitor, opponent, adversary, or recognized enemy) government or NGOs, companies, businesses, corporations, consortiums, groups, agencies, cells or persons, terrorists, insurgents, guerrilla entities, and persons whose demonstrated actions, views, or opinions are inimical to US interests.

11-3. CI elements are instrumental in contributing to situational awareness in the AOI. CI elements may corroborate other intelligence discipline information as well as cue other intelligence assets through the CI core competencies and through CI technical services. The CI core competencies are collection, investigations of national security crimes within the purview of CI, operations, and analysis and production. CI technical services include computer network operations, technical surveillance countermeasures (TSCM), and polygraph. CI focuses on combating adversary intelligence activities targeting Army personnel, plans, operations, activities, technologies, and other critical information and infrastructure.

COUNTERINTELLIGENCE FUNCTIONS

11-4. CI functions are interrelated, mutually supporting, and can be derived from one another. No single function or technical capability can defeat adversary intelligence efforts to target US interests. CI functions are discussed below.

CI INVESTIGATIONS

11-5. Investigative activity is essential to countering the adversary intelligence threat to Army interests. CI places emphasis on investigative activity to support force, infrastructure and technology protection, homeland defense, information assurance, and security programs. CI investigations focus on resolving allegations of known or suspected acts that may constitute National Security Crimes under US law. The primary objective in any CI investigation is the detection, identification, exploitation, and/or neutralization of adversary intelligence threats directed against the US Army. CI investigations are also conducted to identify systemic security problems that may have damaging repercussions to Army operations and national security interests. All CI investigations are conducted within guidelines established in AR 381-10, AR 381-12, AR 381-20, applicable DOD policy and directives, and US laws.

CI OPERATIONS

11-6. CI operations are characterized as those activities that are not solely associated with investigative, collection, analysis, or production functions. CI operations can be either offensive or defensive in nature; they are derived from or transition to a collection or investigative activity depending on the scope, objective, or continued possibility for operational exploitation. CI operations fall into two categories: CI support operations and CI sensitive operations.

- **CI Support Operations**. These are defensive operations used to support ARFOR and technology protection, security projects, and programs. They include technical services support, support to acquisition, FP, special access, international security, foreign visitor or contact, treaty verification, information assurance, homeland defense, and other approved projects and programs.
- **CI Sensitive Operations**. These operations are generally offensive in nature and involve direct or indirect operations against a known adversary intelligence threat. These operations include counter-espionage (CE) and CI projects and are conducted by designated units.

CI COLLECTION

11-7. **Collection Activities**. CI elements conduct collection activities focused on adversary intelligence threats that target US and multinational interests. CI collection is conducted through the use of sources, assets, official contacts, and other human or multimedia sources to obtain information that impacts the supported unit. CI will not be used as a substitute for Army HUMINT collection. These activities are designed to collect specific information or develop leads concerning adversary intelligence collection requirements, capabilities, efforts, operations, structure, personalities, and methods of operation targeting US and multinational interests. CI collection can result from ongoing CI investigations or operations or serve to initiate CI investigations and/or operations.

11-8. **Liaison**. CI elements conduct liaison with US, multinational, and HN military and civilian agencies, to include NGO, for the purpose of obtaining information of CI interest and coordinating or deconflicting CI activities. Liaison activities are designed to ensure a cooperative operating environment for CI elements and to develop CI leads for further exploitation.

11-9. **CI Collection Activities and Source Operations (CASO)**. CASO is used to collect information on direct threats to US Army installations, organizations, activities, and personnel. The CASO program is not intended to be used as a substitute for tactical HUMINT contact operations. CASO can be used to initiate CI investigations, identify potential leads for offensive operations, or develop additional CASO leads. **Only designated units, in accordance with applicable policy, will be authorized to pursue investigative and offensive operational leads from ongoing CASO operations.**

11-10. **Screening**. CI Special Agents work jointly with HUMINT Collectors during screening operations to identify civilians on the battlefield, EPWs, detainees, and other noncombatants who may have information of CI interest or to develop CI leads. CI screening is also conducted during the process of hiring HN citizens for Army and DOD employment. Information obtained during screening operations may be used to initiate CI investigations and operations or to cue other intelligence collection disciplines such as HUMINT, IMINT, SIGINT, and MASINT.

11-11. **Debriefing**. CI Special Agents conduct debriefings of friendly force, HN, or the local population who may have information of CI interest regarding adversary intelligence collection or targeting efforts focused on US and multinational interests.

11-12. **Functional Services**. CI elements conduct functional services to assist traditional CI activities of investigations, collection, operations, analysis, and production; they also provide tailored support to US, DOD, and Army protection and security programs for commanders at all echelons. CI elements may use one or several of the functional services simultaneously to provide tailored support to a particular CI mission or supported program. CI functional services consist of—

- CI threat vulnerability assessments (TVAs).
- Adversary intelligence simulation (Red Team Evaluation).
- Covering agent support.

11-13. **CI Technical Services**. CI technical services are used to assist the CI functions of investigations, collections, and operations or to support functional services conducted by CI elements. For additional information on CI technical services, refer to AR 381-20 (SECRET/NOFORN) and FM 34-5. CI technical services consist of—

- Surveillance.
- Intelligence polygraphs.
- TSCM.
- Computer Network Operations (CNO).
- IO.
- Counter-Signals Intelligence (C-SIGINT).

ANALYSIS

11-14. IPB, all-source, and single-source analysis are used to template adversary intelligence activities. Analysis is also used to make recommendations to the commander on how to counter adversary intelligence efforts and to refine CI activities to potentially neutralize and/or exploit those efforts and to continually focus the efforts of CI teams. CI analysis focuses on the multidiscipline adversary intelligence collection threat targeting information on US multinational personnel, operations, activities, technology, and intentions. Raw information, open-source material, and finished intelligence products are analyzed in response to local and national requirements. Analysis occurs at all levels from tactical to strategic.

- At the tactical level, CI teams focus their efforts on supporting mission requirements and contributing to the all-source COP.
- Operational analysis is used to assess how adversary intelligence views and targets US interests; identify US vulnerabilities that could be exploited or targeted; and to determine adversary intelligence targeting methods of operation (MOs).
- CI collection priorities; assessing adversary intelligence technical options for countering US weapons and intelligence systems; and

assessing the impact of technology transfer activities on the US military's technological overmatch.

PRODUCTION

11-15. CI products consist of, but are not limited to, target nomination, CI input to TVAs, CI estimates, and investigative and intelligence information reports. Finalized intelligence derived from CI activities are incorporated into joint and national intelligence databases, assessments, and analysis products. CI products are also incorporated into the COP to support battlefield situational awareness. CI production takes place at all levels.

- Operational and tactical production includes tactical alerts, spot reports, and current intelligence; CI threat and/or vulnerability assessments tailored to specific activities, units, installations, programs, or geographic areas; CI studies to support contingency planning and major exercises; studies of adversary intelligence organization, MO, personnel, activities, and intentions that pose a current or potential threat to the supported command.
- Strategic products include assessments supporting national and Army programs including SAPs and acquisition programs; worldwide assessments of the organization, location, funding, training, operations capabilities, and intentions of terrorist organizations; global trends in adversary intelligence MO; after-action studies of individual espionage cases; analyses of the intelligence collection capabilities of international narcotics trafficking organizations; and multimedia threat products to support Army CI awareness programs.

TECHNICAL SERVICES

11-16. CI organizations with technically trained CI Special Agents are chartered with providing unique technical capabilities to augment CI investigations, collection, and operations. These technical capabilities are not used as substitutes for CI activities, but support traditional CI techniques employed to counter and neutralize adversary intelligence activities targeting US interests.

11-17. Selected CI Special Agents are trained to identify human deception indicators during the course of investigations, operations, and collection missions. Required training and skills include the use of polygraph and emerging biometric technologies that can recognize indications of deception by human sources, contacts, and subjects of investigations, as well as analysis and reporting of results.

11-18. TSCM CI Special Agents are trained in sophisticated electronic and sensing equipment to identify technical collection activities carried out by adversary intelligence entities. The use of TSCM is critical to ensuring that sensitive and/or restricted areas are clear of any adversary intelligence-placed active or passive electronic sensing, or eavesdropping or collection devices. These areas include SCIF, secure working and planning areas, and C2 facilities.

11-19. CNO consist of computer network attack (CNA), computer network defense (CND), and computer network exploitation (CNE). CI Special Agents are specially trained in the areas of computer operation, network theory and administration, and forensics, along with IO in order to ensure US information dominance. The reliance on networked systems will result in greater emphasis being placed on information assurance. Specially selected CI agents will be trained in CNO in order to assist in protecting US information and information systems while exploiting and/or attacking adversary information and information systems.

11-20. For more information on technical services, refer AR 381-20 and FM 34-5.

OPERATIONAL EMPLOYMENT

11-21. Army CI supports the full spectrum of military operations. The Army requires a well-trained CI force consisting of AC, RC, civilian government employee, and contractor personnel. CI elements are focused on and dedicated to detecting, identifying, neutralizing, and/or exploiting adversary intelligence elements attempting to collect information on US forces. Effective employment of Army CI elements in all phases of operations and at all levels from tactical to strategic is paramount to countering any adversary intelligence threat to US interests and resources. Moreover, Army CI protects special programs that provide R&D, acquisition, and integration of technologies leading to future technological overmatch. CI elements have the capability and authority to conduct complex, nontraditional operations in this Information Age to CE, to both protect critical technologies and satisfy Army, DOD, and national level CI objectives.

LEVELS OF EMPLOYMENT

Strategic and Departmental

11-22. Strategic and departmental operations will be conducted by CI elements supporting national, DOD, and DA required missions (for example, support to NATO and special operations and missions). Strategic and departmental CI will conduct compartmented investigations and operations to affect the knowledge of adversary intelligence regarding CONOPS and defense information. Army CI will execute the full range of CI functions and missions at the strategic and departmental level including CI investigations and operations, CE, technology protection, SAP support, treaty verification, and technical CI services (polygraph, TSCM, and computer forensics). Strategic and departmental CI will also support SOF and special mission units (SMUs) within the scope of applicable national, DOD, and DA security policies and regulations.

Operational

11-23. Operational missions of CI elements will support combatant commanders, generally in geographic AOIR. Operational CI elements will focus on threat identification and countering regional adversary intelligence threats. Operational level CI activities and functions include CI investigations and operations, CE, technology protection, SAP support, treaty verification, and technical CI services (polygraph, TSCM, and computer forensics). CI elements must be capable of quickly transitioning from a peacetime mission to crisis operations to support combatant commander requirements. Theater CI assets will conduct unilateral to multinational operations in designated theaters. Operational elements may also be deployed to support or reinforce tactical forces in CONOPS.

Tactical

11-24. CI teams will conduct operations throughout the battlespace during CONOPS. CONOPS support activities include conduct of CASO, limited CI investigative capability, personnel security investigations, screenings, and debriefings. CI activities in CONOPS focus on countering the adversary intelligence threat and assisting in conducting TVAs. TVAs will be conducted in conjunction with MPs, Engineers, and Medical Service personnel to provide the commander with a comprehensive FP assessment. During peacetime, organic tactical CI teams conduct activities in accordance with approved regulations and command guidance.

SUPPORT TO CONTINGENCY OPERATIONS

11-25. The initial phase of operations from PME to MTW lays the foundation of future team operations. In general, the priority of effort focuses inward on security of operating bases, areas of troop concentration, and C2 nodes to identify the collection threat to US forces that could be used by adversary elements to plan hostile acts against US activities and locations.

11-26. Once security of the operating bases has been established, the operational focus of CI teams shifts outside the operating base to continue to detect, identify, and neutralize the collection threat to US forces as well as to provide I&W of hostile acts targeting US activities. The CI team uses several collection methods, to include CASO, elicitation, and liaison, to answer the supported commander's requirements. This is referred to as the continuation phase. The CI team conducts CI investigations to identify, neutralize, and exploit reported threat intelligence collection efforts.

11-27. A key element to the CI team's success is the opportunity to spot, assess, and develop relationships with potential sources of information. Operating as independent teams, without being tied to ISR or combat assets, enables the CI team's maximum interaction with the local population, thereby maximizing the pool of potential sources of information. Along with the opportunity to spot, assess, and interact with potential sources of information, a second key element of a CI team's success is its approachability to the local population. A soft posture enables a CI team to appear as non-threatening as possible. Experience has shown that the local population in general is apprehensive of fully and openly armed patrols and soldiers moving around population centers.

11-28. During some operations, civilian attire or nontactical vehicles may be used to lower the CI team profile. In some special situations, these measures are taken to make the operation less visible to the casual observer. Also, in some cultures, sharing food and beverages among friends is expected; exceptions to restrictions or general orders should be considered to facilitate successful CI team operations, many of which are geared towards developing relationships with potential sources of information.

SUPPORT TO INSTALLATIONS AND OPERATING BASES

11-29. CI teams, as part of a multi-agency team consisting of MPs, CA, medical, and EOD, support the conduct of TVAs of installations and operating bases to identify the intelligence threat to the operating locations. Detailed TVAs identify weaknesses in operational and physical security procedures and recommend countermeasures to mitigate intelligence collection on friendly forces limiting the ability to plan hostile acts on US activities and locations. CI activities supporting installations and operating bases include—

- Interviewing walk-in sources and locally employed personnel.
- Screening local national (LN) hires. Commanders, staff planners, and SIOs should always provide input to personnel assigned to establish and negotiate contracts using local national (LN) hires. This requirement ensures that LN hires can be screened, interviewed, and in some instances used as CI sources or assets in order to provide intelligence information that impacts the security of the base camp.
- Debriefing friendly force personnel who are in contact with the local population, such as—
  - ISR patrols.
  - MP patrols.
  - Combat patrols.
  - Liaison personnel.
  - CA and PSYOP teams.
- Conducting limited local open-source information collection.
- Providing support to TVAs of the base camp.

TACTICS, TECHNIQUES, AND PROCEDURES

11-30. At the CI team level, team members conduct mission analysis and planning specific to their AO. Backwards planning and source profiling are used extensively to choose CI targets. To verify adequate area coverage, the CI team may periodically develop and use CI target overlays and other CI analytical tools that illustrate the CI situation, identify CI gaps, and help refocus the collection effort.

11-31. The CI team is also in constant contact with the supported S2 and the other ISR assets (Scouts, PSYOP, CA, and MP) in order to coordinate and deconflict operations and to cross-check collected information. The supported unit S2, with the help of the CI team, regularly and systematically debriefs all ISR assets.

11-32. The CI team must be integrated into the supported unit's ISR plan. The CI OMT chief will advise the supported unit on the specific capabilities and requirements of the team to maximize mission success.

OPERATIONAL RISK MITIGATION

11-33. The employment of CI teams includes varying degrees of contact with the local population. As the degree of contact with the population increases, both the quantity and quality of CI collection increases. In many instances, however, there is a risk to the CI team inherent with increased exposure to the local population. The decision at what level to employ a CI team is METT-TC dependent. The risk to the CI assets must be balanced with the need to collect priority information and to protect the force as a whole. ROE, SOFA, direction from higher headquarters, and the overall threat level may also restrict the deployment and use of CI teams. The commander should consider exceptions to the ROE to facilitate CI collection.

11-34. Risks are minimized through the situational awareness of CI team members. They plan and rehearse to readily react to any situation and carry the necessary firepower to disengage from difficult situations. If it becomes necessary to call for assistance, adequate and redundant communications equipment is critical. These scenarios and actions should be trained prior to deployment into a contingency area and rehearsed continuously throughout the deployment.

11-35. A supported unit commander is often tempted to keep the CI team "inside the wire" when the THREATCON level increases. The supported commander must weigh the risk versus potential information gain when establishing operational parameters of supporting CI teams. This is necessary especially during high THREATCON levels when the supported unit commander needs as complete a picture as possible of the threat arrayed against US and multinational forces.

11-36. When it is not expedient to deploy the CI team independently due to threat levels or other restrictions, the team can be integrated into other ongoing operations. The CI team may be employed as part of a combat, ISR, or MP patrol or used to support CA, PSYOP, engineer, or other operations. This method reduces the risk to the team while allowing a limited ability to collect information. It has the advantage of placing the team in contact with the local population and allowing it to spot, assess, and interact with potential sources of information. However, this deployment method restricts collection by subordinating the team's efforts to the requirements, locations, and timetables of the unit or operation into which it is integrated and does not allow for the conduct of sensitive source operations. **This method of employment should be considered a last resort.**

COUNTERINTELLIGENCE EQUIPMENT

11-37. Basic C2, transportation, and weapons requirements do not differ significantly from most soldier requirements and are available as unit issue items. However, CI teams have unique communications, collection, processing, and mission-specific requirements.

COMMUNICATIONS

11-38. **Dedicated and Secure Long-Range Communications.** These are keys to the success of the CI team mission. CI team operations require a secure, three-tiered communications architecture consisting of inter/intra-team radios, vehicle-based communications, and a CI and HUMINT base station.

11-39. **Communications Network.** The CI team must have access to existing communications networks such as the tactical LAN. The CI team must also be equipped with its own COMSEC devices. It is imperative that the CI team acquire access to the public communication system of the HN. This can be in the form of either landlines or cellular telephones. Such access enables the CI team to develop leads which can provide early indicators to US forces.

11-40. **Interoperability.** Communications systems must be equipped with an open-ended architecture to allow for expansion and compatibility with other service elements, government organizations, NGOs, and multinational elements to effectively communicate during CONOPS. All ISR systems must be vertically and horizontally integrated to be compatible across all BOSs and with Legacy and Interim Force elements.

11-41. **SOTM.** To provide real-time and NRT information reporting, CI elements must have the capability to transmit voice, data, imagery, and video while on the move. CI teams must be able to transmit while geographically separated from their parent unit while operating remotely. This broadband requirement can only be achieved through a SATCOM capability and must be achievable while mobile.

CI COLLECTION AND PROCESSING SYSTEMS

11-42. The CI team must rely on automation to achieve and maintain information dominance in a given operation. With time, effective collection planning and management at all echelons, the CI team can collect a wealth of information. The sorting and analysis of this information in a timely and efficient manner is crucial to operations. Automation helps the CI team to report, database, analyze, and evaluate the collected information quickly and to provide the supported unit with accurate data in the form of timely, relevant, accurate, and predictive intelligence.

11-43. Automation hardware and software must be user friendly as well as interoperable among different echelons and services. They must interface with the communications equipment of the CI team as well as facilitate the interface of audiovisual devices. Technical support for hardware and software must be available and responsive.

11-44. The demand for accurate and timely CI reporting, DOCEX, and open-source information has grown tremendously. Biometric (physiological, neuro-logical, thermal analysis, facial and fingerprint recognition) technologies will allow rapid identification, coding, and tracking of adversaries and human sources; as well as cataloging of information concerning EPWs, detainees, and civilians of CI interest on the battlefield. Biometrics will also provide secure authentication of individuals seeking network or facility access.

11-45. CI teams work with multinational forces and other foreign nationals and require the ability to communicate in their respective languages. Often CI personnel have little or no training in the target language, and lack of skilled interpreters can hinder CI activities. CI teams require textual and voice translation devices, source verification, and deception detection machines (biometrics) to improve collection capability and accuracy.

11-46. CI teams require dynamic MLT tools that provide both non-linguists and those with limited linguist skills a comprehensive, accurate means to conduct initial CI screenings and basic interviews in a variety of situations. CI elements will focus on in-depth interviews and communications with persons of higher priority. MLT tools minimize reliance on contract linguists and allow soldiers to concentrate on mission accomplishment.

MISSION SPECIFIC

11-47. The CI team may conduct night operations and must be equipped with NVDs for its members, and photographic and weapons systems. The CI team also may operate in urban and rural areas, where the threat level can vary from semi-hostile to hostile. The safety of the CI team can be enhanced with equipment that can detect, locate, suppress, illuminate, and designate hostile optical and E-O devices. In addition, high power, gyro-stabilized binoculars, which can be used from a moving vehicle, increases the survivability of the CI team and also gives the team another surveillance and collection device.

11-48. Some of the CI team missions may require the documentation of incidents. The CI teams can use the following equipment in their open-source collection efforts.

- Small, rugged, battery-operated digital camcorders and cameras which are able to interface with the collection and processing systems as well as communication devices.
- GPSs that can be mounted and dismounted to move in the AO efficiently.
- Short-range multichannel RF scanning devices that can also identify frequencies which enhance their security.
- In some cases CI teams require a stand-off, high resolution optical surveillance and recording capability that can provide target identification at extended ranges to protect the intelligence collector while avoiding detection by the adversary target. An advanced optical capability provides intelligence collectors the ability to locate and track adversary targets (passive and hostile) for identification, collection, and target exploitations.

INTEGRATION OF LINGUISTS

11-49. Integrating linguists into the CI team should take place as soon as possible. Security clearances and contractual agreements will help the team determine the level of integration.

11-50. Along with the basic briefing of what is expected of the civilian linguists as interpreters, CI teams should be informed about the civilians' chain of command and the scope of their duties beyond interpreting. The CI team leader must ensure that linguists are trained and capable of completing all tasks expected of them.

BATTLE HAND-OFF

11-51. CI teams are always engaged. A good battle hand-off is critical to smooth transition and mission success. The battle hand-off can directly contribute to mission success or failure of the outgoing team, but especially of the incoming team. The battle hand-off begins the first day the CI team begins to operate in an AO. Regardless of how long the team believes it will operate within the AO, it must ensure there is a seamless transition to an incoming team, other US unit, or agency. The CI team accomplishes this transition by establishing procedures for source administration, database maintenance, and report files.

11-52. Teams must plan and implement a logical and systematic sequence of tasks that enables an incoming team to assume the operations in the AO. Adequate time must be allotted for an effective battle hand-off. In some environments, a few weeks may be necessary to accomplish an effective battle hand-off. Introductions to sources of information, especially CASO sources, are critical, and teams must prioritize their time. During this time the outgoing CI team must familiarize the new CI team with all aspects of the operation, which include past, present, and planned activities within the AO. Area orientation is critical. These include major routes, population centers, potential hot spots, and other points of interest (such as police stations, political centers, and social centers).

ORGANIZATION

11-53. CI activities require a complex C2 relationship to ensure that the requirements of the supported commander are fulfilled while balancing the need for strict integrity and legality of CI operations. This complex relationship balances the role of the SIO as the requirements manager and the 2X as the mission manager with the MI commander as the asset manager.

COMMAND VERSUS CONTROL

11-54. ARFOR will normally deploy as part of a joint, multinational, and/or combined operation. In all cases, commanders at each echelon will exercise command over the forces assigned to their organization. Command includes the authority and responsibility for effectively using resources, planning for and employment of forces, and ensuring that forces accomplish assigned missions. Leaders and staffs exercise control to facilitate mission accomplishment.

11-55. While the MI commander supervises subordinates and produces reports, the *2X synchronizes activities between intelligence units and provides single-source processing and limited analysis. **(*2X " indicates 2X functions at all levels.)** While the MI commander takes care of the operators executing missions, the *2X obtains the data and reports from higher echelons required to execute the missions.

STAFF RESPONSIBILITIES AND FUNCTIONS

11-56. The *2X staff is responsible for the integration, correlation, and fusion of all Human Sensor information into the Intelligence BOS within the *2X AOIR. The *2X is also responsible for analyzing adversary intelligence collection, terrorist and sabotage activities, developing countermeasures to defeat threat collection activities, identifying and submitting collection requirements to fill CI collection gaps, and providing input to the all-source picture regarding adversary intelligence activities.

11-57. The *2X Staff Officer provides CI and HUMINT collection expertise. The *2X—

- Is the single focal point for all matters associated with CI and HUMINT in the AOIR.
- Is the CI and HUMINT advisor to the G2 and commander.
- Is an extension of the collection manager and ensures that the best asset or combinations of assets are used to satisfy information requirements.
- Along with his subordinate elements—CICA, HOC, OSC, CIAC, and HAC—exercises technical control over his assigned Army CI and HUMINT elements in the designated AOIR.
- Is the principal representative of the G2 and the commander when coordinating and deconflicting CI and HUMINT activities with national or theater agencies operating in the AOIR.
- Supports specific RM efforts in conjunction with the requirements manager through the planning and coordination of CI and HUMINT operations; the review and validation of CI requirements; the recommendation for assignment of tasks to specific collectors; and the conduct of liaison with non-organic HUMINT collection. This liaison includes national level and multinational force assets for source deconfliction and special activities outside the *2X AOIR.
- Will provide OMTs with capability to reach back to current database information, technical information and guidance, and source deconfliction necessary to monitor the collection activities of the CI teams.
  - *CICA.* The CICA is responsible for coordinating and synchronizing all CI activities in the designated AOIR. The CICA exercises technical control over all CI entities in the designated AOIR and deconflicts CI activities with higher, lower, and adjacent CI elements. The CICA accomplishes all responsibilities through coordination with the operational units and other *2X staff elements.
  - *OSC.* The OSC in the *2X staff maintains the source registry for all CI activities in the designated AOIR. The OSC provides management of intelligence property book operations, source incentive programs, and ICFs.
  - *CIAC.* The CIAC analyzes adversary intelligence collection capabilities. The CIAC leverages all intelligence discipline reporting and analysis to counter threat collection capabilities against the deployed force. CIAC analysis provides information and analysis to the COP.

11-58. The ACE and JISE CI analysis team analyzes threat intelligence collection and the intelligence collection efforts of foreign organizations involved in terrorism and sabotage in order to develop countermeasures against them. CI analysis cross-cues IMINT, SIGINT, MASINT, and TECHINT resources in addition to CI-related HUMINT reporting and analysis to counter threat collection capabilities against the deployed force. While the HAC supports the positive collection efforts of the force, the CI analysis team supports the "defend" aspects of the commander's FP program.

11-59. CI analysis is the analysis of the adversary's HUMINT, IMINT, SIGINT, and MASINT capabilities in support of intelligence collection, terrorism, and sabotage in order to develop countermeasures against them. It involves a reverse IPB process in which the analyst looks at US forces and operations from the threat's perspective. CI analytical products are an important tool in the COA development in the MDMP. This analytical tool supports the commander's FP program and facilitates the nomination of CI targets for neutralization or exploitation. (See FM 2-01.2 (FM 34-60) for more information on CI analysis.) Specifically, CI analysis—

- Produces and disseminates CI products and provides input to INTSUMs.
- Provides collection requirements input to the CICA.
- Analyzes source reliability and credibility as reflected in reporting and communicating that analysis to the collector.
- Nominates CI targets for neutralization or exploitation.
- Identifies and submits CI-related requirements to fill collection gaps.
- Assists HAC personnel in focusing the CI aspects of the HUMINT collection program.
- Presents CI analysis products such as CI estimates, target lists, reports, and graphics that support the commander.

11-60. For intelligence reach operations, CI products are available and disseminated in a variety of forms. It is incumbent on the requestor to ensure the CI product can be transmitted over the available communications systems. This includes verifying the appropriate security level of the communications systems.

CI TEAM STRUCTURE

11-61. **OMT.** The OMT is a four-person team consisting of a WO, two NCOs, and a junior enlisted soldier. (Civilians may be inserted into this structure as appropriate.) Rank structure and standards of grade for OMTs will vary depending upon the skill sets required and mission focus. CI OMTs will provide operational guidance for 1 to 4 CI teams, depending on mission focus and operational tempo. When two or more CI teams are deployed in a DS role, an OMT is also deployed to provide technical control. The OMT works closely with the supported S2 and ACT to furnish current threat information and to answer the supported commander's PIRs and IRs. OMTs coordinate with the supported 2X and manage subordinate CI teams to—

- Provide guidance and technical control of operational activity.
- Provide the collection and operational focus for CI teams.

- Provide quality control and dissemination of reports for subordinate CI teams.
- Conduct single-discipline CI analysis and assist in mission analysis for the supported commander.
- Act as a conduit between subordinate CI teams, the CICA, and supported unit headquarters.
- Provide administrative support for subordinate CI teams to include reporting mission and equipment status to the CICA and the supported unit headquarters.
- Educate the supported commander on the capabilities of the CI teams.
- Integrate the CI teams directly into the maneuver commander's ISR planning.

11-62. **CI Team.** The CI team is a four-person team consisting of two NCOs and two junior enlisted personnel. Rank structure and standards of grade for CI teams will vary depending upon the skill sets required and mission focus. CI teams are trained to execute the full range of CI functions; however, they may be assigned to mission-focused elements (for example, CE, CI projects). Assignment to a TSCM, polygraph, or information warfare team requires additional, specialized technical training.

Appendix A

Intelligence and Information Operations

THE INFORMATION ENVIRONMENT

A-1. The information environment is the aggregate of individuals, organizations, and systems that collect, process, store, display, and disseminate information; also included is the information itself. (FM 3-0) The information environment, it should be noted, is not an exclusively military one; in fact, the military applications of information are almost obscured in today's universal usage of the information spectrum by national, international, and non-state players.

THE COMMANDER AND INFORMATION

A-2. Information is facts, data, or instructions in any medium or form; the meaning that a human assigns to data by means of the known conventions used in their representation. (JP 1-02) Information provides the key to battlefield success in the 21st century. Commanders must have detailed information to command. Information is the medium that allows the commander's decisionmaking and execution cycle to function. Information gives direction to actions by the force, identifies the enemy's centers of gravity, provides COAs for force activity, and enables the force to accomplish its operational mission.

INFORMATION SUPERIORITY

A-3. Information superiority is the operational advantage derived from the ability to collect, process and disseminate an uninterrupted flow of information while exploiting or denying an adversary's ability to do the same. (FM 3-0) Relevant information drawn from intelligence supports the creation and development of situational understanding that contributes directly to information superiority during decisive operations. The requirement for information superiority is not new; what is new is that today's information technologies are creating a base of knowledge for military planning and execution that is unprecedented in scope, volume, accuracy, and timeliness. This means commanders receive accurate, timely information that enables them to make better decisions and act faster than their adversaries.

A-4. Information superiority, however, is neither a staple in today's battlespace nor is it necessary in military operations as a constant condition; rather, at specific times during operations, information superiority becomes a key enabler for assuring military success. At the operational level of predominant interest to the land component commander (LCC), information superiority is realized through the integration of operational level, interdependent ISR, information management, and IO to gain and maintain operational initiative and to achieve an operational advantage.

INFORMATION OPERATIONS

A-5. IO are the employment of core capabilities of electronic warfare, computer network operations, PSYOP, military deception, and operations security, in concert with specified supporting and related capabilities to affect or defend information and information systems, and to influence decisionmaking.... Information systems are...the equipment and facilities that collect, process, store, display, disseminate information. These include computers, hardware, software, and communications, as well as policies and procedures for their use. (FM 3-0). Offensive IO are capable of degrading an adversary's will to resist and ability to fight. Defensive IO measures passively and actively protect friendly information and C2 systems and limit their vulnerability.

A-6. Offensive and defensive IO are conducted as a fully coordinated effort to ensure the complementary, asymmetric, and reinforcing effects to attack enemy forces, influence others, and protect friendly forces. Relevant information assures that the right person has the right information at the right time for decisionmaking and execution.

THE ELEMENTS OF IO

A-7. Full-spectrum IO incorporates, integrates, and synchronizes traditionally independent capabilities and activities in support of the commander's mission. (See FM 3-0 and FM 3-13.) The Army's doctrinal view of full-spectrum IO as evolving core capabilities are—

- PSYOP.
- OPSEC.
- EW.
- Military deception.
- CNO.
  - CNA.
  - CND.
  - CNE.

A-8. Supporting IO capabilities are—

- Physical destruction.
- Physical security.
- Information assurance.
- Counterpropaganda.
- Counterdeception.
- Counterintelligence.

IO-RELATED ELEMENTS

A-9. The IO-related activities of PA and CMO are closely associated and integrated with the elements of IO as key contributors to information superiority.

INTELLIGENCE SUPPORT TO IO

A-10. All-source intelligence support, encompassing all of its forms, is the principal enabler for successful IO. The commander's capability to synchronize military operations can be heavily influenced by the ability to identify the threat and understand the adversary's capabilities and intentions in the information environment.

A-11. It is essential that the elements of the command's IO capabilities and vulnerabilities are integrated into the command intelligence plan. This leads to COAs that synchronize the elements of IO and IO-related activities into the commander's warfighting plans. Integration of the full-spectrum aspects of IO ensures that the relative importance of information is recognized in the development of a synchronized OPLAN. Further, integration of IO into the planning process provides a methodology for analyzing the threat from a knowledge base that enhances protection of friendly systems and assets while exposing windows of opportunity for attack or exploitation. IPB is the best process we have for understanding the battlefield and the options it presents to friendly and threat forces.

A-12. Provide Intelligence support to IO includes, but is not limited to, the following:

- **Provide Intelligence Support to Offensive IO**. The Intelligence BOS supports offensive IO by providing information to identify critical enemy C2 nodes. Intelligence also helps to identify enemy systems and procedures that may be vulnerable for offensive IO. Additionally, intelligence plays a key role in evaluating and assessing the effectiveness of offensive IO.
  - *Provide Intelligence Support to PSYOP.* This task identifies the cultural, social, economic, and political environment of the AO. It identifies target groups and subgroups and their location, conditions, vulnerabilities, susceptibilities, cultures, attitudes, and behaviors. PSYOP influence foreign target audiences in the AO to support achieving the commander's goals in the AO.
    - Identify the cultural, social, economic, and political environment of the AOI; for example, adversary mechanisms for political control, adversary communication and broadcast systems used to elicit support from the populace, and current and past adversary propaganda activities and their effectiveness.
    - Identify target groups and subgroups and their location, conditions, vulnerabilities, susceptibilities, cultures, attitudes, and behaviors. This includes determining the audience demographics of popular radio and television programs and periodicals; groups influenced by media personalities and political cartoons.
    - Identify impact of planned PSYOP on individuals outside the targeted group (for example, multinational partners, neighboring populations).
  - *Provide Intelligence Support to Military Deception.* This task identifies the capabilities and limitations of the adversary's

intelligence-gathering systems and identifies adversary biases and perceptions.

  - Profiles of key adversary leaders.
  - Cultural, religious, social, and political characteristics of the country and region.
  - Sources of military, economic, or political support.
  - Adversary decisionmaking processes, patterns, and biases.
  - Adversary perceptions of the military situation in the AO.
  - Capabilities and limitations of adversary CI and security services.
- *Provide Intelligence Support to Electronic Attack.* This task supports electronic attack employing jamming, electromagnetic energy, or directed energy against personnel, facilities, or equipment. It identifies critical adversary information systems and C2 nodes. It includes determining and presenting the adversary's electronic OB and their information system infrastructure. The enemy's C2 system vulnerabilities and the means they use to protect their C2 systems is part of the electronic OB.

- **Provide Intelligence Support to Defensive IO**. The Intelligence BOS supports defensive IO by providing information to identify threat IO capabilities and tactics. Intelligence provides information relating to CND, physical security, OPSEC, counterdeception, and counterpropaganda. The Intelligence BOS supports defensive IO by providing information to identify threat IO capabilities and tactics. Intelligence provides information relating to CND. Provide Intelligence Support to OPSEC—identify capabilities and limitations of the adversary's intelligence system to include adversary intelligence objectives and the means, methods, and facilities used by the threat to collect, process, and analyze information—supports the identification of indicators that could be interpreted or pieced together to penetrate EEFI in time to be useful to adversaries.
- **Provide Intelligence Support to Activities Related to IO**. The Intelligence BOS when operating outside US territories supports activities related to IO under some circumstances.
  - *Provide Intelligence Support to CMO.* This task allows military intelligence organizations to collect and provide information and intelligence products concerning foreign cultural, social, economic, and political elements within an IO in support of CMO. Identify cultural, social, economic, and political environment of the AOI, including—
    - Population demographics.
    - Civilian populace attitudes, alliances, and behavior.
    - Availability of basic necessities (food, clothing, water, shelter, medical care) and the ability of the populace to care for itself.
    - Access to medical care.
    - Locations and potential routes, destinations, and assembly areas or sites of displaced persons.
    - Local government type, status, organization, and capabilities.

- Availability of local material and personnel to support military operations.
- NGOs in the AOI, their agenda, resources, and capabilities.

- *Provide Intelligence Support to Public Affairs.* This task identifies the multinational and foreign public physical and social environment, as well as world, HN national, and HN local public opinion, in addition to the propaganda and misinformation capabilities, activities, targets, themes, and dissemination means of the adversary. Identify world, national, and local public opinion (location, biases or predispositions, and agenda of national and international media representatives in the AOI, and trends reflected by the national and international media).

Appendix B

Linguist Support

ROLE OF LINGUISTS

B-1. Military operations are highly dependent on foreign language support. The requirement to communicate with and serve on multinational staffs, communicate with local populations, and exploit enemy forces necessitates the use of linguists. The growing focus on multinational operations increases the competition for limited linguist resources that are vital for mission success. This appendix establishes the framework and process to access, prioritize, and employ the Army's limited organic linguist resources.

LINGUISTIC SUPPORT CATEGORIES

B-2. Foreign language support requirements of US Armed Forces typically fall into one of four broad categories:

- Intelligence and Information Gathering. This category includes the traditional SIGINT and HUMINT disciplines, as well as foreign language support to FP and exploitation of open-source information.
- CMO. This category encompasses all functions relating to military interaction with the civilian population. Foreign language support is critical to CMO in areas such as government liaison, legal agreements, medical support and operations, law enforcement, engineering projects, public safety, security and population control, CA, and PSYOP.
- Logistics. This category consists of foreign language support to sustainment or transportation functions. These include logistical contracting, port, railhead, airhead, or transshipment operations and convoy operations.
- Multinational Operations and Liaison. This category includes the coordination of military operations and liaison with multinational partners, previously unaffiliated nations, and at times adversary or former adversary nations. Multinational operations are becoming more common and increasingly important.

DETERMINING LINGUIST REQUIREMENTS

B-3. To identify linguist requirements, the staff conducts mission analysis and identifies specified or implied tasks requiring foreign language support. Other critical factors are the organization or echelon of command and the location of the mission. The staff uses these criteria to determine the allocation of linguists, such as one linguist team per echelon of command, one linguist per piece of equipment, or one linguist team per location where the function is to be performed. The staff then applies task organization and scheme of maneuver to determine the number of linguists needed for an operation.

B-4. The staff must analyze each linguist assignment to determine the minimum level of foreign language proficiency needed. While interpretation for a peace negotiation requires not only outstanding linguistic capability but also cultural acumen, the translation of routine documents (with the aid of a dictionary) requires a much different skill set. Poor identification of linguist proficiency requirements can tie up the best linguists in less effective roles, creating linguist shortfalls in other areas.

B-5. The relative importance of each of the four linguist support categories is mission dependent. For example, during a NEO civil and military coordination would probably not be as critical as intelligence and information gathering. However, the situation is reversed for a humanitarian assistance mission in which CMOs have a significant impact on mission success. Identifying these "dynamics" helps the commander and staff prioritize linguist requirements.

B-6. Determining linguist requirements for any operation can be difficult because each operation is unique. However, commanders and staffs with a basic knowledge of organic Army linguistic assets, foreign language resource alternatives, and MI skills can successfully assess, prioritize, and employ linguists in support of their military operations.

PLANNING AND MANAGING LINGUIST SUPPORT

B-7. Commanders must consider the linguist requirements as part of their MDMP for every CONPLAN and OPLAN assigned to their commands. Prior staff planning and identification of linguist requirements should prompt commanders to initiate linguist support requests and identify command relationships prior to actual operations. If the mission analysis reveals requirements for linguistic support, the commander must identify what foreign languages are needed, the foreign language proficiency levels needed for each assignment, and the best source of linguists. In addition, if the mission includes intelligence and information collection, the commander must identify MI collection skills required. During mission analysis, the commander should consider linguist requirements for every CONPLAN and OPLAN assigned to his command.

LINGUIST CATEGORIES

B-8. The commander and staff must identify linguist requirements by category:

- Category I – Have native proficiency in the target language (level 4-5) and an advanced working proficiency (Interagency Language Round Table [ILRT] level 2+) in English. They may be locally hired or from a region outside the AO. They do not require a security clearance. They must be screened by the Army CI support team.
- Category II – Are US citizens screened by Army CI personnel and are granted access to SECRET by the designated US government personnel security authority. Have native proficiency in the target language (level 4-5) and an advanced working proficiency (ILRT 2+) in English.
- Category III – Are US citizens screened by Army CI personnel and are granted either TS/SCI clearance or an interim TS/SCI clearance by the

designated US government personnel security authority. Meet a minimum requirement of ILRT level 3. They are capable of understanding the essentials of all speech in a standard dialect. They must be able to follow accurately the essentials of conversation, make and answer phone calls, understand radio broadcasts and news stories, and oral reports (both of a technical and non-technical nature).

PRIMARY STAFF RESPONSIBILITIES

B-9. Primary staff at each echelon has responsibilities for evaluating requirements and managing linguist support. The responsibilities include but are not limited to those discussed below. In addition, each staff section is responsible for determining its linguist support required to meet its operational missions.

Assistant Chief of Staff, G1 (S1):

- Identify linguist requirements needed to support G1/S1 functions in all contingency areas. G1/S1 requirements for linguist support include but are not limited to the following:
  - Coordinate with local authorities on matters of civilian hire, finance, and recordkeeping.
  - Contract for local hire personnel.
  - Coordinate for local morale support and community activities.
  - Coordinate with local authorities for postal operations.
  - Support for administration, counseling, personal affairs, and leave for LN and third-country national (TCN) personnel.
  - Coordinate for local medical support.
  - Liaison with multinational counterparts.
- Linguist staffing and linguist replacement management.
- Identify foreign language skill identifiers for all assigned, attached, or OPCON Army linguists.
- Identify all Army foreign language skilled soldiers not identified on electronic Military Personnel Office System (eMILPO) and Defense Integrated Management Human Resource System (DIMHRS). The Standard Installation Division Personnel System (SIDPERS) was replaced by eMILPO.
- Deploy and provide administrative support of DA and DOD civilian linguists.
- Hire, contract for, and provide administrative support of LN linguists.
- Procure Army foreign language support personnel for screening local labor resources.

Assistant Chief of Staff, G2 (S2):

- Identify linguist requirements needed to support G2/S2 functions in all contingency areas. G2/S2 requirements for linguist support include but are not limited to—
  - Evaluate and/or use local maps and terrain products in operations.
  - Process for MI purposes material taken from EPWs or civilian internees.

  - At lower echelons, conduct tactical questioning of refugees, detainees, and EPWs.
  - Assess local open-source information for intelligence value.
  - Coordinate intelligence and liaison with multinational and HN counterpart.
- Determine, during the initial IPB, all foreign languages (spoken and written) and dialects needed for mission accomplishment.
- Collect, process, produce, and disseminate information derived from linguist sources.
- Provide intelligence training for MI linguists employed in AOs.
- Coordinate for security investigations, as necessary, for local hire linguists.
- Provide support to CI screening of contracted linguists and LN labor force.

Assistant Chief of Staff, G3 (S3):

- Identify linguist requirements needed to support G3/S3 functions in all contingency areas. G3/S3 requirements for linguist support include but are not limited to—
  - Operational coordination and liaison with multinational and HN counterparts.
  - Translate OPORDs and OPLANs for use by multinational counterparts.
- Consolidate unit linguistic requirements and establish priorities.
- Develop linguist deployment and employment plans.
- Develop plans to train linguists and to use linguists for training the force in AO's foreign language survival skills. In addition to global language skills, linguists must have training in specific vocabulary used in the AO; for example, terms used for military, paramilitary, civilian or terrorist organizations, and ethnic groups within the area, nomenclatures of equipment used, and other military or technical vocabulary. Training in the specific dialect used in the AO would also be beneficial.
- Assign, attach, and detach linguists and linguist teams.
- Integrate additional or replacement linguists through operational channels.
- Recommend modernization and development of linguist systems and methods.
- Coordinate mobilization and demobilization of RC linguist support.
- Plan linguist usage for deception operations.
- Plan linguist support for movement of EPWs, detainees, and refugees.
- Coordinate evaluation of linguist support by all staff elements.

Assistant Chief of Staff, G4 (S4):

- Identify linguist requirements needed to support G4/S4 functions in all contingency areas. G4/S4 linguist requirements for linguist support include but are not limited to—

    - Procure local supply, maintenance, transportation, and services.
    - Coordinate logistics at air and seaports of debarkation.
    - Contract with local governments, agencies, and individuals for sites and storage.
- Provide logistical, supply, maintenance, and transportation support to attached linguists.

Assistant Chief of Staff, G5 (S5):

- Identify linguist requirements needed to support G5/S5 functions in all contingency areas. G5/S5 linguist requirements for linguist support include but are not limited to—
    - Determine civilian impact on military operations.
    - Minimize civilian interference with combat operations.
    - Inform civilians of curfews, movement restrictions, and relocations.
    - Provide assistance to liaison with HN and multinational agencies, dignitaries, and authorities.
    - Promote positive community programs to win over support.
    - Determine if multinational operations PSYOP efforts are mutually planned and synchronized.
    - Interpret support to assist resolution of civilian claims against the US Government.
    - Solicit linguistic and cultural knowledge support to protect culturally significant sites.
    - Use linguistic and cultural support to identify cultural and religious customs.
- Assist the G1 in the contracting of local hire linguists.
- Identify foreign language requirements for CMOs.

Assistant Chief of Staff, G6 (S6):

- Identify linguist requirements needed to support G6/S6 functions in all contingency areas. G6/S6 linguist requirements for linguist support include but are not limited to—
    - Coordinate suitable commercial information systems and services.
    - Coordinate with multinational forces on command frequency lists.
    - Coordinate signal support interfaces with HN and multinational forces.
- Manage RF assignments for supporting SIGINT linguist elements.
- Support linguist operations with internal document reproduction, distribution, and message services.
- Integrate automation management systems of linguist units.

SPECIAL STAFF OFFICER RESPONSIBILITIES

B-10. If no special staff officer is assigned the duties below, the corresponding coordinating staff officer should assume those responsibilities. Linguist requirements for special staff officers include but are not limited to the following staff officers.

Liaison Officer:

- Should speak the required foreign language. If not, he requires a translator or interpreter for all aspects of his duties.
- Request interpreters to assist when representing the multinational operations.
- Translate orders, maps, traces, overlays, and documents into multinational foreign languages.

Civilian Personnel Officer:

- Recruit, interview for suitability, and hire civilian local labor force if required.
- Negotiate host country on labor agreements.

Dental Surgeon:

- Administer dental care to support humanitarian mission requirements.
- Rehabilitate, construct, and gain usage of existing dental facilities as required.

Finance Officer:

- Support the procurement process of local goods and services not readily available through normal logistical channels.
- Ensure limited non-US and US pay functions to foreign national, HN, civilian internees, and EPWs are provided.
- Ensure all necessary banking functions are performed in theater.

Surgeon:

- Support medical humanitarian assistance and disaster relief operations.
- Provide medical care of EPWs and civilians within the command's AO.
- Coordinate medical laboratory access in AO.
- Determine the nature of local health threats to the force through populace interviews.
- Determine the identity of local or captured medical supplies.

Veterinary Officer:

- Determine source and suitability of local foods.
- Assist the local population with veterinary service needs.

Chemical Officer:

- Identify enemy force chemical weapons and equipment.
- Communicate NBC risks to supported populations.

Engineer Coordinator:

- Procure proper local materials to support engineering missions.
- Communicate engineering project requirements to contracted local work force.

- Communicate engineering project impact on local landowners and other affected parties.
- Determine, in coordination with G2/S2, suitability of local topographic maps and terrain products.
- Assess environmental concerns of HN and local populations in combined operations.

Provost Marshal:

- Support dislocated and civilian straggler control activities.
- Support internment and resettlement operations, to include displaced civilians.
- Support weapons buy-back programs, as required, and work closely with civil-military liaisons for payments to local officials.
- Support counter-drug and customs activities.
- When authorized, help foreign civil authorities maintain control.
- Conduct liaison with local LEAs.

PSYOP Officer:

- Produce approved PSYOP propaganda and counter-propaganda media.
- Evaluate PSYOP impact on target audience.

Air Defense Coordinator:

- Identify enemy air defense artillery (ADA) weapons and radars.
- Communicate air defense warnings to supported populations.
- Communicate air defense project requirements to contracted local work force.

Safety Officer:

- Provide safety training to local labor force.
- Communicate warnings of dangerous military operations and other hazards to local populace.

Transportation Officer:

- Coordinate commercial and local transportation needs.
- Coordinate movement scheduling and routes with multinational forces and/or HN.

PERSONAL STAFF OFFICER RESPONSIBILITIES

B-11. Personal staff officers are under immediate control of the commander and have direct access to the commander. Most personal staff officers also perform special staff officer duties, working with a coordinating staff officer. These assignments are on a case-by-case basis, depending on the commander's guidance and the nature of the mission; they are very common in stability operations and support operations. Linguist requirements for special staff officers include but are not limited to the following staff officers.

Chaplain:

- Coordinate religious support with multinational partners.
- Determine the impact of local population religious group faiths and practices on military operations.
- Provide religious support to the community to include hospital patients, EPWs, refugees, and civilian detainees.
- Conduct liaison with local population religious leaders in close coordination with the G5.

Public Affairs Officer:

- Act as the commander's spokesman for all communication with external media.
- Assess the accuracy of foreign media interpretation of Public Affairs Office (PAO) releases.
- Assess and recommend news, entertainment, and other information (assisting G5) for contracted services foreign nationals.

Staff Judge Advocate:

- Translate and interpret foreign legal codes, SOFAs, and international laws.
- Determine local environmental laws and treaties through translation services.
- Assess the treatment of EPWs and civilian internees.
- Translate documents to support G4 in local contracts.

SOURCES OF LINGUISTS

B-12. There are various sources that a commander can use to obtain the linguists necessary to support operations. It is vital to know the advantages and disadvantages of each type of linguist and to carefully match the available linguists to the various aspects of the operation.

ARMY LANGUAGE-QUALIFIED MOSs

B-13. The AC MI language-dependent military occupational specialities (MOSs) are 98G with a skill qualification identifier (SQI) of L/352G (Cryptologic Communications Interceptor/ Locator), and their related WO fields. Some soldiers in MOS 96B (All-Source Intelligence Analyst), MOS 97B (CI Agent), MOS 97E/351E (HUMINT Collector) and MOS 98C (SIGINT Analyst), and their related WO fields are trained in foreign languages. Using soldiers in the MOSs mentioned above has many advantages. They are already trained in the military system, are not subject to deployment restrictions (a limiting factor with civilian linguists), have a security clearance and, as US personnel, support the command's interests. The major disadvantage to utilizing these individuals for general foreign language support is that in doing so, they are removed from their primary MI functions. They should be used only in linguistic duties that include intelligence potential. For example, a HUMINT collector (97E) provides linguist support to a medical assistance team as a method to provide access to the local population to determine their attitudes toward US Forces.

B-14. Non-MI Army language qualified MOSs include some enlisted and WOs in career management fields 18 (Special Forces), 37 (PSYOP), 180A (Special Forces); and commissioned officers with a branch code 18 (Special Forces); and functional areas 39 (PSYOP and CA) and 48 (Foreign Area Officer). Particular attention must be paid to the recorded language proficiency and test date of these individuals since the standards vary by field. The same advantages and disadvantages apply as with the AC MI linguists.

B-15. RC language-dependent MOSs include those listed above in the AC. RC linguists have the same set of advantages and disadvantages as listed above for AC language-dependent MOSs. The RC also includes linguists in MOS 97L (translator/interpreter). The 97Ls are specifically trained to be a translator and interpreter. They have the same advantages as the AC linguists. An added advantage is that since their sole job is translation and interpretation, they do not have to be removed from another job in order to be used as a linguist. Their major disadvantage is that they have no additional skill that gives them dual functionality.

Army Linguists Not DOD Trained

B-16. The Army also includes numerous soldiers of all grades who are proficient in a foreign language and are receiving Foreign Language Proficiency Pay (FLPP) but whose primary duties do not require foreign language proficiency. They may have attended a civilian school to learn a foreign language, or they may have acquired proficiency through their heritage. They have the advantage of being trained soldiers and are therefore readily deployable to all areas of the battlefield.

B-17. These soldiers may have the specific vocabulary and military skill knowledge for certain linguist support missions. For example, a supply sergeant who speaks the local language would be an invaluable asset to the G4. There are disadvantages in that they already have another job and units are reluctant to give up personnel especially if they are in key positions. Their capabilities are difficult to assess. Since they are not required to take the Defense Language Proficiency Test (DLPT) if they are not receiving FLPP, it is often difficult for the G1/S1 to identify them as a linguist or for a non-linguist to judge the level of their foreign language capability.

Other Service Linguists

B-18. Other service linguists have the advantage of deployability, loyalty, and clearance, but must often learn the Army system and specific Army vocabulary. They are also difficult to obtain since their parent service probably also lacks a sufficient number of trained linguists. Other service linguists, however, will be valuable in joint operation centers and joint activities. When serving the JTF headquarters, Army commanders and staffs must be aware of the linguists in the other services in order to plan for the participation and optimize their employment.

US Contract Linguists

B-19. US civilians can be contracted to provide linguist support. They have an advantage over LN hires in that their loyalty to the US is more readily evaluated, and it is easier for them to be granted the necessary security clearance. However, there are usually severe limitations on the deployment and use of civilians. A careful assessment of their language ability is important because, in many cases, they use "old fashioned" terms, or interject US idioms. If the linguists are recent émigrés, the use of the language in their country of origin could be dangerous to them, or their loyalty may reside with their own country when at odds with US interests.

Multinational Linguists

B-20. Multinational linguists have their own set of advantages and disadvantages. These linguists may be unfamiliar with the US military system unless they have previously participated in a multinational operation with US forces. They may have a security clearance, but clearances are not necessarily equal or reciprocal, automatically guaranteeing access to classified or sensitive information between nations. They support the command's interest but may have differing priorities or responsibilities within their assigned AOs. These linguists also are already fulfilling specific duties for their own nation, which may also have a shortage of linguists. The major disadvantage to acquiring and maintaining multinational linguist support is that they are outside the C2 (via military authority or military contract) of the US forces. These linguists will be valuable in multinational operations centers and activities.

Local National Contract Linguists

B-21. LN hires will provide the bulk of your linguist support. They are usually less expensive to hire than US civilians and will know the local dialect, idioms, and culture. The expertise of these linguists in particular areas or subject matters can be an asset. However, there are several potential problems with using LN hires, to include limited English skills and loyalty considerations. Therefore, a screening interview or test is necessary to determine their proficiency in English. These individuals must also be carefully selected and screened by CI personnel (with US linguist support) initially and periodically throughout their employment. Their loyalty is always questionable. Local prejudices may influence them, and they may place their own interests above those of the US.

EVALUATING LINGUIST PROFICIENCY

B-22. Commanders and staffs must understand the Army linguist proficiency evaluation system in order to effectively plan for and employ linguists. Evaluation and reevaluation of linguist proficiency is covered in detail in AR 611-6, Section III. Language testing is required for all Army personnel in a language-dependent MOS, who have received foreign language training at government expense, who are receiving FLPP, or who are in a language-required position regardless of MOS. Other Army personnel who have knowledge of a foreign language are encouraged to take the proficiency test and may work as linguists.

B-23. The Army uses the DLPT to determine foreign language proficiency levels. DLPTs are listed by foreign language in DA Pam 611-16. In foreign languages where no printed or recorded test exists, oral interview tests are arranged. The DLPT is an indication of foreign language capability, but it is not the definitive evaluation of an individual's ability to perform linguist support.

B-24. AR 611-6, Appendix D, Sections 1 through 4, describes the proficiency levels for the skills of speaking, listening, reading, and writing a foreign language based on the interagency roundtable descriptions. The plus-level designators, shown as a "+" symbol, are used to designate when a linguist is above a base level, but not yet to the capability of the next level. For example, 2+ would indicate a better than limited working proficiency in the foreign language. The six "base levels" of proficiency, as established by DLPT and/or oral exam, are—

- Level 0 (No proficiency). The soldier has no functional foreign language ability. Level 0+. The minimum standard for Special Forces personnel indicates a memorized proficiency only.
- Level 1 (Elementary proficiency). The soldier has limited control of the foreign language skill area to meet limited practical needs and elementary foreign language requirements.
- Level 2 (Limited working proficiency). The linguist is sufficiently skilled to be able to satisfy routine foreign language demands and limited work requirements.
- Level 3 (General professional proficiency). The linguist is capable of performing most general, technical, formal, and informal foreign language tasks on a practical, social, and professional level.
- Level 4 (Advanced professional proficiency). The linguist is capable of performing advanced professional foreign language tasks fluently and accurately on all levels.
- Level 5 (Functionally native proficiency). The linguist is functionally equivalent to an articulate and well-educated native in all foreign language skills; and reflects the cultural standards of the country where the foreign language is natively spoken.

B-25. The above proficiency base levels designate proficiency in any of the four language skills: listening, reading, speaking, and writing. The most evaluated skills on the DLPT are reading and listening. These tests are not for use in evaluating linguists above the 3 proficiency level. Most Army linguist DLPT scores show only two skill levels: listening and reading (for example, 2+/3, or 3/1+).

SUSTAINING MILITARY LINGUIST PROFICIENCY

B-26. Language proficiency diminishes with lack of use and absence of exposure to the foreign language. To ensure combat readiness, commanders should require all military linguists receive periodic language training. In-country language immersion training, in-garrison contracted language instructors, on-line foreign newspapers, and foreign radio broadcasts are all examples of language training resources. Funding for language training is available through MACOM language training program funds.

Glossary

The glossary lists acronyms and terms with Army or joint definitions, and other selected terms. Where Army and joint definitions are different, (Army) follows the term. Terms for which FM 2-0 is the proponent FM (authority) are marked with an asterisk (*) and followed by the number of the paragraph (¶) where they are discussed. For other terms, the number of the proponent FM follows the definition. JP 1-02 and FM 1-02 are posted in the Joint Electronic Library, which is available online and on CD-ROM.

- Use this URL to access JP 1-02 online: http://www.dtic.mil/doctrine/jel/doddict/.
- Use this URL to access FM 1-02 online: http://www.dtic.mil/doctrine/jel/service_pubs/101_5_1.pdf.
- Follow this path to access JP 1-02 on the Joint Electronic Library CD-ROM: Mainmenu>Joint Electronic Library>DOD Dictionary.
- Follow this path to access FM 1-02 on the Joint Electronic Library CD-ROM: Mainmenu>Joint Electronic Library>Service Publications>Multiservice Pubs> FM 101-5-1.

A2C2	Army Airspace Command and Control
AA	avenue of approach
AAMDC	Area Air and Missile Defense Command
ABCS	Army Battle Command System
AC	Active Component
ACE	analysis and control element
ACINT	acoustic intelligence (water)
ACOUSTINT	acoustic intelligence (atmosphere)
ACR	armored cavalry regiment
ACT	analysis and control team
ADA	air defense artillery
ADCON	administrative control
ADDO	assistant deputy director for operations
ADO	air defense officer
ADP	automated data processing
AFACSI	Air Force Assistant Chief of Staff for Intelligence
AFIWC	Air Force Information Warfare Center
AFMIC	Armed Forces Medical Intelligence Center

AFOSI	Air Force Office of Special Investigation
AGM	attack guidance matrix
AIA	Air Force Intelligence Agency
AITR	aided target recognition
ALOC	administrative and logistics operations center
AMC	US Army Materiel Command
AMD	Air Missile Defense
AMHS	Automated Message Handling System
AMIB	Allied Military Intelligence Battalion
***analysis**	Determination of the significance of the information, relative to information and intelligence already known, and drawing deductions about the probable meaning of the evaluated information. (¶1-58)
AO	area of operations
AOC	area of concentration
AOI	area of interest
AOIR	area of intelligence responsibility
AOR	area of responsibility
AR	Army Regulation
ARAT	Army Reprogramming Analysis Team
ARCENT	US Army Central Command
ARFOR	Army forces
ARISC	Army Reserve Intelligence Support Center
ARL	airborne reconnaissance low
ARNG	US Army National Guard
ARSOF	Army Special Operations Forces
ASCC	Army Service Component Commander
ASPO	Army Space Program Office
AT	antiterrorism
ATCAE	Army Technical Control Analysis Element
ATCCS	Army Tactical Command and Control System
ATR	automatic target recognition
ART	Army tactical task
BDA	battle damage assessment
BDU	battle-dress uniform

BFACS	Battlefield Functional Area Control System
C/J2	combined J2
C2	command and control
CA	civil affairs
CASO	Collection Activities and Source Operations
CAT	crisis action team
CBRN	chemical, biological, radiological, and nuclear
CBRNE	chemical, biological, radiological, nuclear, and high-yield explosive
CCD	coherent change detection
CCIR	commander's critical information requirement
CD-ROM	compact disc-read only memory
CE	counterespionage
CECOM	Communications-Electronics Command
CED	captured enemy document
CEE	captured enemy equipment
CEM	captured enemy materiel
CFSO	counterintelligence force protection source operations
CGS	common ground station
CI	counterintelligence
CIA	Central Intelligence Agency
CIAC	counterintelligence analysis cell
CICA	Counterintelligence Coordinating Authority
CITO	Central Imagery Tasking Office
CJCS	Chairman, Joint Chiefs of Staff
CMEC	Captured Materiel Exploitation Center
CMF	career management field
CMISE	Corps Military Intelligence Support Element
CMO	civil-military operations
CNA	computer network attack
CND	computer network defense
CNE	computer network exploitation
CNN	Cable News Network
CNO	computer network operations

COA	course of action
COCOM	combatant command (command authority)
COG	center of gravity
COMINT	communications intelligence
COMSEC	communications security
CONOPS	contingency operations
CONPLAN	contingency plan
CONUS	continental United States
COP	common operational picture
COSCOM	Corps Support Command
***counterintelligence**	**1.** Information gathered and activities conducted to protect against espionage, other intelligence activities, sabotage, or assassinations conducted by or on behalf of foreign governments or elements thereof, foreign organizations, or foreign persons, or international terrorist activities. (JP 1-02) **2.** (Army) Counterintelligence counters or neutralizes intelligence collection efforts through collection, CI investigations, operations, analysis and production, and functional and technical services. CI includes all actions taken to detect, identify, exploit, and neutralize the multidiscipline intelligence activities of friends, competitors, opponents, adversaries, and enemies; and is the key intelligence community contributor to protect US interests and equities. (FM 2-0) (¶1-111)
CP	command post
CRN	command radio net
CS	combat support
CSG	cryptologic support group
CSS	combat service support
CSSCS	Combat Service Support Control System
CT	Counterterrorism
***cueing**	The use of one or more sensor systems to provide data that directs collection by other systems. (¶2-53)
CWT	combat weather team
DA	Department of the Army
DAC	Department of the Army Civilian
DAPDM	Department of the Army Production and Dissemination Management
DCI	Director of Central Intelligence
DCS, G2	Deputy Chief of Staff, Intelligence

***debriefing** The systematic questioning of individuals to procure information to answer specific collection requirements by direct and indirect questioning techniques. (¶6-4)

DF direction finding

DHS Defense HUMINT Service

DIA Defense Intelligence Agency

DIAC Defense Intelligence Analysis Center

DIMHRS Defense Integrated Management Human Resource System

DIO Defensive Information Operations

DISCOM Division Support Command

DISUM daily intelligence summary

DLPT Defense Language Proficiency Test

DMS Director of Military Support

DOCEX document exploitation

***document exploitation** The systematic extraction of information from all media formats in response to collection requirements. (¶6-4)

DOD Department of Defense

DOJ Department of Justice

DOT Department of Transportation

DP decision point

DPM dissemination program manager

DS direct support

DSM decision support matrix

DST decision support template

EA electronic attack

EAC echelons above corps

ECOA enemy course of action

EEFI essential elements of friendly information

ELINT electronic intelligence

EM electromagnetic

eMILPO electronic Military Personnel Office System

EMS electromagnetic spectrum

E-O electro-optical

EOB electronic order of battle

EOD explosive ordinance disposal

EP	electronic protection
EPW	enemy prisoner of war
ES	electronic warfare support (Army)
***evaluating**	**1.** Comparing relevant information (RI) on the situation or operation against criteria to determine success or progress. (FM 6-0) **2.** In intelligence usage, appraisal of an item of information in terms of credibility, reliability, pertinence, and accuracy. (FM 2-0) (¶1-26 and 1-98)
EW	electronic warfare
EWS	electronic warfare support (joint)
FBI	Federal Bureau of Investigation
FDA	functional damage assessment
FDU	Force Design Update
FEMA	Federal Emergency Management Agency
FFIR	friendly force information requirement
FISINT	foreign instrumentation signals intelligence
FISS	Foreign Intelligence Security Service
FLPP	Foreign Language Proficiency Pay
FM	Field Manual
FMA	foreign materiel acquisition
FME	foreign materiel exploitation
FMEP	Foreign Materiel Exploitation Program
FMP	Foreign Materiel Program
FMT	Foreign Materiel for Training
FP	force protection
FRAGO	fragmentary order
FS	fire support
FSAC	Fire Support Armaments Center
FSE	fire support element
FSO	fire support officer
G1	Assistant Chief of Staff, Personnel
G2	Assistant Chief of Staff, Intelligence
G3	Assistant Chief of Staff, Operations
G4	Assistant Chief of Staff, Logistics
G5	Assistant Chief of Staff, Civil/Military Affairs

G6 Assistant Chief of Staff, Command, Control, Communications, and Computer Operations

G7 Assistant Chief of Staff, Information Operations

GEOINT geospatial intelligence

GMI general military intelligence

GPS Global Positioning System

GS general support

GS-R general support-reinforcing

HN host nation

HOC HUMINT operations cell

HPT high-payoff target

HQ headquarters

HQDA Headquarters, Department of the Army

HSI hyperspectral imaging

HUMINT human intelligence

***human intelligence** Collection by a trained HUMINT collector of foreign information from people and multimedia to identify elements, intentions, composition, strength, dispositions, tactics, equipment, personnel, and capabilities. It uses human sources and a variety of collection methods, both passively and actively, to gather information to satisfy the commander's intelligence requirements and cross-cue other intelligence disciplines. (¶1-106 and ¶6-1)

HVT high-value target

I/R internment and resettlement (operations)

I&W indications and warnings

ICC Intelligence Coordination Center (US Coast Guard)

ICF intelligence contingency fund

ICL intelligence coordination line

IDC Information Dominance Center

IIR intelligence information report

IM information management

IMA individual mobilization augmentee

IMEP International Materiel Evaluation Program

IMETS Integrated Meteorological System

IMINT imagery intelligence

***indications and warning** (joint) Those intelligence activities intended to detect and report time-sensitive intelligence information on foreign developments that could involve a threat to the United States or allied and/or coalition military, political, or economic interests or to US citizens abroad. It includes forewarning of enemy actions or intentions; the imminence of hostilities; insurgency; nuclear or non-nuclear attack on the United States, its overseas forces, or allied and/or coalition nations; hostile reactions to US reconnaissance activities; terrorists attacks; and other similar events. (¶2-19)

***indicator** (Army) Positive or negative evidence of threat activity or any characteristic of the AO which points toward threat vulnerabilities or the adoption or rejection by the threat of a particular capability, or which may influence the commander's selection of a COA. Indicators may result from previous actions or from threat failure to take action. (FM 2-0) (¶5-14)

INFOSEC information security

***infrastructure** **1.** (joint) All building and permanent installations necessary for support, redeployment, and military forces operations (e.g., barracks, headquarters, airfields, communications facilities, stores, port installations, and maintenance stations). (JP 4-01.8) **2.** (Army) In intelligence usage, the basic underlying framework or feature of a thing; in economics, basic resources, communications, industries, and so forth, upon which others depend; in insurgency, the organization (usually hidden) of insurgent leadership. (FM 2-0) (¶5-8)

INR Bureau of Intelligence and Research, Department of State

INS Immigration and Naturalization Service

INSCOM US Army Intelligence and Security Command

intelligence **1.** (joint) The product resulting from the collection, processing, integration, analysis, evaluation, and interpretation of available information concerning foreign countries or areas. **2.** Information and knowledge about an adversary obtained through observation, investigation, analysis, or understanding. (¶1-3)

***intelligence coordination line** A line that designates the boundary between AOIRs. The G2/S2 establishes ICLs to facilitate coordination between higher, lateral, and subordinate units; coordinates with the G3/S3 to direct subordinates to track enemy units and HPTs in their areas; and hands over intelligence responsibility for areas of the battlefield. The establishment of ICLs ensures that there are no gaps in the collection effort; that all echelons are aware of the location, mission, and capabilities of other assets; facilitates asset cueing, and provides timely exchange of information between assets. The G2/S2 keeps abreast of collection activities in progress (all echelons) and battlefield developments through the ICLs. (¶3-13)

intelligence discipline (joint) A well-defined area of intelligence collection, processing, exploitation, and reporting using a specific category of technical or human resources. There are seven major disciplines: human intelligence, imagery intelligence, measurement and signature intelligence, signals intelligence (communications intelligence, electronic intelligence, and foreign instrumentation signals intelligence), all-source intelligence, technical intelligence, and counterintelligence. (¶1-104)

***intelligence process** A theoretical model used to describe intelligence operations. The intelligence process is not a framework for actual operations because the functions and common tasks occur in parallel as opposed to sequentially. (¶1-91)

***intelligence products** Intelligence products are generally placed in one of six categories: I&W, current, general military, target, S&T, and CI. The categories are distinguished from each other primarily by the purpose for which the intelligence was produced. The categories often overlap, and the same intelligence data can be used in each of the categories. (¶1-5)

INTREP intelligence report

INTSUM intelligence summary

***intelligence requirements** (Army) Those requirements generated from the staff's IRs regarding the enemy and environment that are not a part of the CCIR (PIR and FFIR). Intelligence requirements require collection and can pro-vide answers in order to identify indicators of enemy actions or intent, which reduce the uncertainties associated with an operation. Significant changes (i.e., branches and sequels) with an operation usually lead to changes in intelligence requirements. (¶2-7)

***intelligence running estimate** The intelligence running estimate is a continuous flow and presentation of relevant information and predictive intelligence that, when combined with the other staff running estimates, enable the decisionmaker's visualization and situational understanding of the AOI in order to achieve information superiority. The intelligence running estimate requires constant verification to support situational understanding of the current situation as well as predictive assessments for future operations. (¶5-30)

***intelligence synchronization** The task that ensures ISR operations are linked to the commander's requirements and respond in time to influence decisions and operations. The intelligence officer, with staff participation, synchronizes the entire collection effort to include all assets the commander controls, assets of lateral units and higher echelons units and organizations, and intelligence reach to answer the commander's CCIR (PIR and FFIR). (¶1-24)

***intelligence synchronization plan**	The plan the intelligence officer uses, with staff input, to synchronize the entire collection effort to include all assets the commander controls, assets of lateral units, and higher echelon units and organizations, and intelligence reach to answer the commander's CCIR (PIR and FFIR). (¶1-38)
intelligence warfighting function	The related tasks and systems that facilitate understanding of the operational environment
IO	information operations
IPB	intelligence preparation of the battlefield
IPIR	Initial Phase Imagery Report
IPTF	international police task force
IR	information requirement
IRINT	infrared intelligence
IRR	Individual Ready Reserve
ISM	intelligence synchronization matrix
ISR	intelligence, surveillance, and reconnaissance
J2	Joint Staff Directorate, Intelligence
J2A	Joint Staff Directorate, Intelligence (Administrative)
J2J	Joint Staff Directorate, Intelligence (Joint)
J2M	Joint Staff Directorate, Intelligence (Management)
J2O	Joint Staff Directorate, Intelligence (Operations)
J2P	Joint Staff Directorate, Intelligence (Assessments, Doctrine)
J2T	Joint Staff Directorate, Intelligence (Targeting)
J2X	Joint Staff Directorate, Intelligence (CI & HUMINT)
J3	Joint Staff Directorate, Operations
J5	Joint Staff Directorate, Civil Affairs
JAC	Joint Analysis Center
JCMEC	Joint Captured Materiel Exploitation Center
JCO	Joint Commission Observers
JCS	Joint Chiefs of Staff
JDEC	Joint Document Exploitation Center
JDISS	Joint Deployable Intelligence Support System
JFC	joint force commander
JIAWG	joint interagency working group
JIC	Joint Intelligence Center

JIF	Joint Interrogation Facility
JISE	joint intelligence support element
JOA	joint operations area
JP	Joint Publication
JSTARS	Joint Surveillance Target Attack Radar System
JTF	joint task force
JTMD	joint table of mobilization and distribution
JWICS	Joint Worldwide Intelligence Communications System
LAN	local area network
LASINT	laser intelligence
LCC	land component commander
LEA	law enforcement agency
LIDAR	laser radar
LN	local national
LNO	liaison officer
LOC	line of communication
LOS	line of sight
LPT	logistics preparation of the theater
LTIOV	latest time information is of value
MACOM	major Army command
MASINT	measurement and signature intelligence
***materiel**	(Army) In intelligence usage, the all-encompassing term for the weapons systems, equipment, apparatus, documents, and supplies of a foreign military force or nonmilitary organization. (¶2-24)
MCIA	Marine Corps Intelligence Agency
MCOO	modified combined obstacle overlay
MCS	Maneuver Control System
MDMP	military decisionmaking process
MEA	munitions effects assessment
METL	mission-essential task list
METT-TC	mission, enemy, terrain and weather, troops and support available, time available, civil considerations
MI	military intelligence
MLT	machine language translation

MNFC	multinational force commander
MO	method of operation
MOPP	mission-oriented protective posture
MOS	military occupational specialty
MP	military police
MSC	major subordinate command
MSE	mobile subscriber equipment
MSI	multi-spectral imagery
MSIC	Missile and Space Intelligence Center
MTI	moving target indicator
MTOE	modified table of organization and equipment
MTW	major theater war
NAI	named area of interest
NAIC	National Air Intelligence Center
NATO	North Atlantic Treaty Organization
NBC	nuclear, biological, and chemical
NCA	National Command Authorities
NCO	noncommissioned officer
NCR	national cryptologic representative
NCTI	non-cooperative target identification
NEO	noncombatant evacuation operation
NFH	National Defense Headquarters, Canada
NGA	National Geospatial-Intelligence Agency
NGO	non-governmental organization
NIC	National Intelligence Center
NIIRS	National Imagery Interpretability Rating Scale
NIMA	National Imagery and Mapping Agency
NIST	National Intelligence Support Team
NLT	not later than
NMCC	National Military Command
NMIC	National Maritime Intelligence Center
NMJIC	National Military Joint Intelligence Center
NRO	National Reconnaissance Office
NRT	near-real time

NSA	National Security Agency
NSC	National Security Council
NUCINT	Nuclear Intelligence
NVD	night vision device
OB	order of battle
OE	operational environment
OMA	Office of Military Affairs
OMB	Office of Management and Budget
OMT	operational management team
ONDCP	Office of National Drug Control Policy
ONI	Office of Naval Intelligence
ONIR	overhead non-imaging infrared
OPCON	operational control
OPLAN	operation plan
OPORD	operation order
OPSEC	operations security
OSC	operations support cell
OSCE	Organization for Security and Cooperation in Europe
OSD	Office of the Secretary of Defense
OSINT	open-source intelligence
PAO	Public Affairs Office
PDA	physical damage assessment
PIAP	Police Information Assessment Program
PIO	police intelligence operations
PIR	priority intelligence requirement
PME	peacetime military engagement
POC	point of contact
POD	port of debarkation
POE	port of embarkation
***predictive intelligence**	Intelligence analysis conclusions, assessments or products that attempt to define, describe, present, or portray the future situation or conditions of the enemy, +terrain, weather and civil considerations. (¶1-3)
***priority intelligence requirements**	(joint) Those intelligence requirements for which a commander has an anticipated and stated priority in his task of planning and decision-making. (FM 2-0) (¶1-32)

***process/ processing** A function of the intelligence process that involves converting collected data, which is not already in a comprehensible form when it is reported, into a form that is understandable and suitable for analysis and production of intelligence. Examples of processing include developing film, enhancing imagery, translating a document from a foreign language, converting electronic data into a standardized report that can be analyzed by a system operator, and correlating dissimilar or jumbled information by assembling like elements before the information is forwarded for analysis. (FM 2-0) (¶1-96)

***production** In intelligence usage, conversion of information into intelligence through the integration, analysis, evaluation, and interpretation of all source data and the preparation of intelligence products in support of known or anticipated user requirements. (¶4-20)

PSYOP psychological operations

QRC quick reaction capability

QRF quick reaction force

R&D research and development

RADINT radar intelligence

RC reserve component

RDEC Research, Development, and Engineering Center

***reconnaissance, surveillance, and target acquisition/intelligence surveillance and reconnaisssance** A full spectrum combined arms mission that integrates ground and air capabilities to provide effective, dynamic, timely, accurate and assured combat information and multi-discipline actionable intelligence for lethal and non-lethal effects/decisions in direct support of the ground tactical commander.

REMBASS Remotely Monitored Battlefield Sensor System

***requirements management** The intelligence task that develops a prioritized list of what information needs to be collected and produced into intelligence, dynamically updates and adjusts those requirements in response to mission adjustments/changes, and places a latest time intelligence is of value to ensure intelligence and information is reported to meet operational requirements. (¶2-83)

RF radio frequency

RFI request for information

RFW radio frequency weapon

RI relevant information

RM requirements management

ROE rules of engagement

ROK Republic of Korea

ROZ restricted operating zone

RSO&I	reception, staging, onward movement, and integration
RSOC	Regional Security Operations Center
RSTA	reconnaissance, surveillance, and target acquisition
RSP	render safe procedure
RTCAE	Regional Technical Control and Analysis Element
running estimate	A staff estimate, continuously updated, based on new information as the operation proceeds. (FM 6-0) (¶5-28)
S&T	scientific and technical
S&TI	scientific and technical intelligence
S1	Staff Officer, Personnel
S2	Staff Officer, Intelligence
S2X	Staff Officer, Intelligence (CI & HUMINT)
S3	Staff Officer, Operations
S4	Staff Officer, Logistics
S5	Staff Officer, Civil Affairs
S6	Staff Officer, Command, Control, Communications, and Computer Operations
SAP	special access program
SAR	synthetic aperture radar
SATCOM	satellite communications
SBCCOM	Chemical and Biological Defense Command
SCI	sensitive compartmented information
SCIF	sensitive compartmented information facility
SIDPERS	Standard Installation Division Personnel System
SIGINT	signals intelligence
SIGO	signal officer
SII	statement of intelligence interest
SIO	senior intelligence officer
SIPRNET	Secret Internet Protocol Router Network
SIR	specific information requirement
SITMAP	situation map

***situation template** (Army) A depiction of a potential adversary course of action as part of a particular adversary operation. Situation templates are developed on the adversary's current situation (for example, training and experience levels, logistic status, losses, and disposition), the environment, and adversary doctrine or patterns of operations. The commander dictates the level to depict the adversary based on the factors of METT-TC (at minimum two levels of command below the friendly force) as a part of his guidance for mission analysis. (FM 2-0) (¶1-58)

SJA staff judge advocate

SMU special mission unit

SOCOM satellite communications on-the-move

SOF special operations forces

SOFA status of forces agreement

SOP standing operating procedure

SOTA SIGINT operational tasking authority

SR special reconnaissance

SRTD signals research and target development

SSO special security office

STANAG standardization agreement (NATO)

STICEUR Scientific and Technology Intelligence Center Europe

SUPR supplemental imagery report

SWO staff weather officer

TA target acquisition

TACON tactical control

***tactical questioning** The expedient initial questioning for information of immediate tactical value. (FM 2-0) (¶6-6)

TAI target area of interest

TCAE technical control and analysis element

TCN third-country national

***technical control** The technical function to ensure adherence to existing policies or regulations and provide technical guidance for MI activities, particularly HUMINT, SIGINT, and CI operations. (FM 2-0) (¶3-23)

TECHINT technical intelligence

TECOM Test and Evaluation Command

TENCAP Tactical Exploitation of National Capabilities

TEU Technical Escort Unit

THREATCON	threat condition
TIB	Theater Intelligence Brigade
TIG	Theater Intelligence Group
TIM	toxic industrial material
TLP	troop-leading procedure
TOC	tactical operations center
TPFDD	Time Phased Force Deployment Data
TRADOC	US Army Training and Doctrine Command
TSA	target system assessment
TSCM	Technical Surveillance Countermeasures
TSE	theater support element
TTP	tactics, techniques, and procedures
TVA	threat vulnerability assessment
UAV	unmanned aerial vehicle
UN	United Nations
US	United States
USAF	United States Air Force
USAR	United States Army Reserve
USC	United States Code
USCENTCOM	United States Central Command
USCG	United States Coast Guard
USCS	United States Cryptologic System
USEUCOM	United States European Command
USJFCOM	United States Joint Forces Command
USMC	United States Marine Corps
USNORTHCOM	United States Northern Command
USPACECOM	United States Space Command
USPACOM	United States Pacific Command
USSID	United States Signal Intelligence Directive
USSOCOM	United States Special Operations Command
USSOUTHCOM	United States Southern Command
USSPACECOM	United States Space Command
USSTRATCOM	United States Strategic Command
USTRANSCOM	United States Transportation Command

VJ2	Vice J2
VTC	video teleconference
WAN	wide area network
WARNO	warning order
WMD	weapons of mass destruction
WO	warrant officer

Bibliography

The bibliography lists field manuals published with new numbers followed by old number.

DOCUMENTS NEEDED

These documents must be available to the intended users of this publication.

JP 1-02. *Department of Defense Dictionary of Military and Associated Terms.* June 1999. [Online] Available http://www.dtic.mil/doctrine/jel/doddict/

FM 1-02/MCRP 5-2A. *Operational Terms and Symbols* (Drag Draft). 21 February 2003.

READINGS RECOMMENDED

These sources contain relevant supplemental information.

ARMY PUBLICATIONS

Most Army doctrinal publications are available online: http://155.217.58.58/atdls.htm

AR 381-10. *US Army Intelligence Activities.* 1 July 1984.

AR 381-12. *Subversion and Espionage Directed Against the U.S. Army (SAEDA).* 15 January 1993.

AR 381-20. *The Army Counterintelligence Program.* 15 November 1993.

AR 611-6. *Army Linguist Management.* 16 February 1996.

AR 715-9. *Contractors Accompanying the Force.* 29 October 1999.

FM 1 (100-1). *The Army.* 14 June 2001.

FM 2-19.402. *STRYKER Brigade Combat Team Intelligence Operations.* 1 March 2003.

FM 2-19.602. *Surveillance Troop.* 1 March 2003.

FM 2-33.5/ST. *Intelligence Reach Operations.* 1 June 2001.

FM 3-0. *Operations.* 14 June 2001.

FM 3-05.102. *Army Special Operations Forces Intelligence.* 31 August 2001 (formerly FM 34-36).

FM 3-05.30. *Psychological Operations.* 19 June 2000.

FM 3-06.11. *Combined Arms Operations in Urban Terrain.* 28 February 2002.

FM 3-13. *Information Operations.* 28 November 2003 (formerly FM 100-6).

FM 3-19. *NBC Reconnaissance.* 19 November 1993.

FM 3-19.1. *Military Police Operations.* 22 March 2001.

FM 3-19.4. *Military Police Leaders' Handbook.* 4 March 2002.

FM 3-19.40. *Military Police Internment/Resettlement Operations.* 30 September 1987.

FM 3-20.98. *Reconnaissance Platoon.* May 2001.

FM 34-2. *Collection Management and Synchronization Planning.* 8 March 1994.

FM 34-52. (FM 2-22-3) *Human Intelligence.* 28 September 1992.

FM 3-90. *Tactics.* 4 July 2001.

FM 3-90.3. *The Mounted Brigade Combat Team.* 1 November 2001.

FM 3-100.4. *Environmental Considerations in Military Operations.* 15 June 2000, Change 1, 11 May 2001.

FM 3-100.21. *Contractors on the Battlefield.* 26 March 2000.

FM 5-170. *Engineer Reconnaissance.* 5 May 1998.

FM 6-0. *Mission Command: Command and Control of Army Forces.* 11 August 2003.

FM 6-20-10. *Tactics. Techniques, and Procedures for the Targeting Process.* 8 May 1996.

FM 6-121. *Tactics, Techniques, and Procedures for Field Artillery Target Acquisition.* 25 September 1990.

FM 7-0. *Training the Force.* 22 October 2002.

FM 7-10. *The Infantry Rifle Company.* 14 December 1990.

FM 7-15. *Army Universal Task List.* 31 August 2003.

FM 7-92. *The Infantry Reconnaissance Platoon and Squad (Airborne, Air Assault, Light Infantry).* 23 December 1992.

FM 7-100. *Opposing Force Doctrinal Framework and Strategy.* 1 May 2003.

FM 17-97. *Cavalry Troop.* 3 October 1995.

FM 17-98. *Scout Platoon.* 10 April 1999.

FM 19-1. *Military Police Support for the Airland Battle.* 23 May 1988.

FM 27-10. *The Law of Land Warfare.* 18 July 1956.

FM 31-20-5. *Special Reconnaissance Tactics, Techniques, and Procedures for Special Forces.* 23 March 1993.

FM 34-2-1. *Tactics, Techniques, and Procedures for Reconnaissance and Surveillance and Intelligence to Counterreconnaissance.* 19 June 1991.

FM 34-3. *Intelligence Analysis.* 15 March 1990.

FM 34-8-2. *Intelligence Officer's Handbook.* 1 May 1998.

FM 34-10. *Division Intelligence and Electronic Warfare Operations.* 25 November 1986.

FM 34-25. *Corps Intelligence and Electronic Warfare Operations.* 30 September 1987.

FM 34-37. *Echelons Above Corps (EAC) Intelligence and Electronic Warfare (IEW) Operations.* 15 January 1991.

FM 34-54. *Technical Intelligence.* 30 January 1998.

FM 34-80. *Brigade and Battalion Intelligence and Electronic Warfare Operations.* 15 April 1986.

FM 34-130. *Intelligence Preparation of the Battlefield.* 8 July 1994.

FM 4-0. *Combat Service Support.* 29 August 2003.

FM 41-10. *Civil Affairs Operations.* 14 February 2000.

FM 100-25. *Doctrine for Army Special Operations Forces.* August 1999.

FM 101-5. *Staff Organization and Operations.* 31 May 1997.

FM 101-5-1. *Operational Terms and Graphics.* 30 September 1997.

FM 101-5-2. *US Army Report and Message Formats.* 29 June 1999.

ST 2-22.7. *Tactical Human Intelligence and Counterintelligence Operations.* April 2002.

ST 2-33.5. *Intelligence Reach Operations.* 1 June 2001.

ST 2-50. *Intelligence and Electronic Warfare Assets.* June 2002.

ST 2-50.4. *Combat Commanders' Handbook on Intelligence.* September 2001 (formerly FM 34-81, 28 September 1992).

JOINT PUBLICATIONS

Most joint publications are available online: http://www.dtic.mil/doctrine/jel/

JP 0-2. *Unified Action Armed Forces (UNAAF).* 10 July 2001.

JP 1. *Joint Warfare for the US Armed Forces.* 14 November 2000.

JP 1-06. *Joint Tactics, Techniques, and Procedures for Financial Management During Joint Operations.* 22 December 1999.

JP 2-0. *Joint Doctrine for Intelligence to Operations.* January 2000.

JP 2-01. *Joint Intelligence Support to Military Operations.* 20 November 1996.

JP 2-01.1. *Joint Tactics, Techniques, and Procedures for Intelligence Support to Targeting.* 9 January 2003.

JP 2-01.3. *Joint Tactics, Techniques, and Procedures for Joint Intelligence Preparation of the Battlespace*, 24 May 2000.

JP 2-02. *National Intelligence to Joint Operations.* 28 September 1998.

JP 2-03. *Joint Tactics, Techniques, and Procedures for Geospatial Information and Services Support to Joint Operations.* 31 March 1999.

JP 3-0. *Doctrine and Joint Operations.* 1 February 1995.

JP 3-07. *Joint Doctrine for Military Operations Other Than War.* 16 June 1995.

JP 3-07.2. *JTTP for Antiterrorism.* 17 March 1998.

JP 3-08. *Interagency Coordination During Joint Operations,* Volume II. 9 October 1996.

JP 3-09. *Doctrine for Joint Fire Support.* 12 May 1998.

JP 3-09.1. *Joint Tactics, Techniques and Procedures for Laser Designation Operations.* 28 May 1999.

JP 3-11. *Joint Doctrine for Operations in Nuclear, Biological, and Chemical (NBC) Environments.* 11 July 2000.

JP 3-13. *Joint Doctrine for Information Operations.* 9 October 1998.

JP 3-18. *Joint Doctrine for Forcible Entry Operations.* 16 July 2001.

JP 3-35. *Joint Deployment and Redeployment Operations.* 7 September 1999.

JP 3-51. *Joint Doctrine for Electronic Warfare.* 7 April 2000.

JP 3-55. *Doctrine for Reconnaissance, Surveillance, and Target Acquisition Support for Joint Operations (RSTA).* 14 April 1993.

JP 3-57. *Joint Doctrine for Civil-Military Operations.* 8 February 2001.

JP 3-60. *Joint Doctrine for Targeting.* 17 January 2002.

JP 4-0. *Doctrine for Logistic Support of Joint Operations.* 6 April 2000.

JP 5-0. *Doctrine for Planning Joint Operations.* 13 April 1995.

JP 5-00.2. *Joint Task Force Planning Guidance and Procedures.* 13 January 1999.

PUBLIC LAWS AND OTHER PUBLICATIONS

The United States Code is available online: http://uscode.house.gov/usc.htm

EO 12333. *United States Intelligence Activities.* 4 December 1981.

CJCS Instruction 1301.01, Policy and Procedures to Assign Individuals to Meet Combatant Command Mission-Related Temporary Duty Requirements. 1 July 2001.

CJCSM 3500.04C. *Universal Joint Task List.* 1 July 2002.

Military Standard 2525B, *DOD Interface Standard, Common Warfighting Symbology.* 30 January 1999.

PUBLICATIONS UNDER DEVELOPMENT

FM 2-01. *Intelligence Synchronization.* When published tentatively in July 2004, FM 2-01 will supersede FM 34-2, 8 March 1994, and FM 34-2-1, 19 June 1991.

FM 2-01.2. *Counterintelligence Analysis.* When published tentatively in April 2005, FM 2-01.2 will supersede FM 34-60, 3 October 1995.

FM 2-01.3. *Intelligence Preparation of the Battlefield.* When published tentatively December 2004, FM 2-01.3 will supersede FM 34-130, 8 July 1994.

Index

Entries are by paragraph number.

J-L

Printed in Dunstable, United Kingdom

91222485R00201